Tight Junctions in Cancer Metastasis

Cancer Metastasis – Biology and Treatment

VOLUME 19

Recent Volumes in this Series

Volume 10: Metastasis of Prostate Cancer
Editors: Richard J. Ablin and Malcolm D. Mason
ISBN 978-1-4020-5846-2

Volume 11: Metastasis of Breast Cancer
Editors: Robert E. Mansel, Oystein Fodstad and Wen G. Jiang
ISBN 978-1-4020-5866-7

Volume 12: Bone Metastases: A Translational and Clinical Approach
Editors: Dimitrios Kardamakis, Vassilios Vassiliou and Edward Chow
ISBN 978-1-4020-9818-5

Volume 13: Lymphangiogenesis in Cancer Metastasis
Editors: Steven A. Stacker and Marc G. Achen
ISBN 978-90-481-2246-2

Volume 14: Metastasis of Colorectal Cancer
Editors: Nicole Beauchemin and Jacques Huot
ISBN 978-90-481-8832-1

Volume 15: Signal Transduction in Cancer Metastasis
Editors: Wen-Sheng Wu and Chi-Tan Hu
ISBN 978-90-481-9521-3

Volume 16: Liver Metastasis: Biology and Clinical Management
Editor: Pnina Brodt
ISBN 978-94-007-0291-2

Volume 17: Electric Cell-Substrate Impedance Sensing and Cancer Metastasis
Editor: Wen G. Jiang
ISBN 978-94-007-4926-9

Volume 18: Central Nervous System Metastasis, the Biological Basis and Clinical Considerations
Editor: Diane Palmieri
ISBN 978-94-007-5290-0

For further volumes:
http://www.springer.com/series/5761

Tracey A. Martin • Wen G. Jiang
Editors

Tight Junctions in Cancer Metastasis

Editors
Tracey A. Martin
Institute of Cancer and Genetics
Cardiff University
Cardiff, UK

Wen G. Jiang
Institute of Cancer and Genetics
Cardiff University
Cardiff, UK

ISSN 1568-2102
ISBN 978-94-007-6027-1 ISBN 978-94-007-6028-8 (eBook)
DOI 10.1007/978-94-007-6028-8
Springer Dordrecht Heidelberg New York London

Library of Congress Control Number: 2013931991

Printed on acid-free paper

Springer is part of Springer Science+Business Media (www.springer.com)

Preface

Overview: Tight junctions, the most apical cellul ... and endothelial cells in the body, are key cellular stru ... ellular permeability in cells and as a result, are key to th ... ace and tissue types in the body. There has been a drama ... of tight junctions in the past decade. The molecular stru ... e cellular functions and the pathophysiological roles of the ... ning clear. Of the most important functions, the role of the cellular stru ... er spreading and drug delivery is increasingly realised. It is now clear that the ... fundamental damage to tight junctions during the process of cancer development. Tight junctions are also critical in the metastatic process of cancer cells. The cellular structure is also critical in drug therapies, namely, the permeability and bioavailability of the drugs, and penetration of barriers such as the blood-brain barrier. There have been few dedicated publications on tight junctions and cancer metastasis. This volume aims to summarise the current knowledge of tight junctions, their role in cancer and cancer metastasis.

Dr. Tracey A. Martin
Professor Wen G. Jiang

Contents

Chapter 1
The Molecular Aspects of Tight Junctions

Elaine A. McSherry, Mark B. Owens, and Ann M. Hopkins

Abstract Tight junctions (TJs) are multi-protein complexes whose principal function is to mediate cell-cell adhesion between epithelial or endothelial cells. While once thought to participate solely as passive effectors of adhesion, it is increasingly being recognised that TJs are dynamic structures which regulate many aspects of cellular function and physiology. Accordingly, dysregulation of TJ-based adhesion or signalling is emerging as an intriguing contributor to several pathophysiologies including cancer. This review will attempt to summarise the current state of knowledge about molecular aspects which regulate, and are regulated by, TJs. The first section will outline selected physiological processes known to influence TJ structure or function, under the headings of cell adhesion/polarity, cell-matrix signalling, ion transport, hormone effects, pro-inflammatory cytokines and hypoxia. The second section will describe selected functional behaviours within the pathophysiology of cancer which TJs have been demonstrated to influence, encompassing cell proliferation and apoptosis, migration and invasion, cell fate and differentiation, metastasis across the blood brain barrier and finally angiogenesis. Collectively, these sections illustrate that a wealth of mechanistic information can be gained from interrogating the contribution of TJs to normal physiology. In turn they highlight how TJ-based disturbances can promote some of the functional behaviours associated with cancer, and thereby offer insight into new TJ-based targets that may offer pharmacological promise in halting tumour progression.

Keywords Tight junction • Epithelium • Barrier function • Cancer • Polarity • Tumour progression • Metastasis • Cell-Cell adhesion • Cell migration

---

E.A. McSherry • M.B. Owens • A.M. Hopkins (✉)
Department of Surgery, RCSI Education and Research Centre,
Beaumont Hospital, Dublin 9, Ireland
e-mail: annhopkins@rcsi.ie

T.A. Martin and W.G. Jiang (eds.), *Tight Junctions in Cancer Metastasis*,
Cancer Metastasis - Biology and Treatment 19, DOI 10.1007/978-94-007-6028-8_1,

1.1 Introduction

Externally, epithelial cells of the skin form a selective physical barrier between an organism and environmental insults including allergens and chemicals. Internally, the epithelial cells lining most visceral organs in conjunction with endothelial cells lining the vasculature also function as barriers to prevent absorption of pathogens and harmful substances from their external surfaces.

Tight junctions (TJs), adhesion complexes which connect the lateral membranes of adjacent epithelial or endothelial cells close to their external surfaces, are responsible for sealing the intracellular space and thereby creating separate apical and basolateral compartments (Schneeberger and Lynch 2004; Tsukita et al. 2001). This intracellular TJ seal performs several crucial functions. Firstly, TJ proteins constitute a molecular barrier, which controls paracellular permeability and transport of ions, solutes and even cells. Secondly, homo- and hetero-typic binding of TJ proteins between neighboring cells aids in the establishment and maintenance of cell polarity by effectively linking polarity complexes with the underlying cytoskeletal structure of individual cells. Finally, TJ proteins can facilitate transmission of signals culminating in processes such as cell differentiation, growth, migration and invasion.

The molecular components of tight junctions can be broadly split into three main groups reviewed in detail in (Brennan et al. 2010): (1) integral transmembrane proteins including occludin, claudins, junctional adhesion molecules (JAMs), crumbs; (2) peripheral or plaque adaptor proteins which generally contain PDZ domains that facilitate protein-protein interactions, such as the Zona Occludens (ZO) family, Par3, Par6, Afadin; (3) associated regulatory/signalling proteins including cingulin and Rho-GTPases.

Deficits in tight junction function which lead to increased paracellular permeability have been linked to several pathologies including blistering skin diseases (Simon et al. 1999) and inflammatory bowel diseases (IBD) such as ulcerative colitis (Schulzke et al. 2009) and Crohn's disease (Hollander 1988) (reviewed in detail in (Marchiando et al. 2010a)). In addition, patients with IBD have a significantly higher risk of developing colitis-associated cancer, suggesting that efficient TJ barrier function may play a crucial role in preventing cancer development. Indeed, a wealth of evidence has recently associated alterations in several TJ proteins with many solid tumours including breast, lung colorectal, and gastric (Martin et al. 2011).

Common putative mechanisms of TJ dysregulation in cancer include aberrant microRNA regulation of gene expression (e.g. JAM-A in breast cancer (Gotte et al. 2010)), aberrant methylation control (e.g. Claudin 6 in breast cancer (Osanai et al. 2007)) and protein mislocalisation (e.g. ZO-1 in pancreatic cancer (Prat et al. 2010)). Furthermore, common dysregulation patterns are evident across multiple cancer types, such as widespread down-regulation of claudin 1 and loss of ZO protein expression and localization (Martin et al. 2011). Intriguingly, genetic classifications of human breast cancer subtypes describe a highly-aggressive subtype, the claudin-low subtype, which is characterized by reduced expression of claudins 3, 4 and 7 and accompanied by increased expression of proteins involved in epithelial to mesenchymal transition (EMT) (Prat et al. 2010). Taken together, this wealth of data suggests that alterations in TJ functions may in fact play a causal role in cancer. The easiest interpretations are that simple downregulation of TJ proteins leading to barrier breakdown and concomitant loss of adhesion would disrupt cell polarity and promote cancer cell dissemination; or that upregulation or mislocalisation of TJ proteins could initiate aberrant signalling cascades that culminate in abnormal differentiation, proliferation, migration, and invasion. However, given the multi-functional roles of TJ proteins in several biological processes, it remains unlikely that a simple paradigm of either TJ protein loss or TJ protein gain will explain the complexities of different cancers.

In this review, we will firstly outline the processes that influence tight junction structure and function at a physiological level; then focus on emerging data describing mechanisms whereby alterations in TJ proteins may facilitate processes critical for tumor formation and progression. Given the huge expansion in the TJ literature over the last number of years, it has not been possible to give an exhaustive overview of all the relevant literature. We therefore chose to focus on some selected aspects, and apologise to those authors whose work we did not have space to include.

1.2 Physiological Processes Which Involve Tight Junction Proteins

The strategic location of TJs at the apical-most interface of the lateral intercellular membrane of polarized epithelia or endothelia facilitates their dynamic regulation by both extracellular and intracellular factors during a variety of physiological processes. Some of these will be summarized in the upcoming paragraphs.

1.2.1 Establishment of Cell Adhesion and Polarity

Tight junction proteins are critically important for the establishment of cellular adhesion and cell polarity through interactions with polarity complex proteins and RhoGTPases. For example, TJ barrier function is strongly regulated through homotypic binding of claudin proteins on adjacent cells, which control permeability through the formation of aqueous pores (Van Itallie and Anderson 2006). However, formation of apico-basal polarity requires the coordinate spatial regulation of the Par, Crumbs, and Scribble core polarity complexes.

Nascent cell adhesions are enriched in the TJ proteins JAM-A and ZO1, and also the adherens junction protein E-cadherin (which activates Rac1 and suppresses

RhoA to facilitate junction formation (Nakagawa et al. 2001)). Upon initiation of cell polarity, Par3 is transported to the apical cortex (Harris and Peifer 2004) where its association with TJ proteins such as JAM-A (Itoh et al. 2001) and PTEN (Feng et al. 2008) facilitates Par3-TIAM1 interactions which mediates stabilization and maturation of junctions (Chen and Macara 2005). Subsequent binding of the small GTPase Cdc42 to Par6 provides the stimulus for recruitment of atypical PKC (aPKC) to the apical surface, which serves to maintain apical domain integrity (Martin-Belmonte et al. 2007) and to recruit the Crumbs complex to the apical cortex via an interaction with PALS1 (Hurd et al. 2003). The basolateral-associated Scribble complex is formed through co-localization of Scribble (Albertson et al. 2004) and Dlg (Dow et al. 2003) proteins at the basolateral cortex; with recent data suggesting that Scribble association with ZO1 may be critical for TJ assembly (Ivanov et al. 2010). This data indicate that the interplay between TJ and polarity proteins is crucial for both maturation of apical junctions and the formation of apico-basal polarity.

1.2.2 Cell-Matrix Signaling

TJ proteins can also influence cell-matrix interactions, transmitting signals to and from the microenvironment to control cell polarity and processes such as migration and invasion. For example, JAM-A-deficient neutrophils show impaired activation of the small GTPase Rap1A (Cera et al. 2009), which is known to promote β1-integrin activation in a Talin-dependent manner (Boettner and Van Aelst 2009). JAM-A knockdown or inhibition using blocking antibodies has also been shown to reduce Rap1-GTPase activity and to decrease cell migration in colonic and breast epithelial cells (McSherry et al. 2011; Severson et al. 2009).

Correct cell-matrix interactions are also critically important for developmental processes involving the generation of multi-dimensional glandular structures such as acini or organoids. For example, activity of the extracellular matrix (ECM)-degrading protein, matrix metalloproteinase MT1-MMP, has been shown to be crucial for normal branching during mammary gland development (Mori et al. 2009). Interestingly, the TJ protein ZO1 has recently been shown to regulate MT1-MMP expression in breast cell lines, suggesting that TJs may participate in modulating cell-matrix interactions during normal morphogenesis (Polette et al. 2005, 2007).

Bidirectionality in the signalling cascades between cell-cell and cell-matrix complexes is also evident, as typified by functional interactions between the cell-matrix protein CD44 and TJs. CD44 binds ECM components such as hyaluronic acid, collagen, fibronectin, laminin, and chondroitin sulfate (Naor et al. 1997). Recent studies have demonstrated that CD44 can regulate TJ assembly and barrier function in keratinocyte epithelial cells (Kirschner et al. 2011). Specifically, CD44 knockout mice exhibited alterations in expression and/or localization of TJ proteins including Claudins-1 and -4, ZO1, and Par3; and a reduction in Rac1 activity culminating in a loss of cell polarity and decreased epidermal barrier function.

1.2.3 Alterations in Ion Transport

A number of epithelial ion channels have emerged as important regulators of TJ function, of which the sodium potassium ATPase (Na^+ K^+ ATPase) is the best studied. This heterodimeric protein exports three sodium ions and imports two of potassium against their concentration gradients in an ATP-dependent reaction to maintain transmembrane ion concentrations (Lingrel and Kuntzweiler 1994; Kaplan 2002; Malik et al. 1996; Sweadner 1989). This serves to maintain transmembrane potential, driving multiple transport mechanisms and controlling cell volume and osmolality.

A functional Na^+ K^+ ATPase plays important roles in the assembly of TJs, establishment of cell polarity and regulation of paracellular permeability. Studies in various epithelial cell types by Rajasekeran et al. (Rajasekaran et al. 2003, 2007, 2008) have shown reversible inhibition of TJ formation upon inhibition of Na^+ K^+ ATPase, either by K^+ depletion or treatment with the inhibitor ouabain, in a sodium-dependent fashion. In addition, expression of Na^+ K^+ ATPase subunits and E-cadherin, in addition to adequate Na^+ K^+ ATPase pump function, have been shown to be necessary for TJ formation and normal epithelial polarization (Rajasekaran et al. 2003, 2007). Accordingly, it has been hypothesised that Na^+ K^+ ATPase and E-cadherin function synergistically in assembling TJs (Rajasekaran et al. 2008).

Interestingly, while treatment with high concentrations of ouabain that inhibit Na^+ K^+ ATPase pump function increase permeability and decrease transepithelial resistance (Rajasekaran et al. 2003; Contreras et al. 1999); treatment with nanomolar ouabain concentrations that do not affect pump function actually *decrease* TJ permeability to both ions and non-ionic molecules (Larre et al. 2010). The latter has been attributed in part to alterations in the expression of claudins -1, -2 and -4 (Larre et al. 2010). It is intriguing to speculate that such profoundly opposing effects on TJ function may in fact be subject to physiological regulation by endogenous hormone-like molecules in addition to exogenous drugs, with reports that an endogenous form of ouabain is synthesized and stored in the mammalian adrenal cortex and hypothalamus (Schoner and Scheiner-Bobis 2007).

Other ion channels which have been implicated in regulatory control of TJ functions include the Na+/glucose co-transporter SGLT-1. Glucose uptake by apically-expressed SGLT-1 in the intestinal brush border has been shown to induce a drop in transepithelial electrical resistance and to increase in the paracellular uptake of small nutrients in vitro (Turner et al. 1997) and in vivo in both rats (Pappenheimer and Reiss 1987) and humans (Turner et al. 2000). This has been associated with TJ strand disruption (Madara and Pappenheimer 1987), dissociation of ZO1 from tight junctions (Atisook and Madara 1991) and phosphorylation of myosin regulatory light chain at the epithelial perijunctional ring (Turner et al. 1997).

Energy-dependent ion channels are not the only ones to have been functionally linked to TJs, with the passive transport chloride transporter ClC-2 also known to localise at TJs in intestinal epithelia. Activation of this channel reportedly stimulates an increase in transepithelial electrical resistance and a concomitant reduction in paracellular permeability (Moeser et al. 2004). Furthermore expression of the chloride channel CFTR has been shown to increase transepithelial resistance (LeSimple et al. 2010); while CLIC-4 co-localises with ZO1 in apical regions of epithelial cells, suggesting a possible but unproven role in regulation of TJs (Berryman and Goldenring 2003). Finally, the transmembrane water channels termed aquaporins are also thought to regulate TJs, with aquaporin-5 in particular being shown to modulate epithelial paracellular permeability (Turner et al. 1997; Murakami et al. 2006).

1.2.4 Hormone Effects

Several hormones from the steroid receptor family and otherwise have been shown to regulate TJs, consistent with the physiological need to actively modulate tissue permeability or other important functions of TJs at different stages of development or hormonal cycles.

Among the most prominent, estrogen has been demonstrated to profoundly affect the TJs of sex hormone-sensitive epithelia ranging from reproductive tissue to the intestinal tract. In cervico-vaginal epithelium, oestrogen can reportedly decrease the resistance of both epithelial TJs and the lateral intercellular space via matrix metalloproteinase 7-induced modulation of occludin, with the net effect of increasing epithelial permeability (Gorodeski 2005, 2007).

Oestrogen receptor-β (ERβ) is expressed in intestinal epithelial cells, where it appears to regulate paracellular permeability in a manner not strictly dependent on the oestrus cycle. In fact both male and female rats that under-express ERβ exhibit greater epithelial permeability and susceptibility to barrier-disruptive injury than their wild type female counterparts (Wada-Hiraike et al. 2006; Looijer-van Langen et al. 2011). Female rats under-expressing ERβ also show ultrastructural evidence of altered TJ and desmosomal morphology (Wada-Hiraike et al. 2006).

Hormonal regulation of mammary epithelial permeability during pregnancy and lactation occurs via not just the complex effects of oestrogen, but rather its interplay with other hormones such as progesterone, glucocorticoids, prolactin and serotonin (5-HT). During pregnancy the mammary gland reaches the expanded alveolar stage of development, however milk synthesis cannot begin until after parturition in conjunction with prolactin and glucocorticoid secretion which dynamically regulate TJ opening to facilitate the delivery of milk proteins during breastfeeding. (Thompson 1996; Zettl et al. 1992; Stelwagen et al. 1999).

The neurotransmitter serotonin (5-HT) also appears to regulate epithelial homeostasis in several organ systems including the mammary gland, where it is locally synthesized (Matsuda et al. 2004). 5-HT regulates the lactation to involution switch, and exhibits biphasic effects on tight junctions in vitro; increasing transepithelial resistance at low concentrations and early time points via protein kinase A while disrupting TJs via p38 MAP kinase signalling following sustained exposure to higher concentrations (Pai and Horseman 2008, 2011).

1.2.5 Pro-inflammatory Cytokines

Prototypic pro-inflammatory cytokines including interleukins-1, -6, -17, -18 (IL-1, -6, -17, -18), tumour necrosis factor-α (TNF-α) and interferon-γ (IFN-γ) are secreted from multiple cellular sources under physiological and pathophysiological circumstances. Among their pleiotrophic effects include profound remodelling of TJs, which often induces endothelial or epithelial barrier disruption and perpetuates inflammation (for a recent review see (Capaldo and Nusrat 2009)). The near-ubiquitous expression of cytokine receptors has fuelled reports of cytokine-induced TJ disruption in most epithelial and endothelial barriers, yet despite functional overlap between different cytokines there is no unifying paradigm of specific alterations which are essential for barrier dysfunction.

IL-1 has been reported to induce a variable combination of reduced transepithelial resistance or increased paracellular permeability in conjunction with occludin and ZO-1 degradation/redistribution in epithelial cells from the intestine (Al-Sadi and Ma 2007), thyroid (Nilsson et al. 1998) and cornea (Kimura et al. 2009) in addition to models of the blood-brain barrier (Bolton et al. 1998), blood-retinal barrier (Bamforth et al. 1997) and blood-testis barrier (Lie et al. 2011). The barrier-disruptive mechanisms associated with IL-1 exposure in both intestinal (Al-Sadi and Ma 2007; Al-Sadi et al. 2008, 2010) and corneal (Kimura et al. 2009) epithelial cells as well as an in vitro model of the blood-brain barrier (Afonso et al. 2007) have been ascribed to canonical NFκB signalling via upstream activators such as MEKK and downstream effectors including myosin light chain kinase.

In addition to phenocopying several noted effects of IL-1 on barrier function and occludin/ZO-1 distribution, TNF-α has been observed to reduce the structural complexity (Schmitz et al. 1999) of claudin-containing TJ strands (Furuse et al. 1998). Also in common with IL-1 signalling mechanisms, TNF-dependent reductions in barrier function have been linked to activation of NFκB in retinal endothelial cells (Aveleira et al. 2010) and corneal epithelial cells (Kimura et al. 2008). Barrier disruption downstream of TNF-α signalling in intestinal epithelial models has alternately been proposed to reflect expressional enhancement of specific micro-RNAs targeting occludin for degradation (Ye et al. 2011) or enhanced removal of occludin from tight junctions via caveolar-mediated endocytosis (Marchiando et al. 2010b).

Observations of IFN-γ-induced epithelial (Madara and Stafford 1989; Youakim and Ahdieh 1999; Adams et al. 1993) or microvasculature endothelial (Oshima et al. 2001) barrier disruption in conjunction with degradation/mislocalization of ZO/occludin proteins might seem to mirror the cellular mechanisms discussed above in response to IL-1 or TNF-α exposure. However several lines of evidence suggest not only overlapping mechanisms but also unique ones whereby IFN-γ disrupts barrier function. One contends that the barrier-disruptive effects of IFN-γ in intestinal epithelia involve PI3-kinase/NFκB cross-talk (Boivin et al. 2009); another that macropinocytotic internalisation of occludin is responsible for induced deficits (Bruewer et al. 2005), while yet another possibility is that IFN-induced protease activation cleaves supporting TJ proteins such as claudin-2 (Willemsen et al. 2005).

Reported alterations in the lipid composition of membrane microdomains following IFN/TNF co-treatment (Li et al. 2008) also offer a novel explanation for putative sub-membranous displacement of occludin and ZO-1 from tight junctions. Accordingly, although synergism between IFN-γ and TNF-α has been reported in many instances (Rodriguez et al. 1995; Wang et al. 2006; Bruewer et al. 2003), it is interesting to note that this can be dissociated from the pro-apoptotic effects of some pro-inflammatory cytokines (Bruewer et al. 2003).

1.2.6 Hypoxia

The adaptive response to reduced oxygen tension, termed hypoxia, also plays an important role in influencing TJ structure and function in physiological and pathophysiological settings. Physiological differences in vascular perfusion between tissues dictate that some body compartments exist in normoxic states (e.g. lung alveoli) while others are relatively hypoxic (e.g. colon). In pathophysiological settings, hypoxia reportedly activates Notch signalling (Chen et al. 2010); which in turn has been implicated in reducing gene expression of TJ components such as occludin and ZO-1 during EMT in airway epithelial cells (Aoyagi-Ikeda et al. 2011). While this implies that Notch activation secondary to hypoxia exerts a negative effect on lung epithelial barrier function, it is interesting to note that *inactivation* of Notch signalling may have a similar net effect in intestinal epithelia (Dahan et al. 2011). Whether this reflects innate differences in sensitivity to hypoxic signalling in tissues with disparate basal oxygen tensions, or merely illustrates the importance of carefully balancing Notch activity levels for barrier function homeostasis in any epithelial tissue is not yet clear.

What is clear is that regulation of epithelial barrier function by hypoxic signalling is complex and multi-factorial. Temporal activation of the transcription factors Slug and Snail during hypoxia (Kurrey et al. 2005) can also trigger junctional disassembly via repression of occludin, ZO-1 and claudin-1 expression (Martinez-Estrada et al. 2006; Ohkubo and Ozawa 2004; Wang et al. 2007). Similarly, reduced expression of occludin and claudin-1 have been demonstrated both in vitro and in vivo in renal epithelial cells deficient in the tumour suppressor gene von Hippel-Lindau (VHL) (Harten et al. 2009). Consequently, VHL inactivation has been associated with loss of barrier function and structural disorder of the renal epithelial phenotype (Calzada et al. 2006). Since a major function of the VHL protein product is to promote proteasomal degradation of hypoxia inducible factors (HIFs) (Maxwell et al. 1999), much interest has focussed on the potential role of HIFs in regulating TJs in various tissues. Recent evidence has suggested that HIF-1α antagonism can temper occludin/ZO-1 redistribution and the associated defects in intestinal epithelial barrier function induced by pro-inflammatory cytokines (Liu et al. 2011). A similar regulatory role has been noted in endothelial tight junctions, with loss of HIF-1α promoting TJ re-sealing in brain microvascular endothelial sheets compromised by prior exposure to either high glucose levels (Yan et al. 2012) or hypoxia-reoxygenation injury (Yeh et al. 2007). Taken together with the links between enhanced HIF activity and tumour progression or metastasis (Liao et al. 2007), the importance of hypoxia as an upstream regulator of tight junctions and barrier function cannot be underestimated.

As described above, TJ proteins are important for the maintenance of cell polarity and for hormonal and cytokine regulation of cellular homeostasis among a myriad of associated functions. Disruptions in cell polarity and tissue architecture are hallmarks of de-differentiation and early features of malignancy (Molitoris and Nelson 1990). In addition, several TJ-associated proteins have recently been shown to be targeted by oncogenes such as ERBB2 (Aranda et al. 2006) and MYC (Zhan et al. 2008) to facilitate malignant transformation. Furthermore, TJ proteins including Scribble (Javier 2008) and ZO2 (Glaunsinger et al. 2001) have been shown to be targeted by oncogenic viruses such as the human papilloma virus. Collectively, these studies provide strong evidence that TJ proteins may indeed act as key regulators of cancer initiation and progression. This will next be addressed.

1.3 Pathophysiological Processes in Cancer Influenced by Tight Junctions

1.3.1 Regulation of Cell Proliferation and Apoptosis

Tumour formation requires the acquisition of alterations that facilitate sustained proliferative capacity, whilst resisting cellular senescence and apoptotic cell death (Hanahan and Weinberg 2011). Several studies have described how alterations in several TJ-associated proteins may upset the delicate balance of growth and death signalling to result in malignant transformation. As noted earlier, however, the complex and tissue-specific regulation of TJ function in various endothelial and epithelial cells makes it unlikely that a single paradigm of simple expressional upregulation or downregulation will emerge to explain the many functional events associated with tumour initiation and progression.

Regardless, members of the largest family of integral membrane TJ proteins, the claudins, are frequently dysregulated in many cancers and appear to have a central role in determining cell fate (Escudero-Esparza et al. 2011). With respect to tumour initiation, Claudin-6 downregulation has been shown to result in increased resistance to apoptosis in vitro (Osanai et al. 2007). Claudin-1 expression, though increased in senescent cells (Swisshelm et al. 1999), has been reported to be decreased throughout several tumour types (Martin et al. 2011). Similarly downregulation of occludin has been correlated with dedifferentiation and progression of several cancers including endometrial (Tobioka et al. 2004a) and lung (Tobioka et al. 2004b). These effects may be due to occludin-mediated regulation of apoptosis, as occludin loss results in decreased expression of pro-apoptotic proteins including Bax (Osanai et al. 2006).

Alterations in proteins of the junctional adhesion molecule (JAM) family have also been shown in breast and renal cancers (McSherry et al. 2011; Gutwein et al. 2009) and melanoma (Langer et al. 2011), where many have demonstrated prognostic value in determining levels of JAM-A expression in patient cohorts. Indeed aberrant expression of microRNA mir145 may be mechanistically responsible for observed overexpression of JAM-A in breast cancer patient tissues which correlate strongly with poor survival outcomes (Gotte et al. 2010; McSherry et al. 2009). Although generally accepted as primarily functioning in adhesive and barrier roles at the TJ, compelling data have recently emerged regarding a role for JAM-A in both apoptosis and proliferation control. Colonic epithelial cells of JAM-A-deficient mice show enhanced crypt proliferation as measured by Ki67 staining (Laukoetter et al. 2007). Specifically, JAM-A appears to control cell proliferation through inhibition of Akt-dependent β-catenin activation (Nava et al. 2011); with Akt inhibition reversing crypt proliferation in JAM-A-deficient mice. Somewhat conversely in the context of cancer, JAM-A deficient mice display significantly reduced tumor growth in a pancreatic tumor model, likely due to decreased angiogenesis and increased immune responses (Murakami et al. 2010). Furthermore, in a breast cancer mouse model, JAM-A deficient mice show significantly decreased tumour growth; with tumour cells displaying increased rates of apoptosis in vivo and in vitro (Murakami et al. 2011). Together these studies suggest that, in contrast to loss of TJ proteins such as claudins and occludin, upregulation of JAM-A may in fact facilitate increased tumor growth and survival by promoting signalling events which protect cells from apoptosis.

As mentioned, JAM-A associates with the peripheral TJ protein Par3 during junctional maturation and establishment of cell polarity (Ebnet et al. 2001). In a mouse model of mammary morphogenesis, Par3 depletion in mammary progenitor cells disrupted mammary development, resulting in ductal hyperplasia. Re-expression of full length Par3 (but not truncated Par3) rescued this defect, demonstrating that Par3/aPKC interaction is essential for normal breast morphogenesis (McCaffrey and Macara 2009). The interaction of another TJ-associated protein and Par polarity complex member, Par 6, with aPKC has also been shown to be required for ErbB2 oncogene-driven evasion of apoptosis and disruption of breast cellular morphogenesis in vitro (Aranda et al. 2006).

Interestingly, association of ZO-1 with the transcription factor ZONAB can directly promote expression of the ErbB2 oncogene (Balda and Matter 2000). Furthermore, ZONAB is a critical determinant of cell cycle progress through effects on cyclin D1 and cdk4 (Balda et al. 2003). Similarly ZO2 can control cell proliferation through sequestration of transcription factors such as Jun and Fos at the TJ in a density-dependent manner (Huerta et al. 2007). Finally, interactions between ZO1 and the polarity complex member Scribble play an important role in normal regulation of cell adhesion (Ivanov et al. 2010). Interestingly, correct localization and expression of Scribble mediates pro-apoptotic signalling critical for both normal mammary gland morphogenesis and resisting MYC–induced transformation (Zhan et al. 2008).

In summary, data suggests that TJ proteins may be critical determinants of cancer initiation through effects on oncogene expression and imbalances in cell proliferative and apoptotic signaling.

1.3.2 Migration and Invasion

Although uncontrolled growth is a fundamental requirement during transformation, cancer cells must acquire both migratory and invasive capabilities in order to successfully disseminate from a primary tumour before seeding metastases at distant sites. Generally, cell migration consists of three main steps: the activation of Rho GTPases extend cell protrusions (through assembly of focal contacts with extracellular matrix proteins), the cell is dragged forward (through myosin II-mediated cell contraction), and finally cell adhesions are disassembled at the rear of the cell. This cyclical process (similar to regulation of apico-basal polarity and establishment of cell adhesion) requires crosstalk between junctional proteins, core polarity regulators (Etienne-Manneville and Hall 2001; Huo et al. 2011), and Rho family GTPases (Etienne-Manneville 2008; Iden and Collard 2008). Malignant cells can hijack these pro-migratory pathways and several TJ associated proteins have been implicated as having a causal role in cancer progression.

Loss of claudin-7 expression has been correlated with increased migration and invasion in lung (Lu et al. 2011), colorectal (Oshima et al. 2008) and oesophageal cancer (Lioni et al. 2007). Specifically, claudin-7 loss or mis-localisation in oesophageal cancer can lead to decreased E-cadherin expression and increased three-dimensional invasion in vitro (Lioni et al. 2007). Furthermore, re-expression of claudin-7 in claudin-7 deficient lung cancer cells resulted in decreased hepatocyte growth factor-mediated in vitro migration and invasion, and decreased in vivo tumour growth via regulation of ERK/MAPK signalling (Lu et al. 2011). Several other claudins have been implicated in regulating invasion through effects on matrix degrading enzymes from the matrix metalloproteinase (MMP) family. Claudin-1 expression in liver cancer cells promotes increased MMP2 activity and migration and invasion through activation of a c-Abl-PKCdelta signaling pathway (Yoon et al. 2010). Conversely, claudin-6 loss has been demonstrated to increase MMP activity and promote invasion of breast cancer cells (Osanai et al. 2007).

JAMs also have established roles in promoting normal leukocyte (Ostermann et al. 2002) and neutrophil migration (Cera et al. 2009), with JAM-A loss in endothelial cells functioning to decrease motility (Bazzoni et al. 2005). With respect to cancer, the majority of studies suggest that JAM proteins signal to increase cancer cell migration and invasion. JAM-A overexpression is associated with increased breast cancer metastasis (McSherry et al. 2009); potentially due to downstream regulation of the migratory protein β1-integrin through AF-6 and Rap1 GTPase adaptor proteins (McSherry et al. 2011; Severson et al. 2009). Furthermore, JAM-C is required for melanoma cell transendothelial migration in vitro (Ghislin et al. 2011); with its increased expression linked to melanoma invasion and metastasis in vivo (Fuse et al. 2007).

JAM proteins interact with several TJ adaptor proteins including AF6 and ZO proteins (Schneeberger and Lynch 2004). Fusion of AF6 and MLL represents the most common alteration in mixed lineage leukemia (MLL), where the Ras association-1 domain of AF6 likely activates the oncogenic potential of the MLL-AF6 protein (Liedtke et al. 2010). Recently, loss of AF6 in breast cancer has also been linked with poor prognosis (Letessier et al. 2007). Further work has demonstrated that AF6 loss dramatically increased heregulin-induced in vitro migration and invasion through activation of RAS/MAPK and Src kinase pathways; as well as significantly increased tumour growth and metastasis in an SKBR3 orthotopic mouse model (Fournier et al. 2011).

Interestingly, ZO1 has been shown to regulate the expression of the matrix metalloproteinase MT1-MMP, with knockdown of ZO1 in breast cancer cell lines reducing MT1-MMP expression and three-dimensional in vitro invasion (Polette et al. 2005). Recently, the TGF-β/Smad pathway (known to target the Par polarity complex (Viloria-Petit et al. 2009)) was demonstrated to induce breast cancer cell invasion through up-regulation of MMPs -2 and -9, reinforcing a potential link between matrix degradation and TJ-associated proteins (Wiercinska et al. 2011).

Furthermore, interactions between the Par complex members Par6 and aPKC lead to Rac GTPase activation in non-small cell lung cancer (NSCLC) cells, which drives anchorage-independent growth and invasion through activation of MMP10 (Frederick et al. 2008). The evidence for an involvement of Par3 in cancer cell migration has also been strengthened by studies demonstrating that Par-3 engages in the spatial regulation of Rac activity. Par3 directly interacts with Tiam1, a Rac1-specific guanine nucleotide exchange factor, to form a complex with aPKC-PAR-6-Cdc42, leading to Rac1 activation (Chen and Macara 2005). Recently, Par3 has been suggested to also be important in regulating squamous cell carcinoma collective cell migration. Recruitment of Par3 by DDR1 reduced actinomyosin contractile activity at cell-cell contacts and antagonized ROCK activity to Rac activation, thus keeping migrating cells clustered together and promoting more efficient collective migration (Hidalgo-Carcedo 2011).

Finally loss or mislocalisation of the ZO1 interacting protein, Scribble, increases migration and invasion of breast cancer cell lines (Zhan et al. 2008; Vaira et al. 2012), and cooperation of Scribble with the Ras oncogene increases MEK-ERK-dependent matrix invasion in a 3D breast acinar morphogenesis model (Dow et al. 2008).

Together, the above studies underline the importance of TJ proteins in mediating pro-migratory and pro-invasive signals and also suggest that targeting these proteins in cancer may be of therapeutic value.

1.3.3 Cell Fate and Differentiation

Recent work has provided evidence that several TJ proteins may regulate cell fate and differentiation during normal development (Balda and Matter 2009; Koch and Nusrat 2009). Expression levels of Claudin-4, ZO1 and ZO2 regulate murine stem cell commitment to hematopoetic or endothelial cell lineages (Stankovich et al. 2011).

In addition, JAM proteins have been shown to be required for maintenance of hematopoietic stem cells in bone marrow (Arcangeli et al. 2011), spermatid differentiation (Gliki et al. 2004) and dendritic cell differentiation (Ogasawara et al. 2009). Furthermore, as mentioned above, TJ proteins interact with polarity complexes such as the Scribble and Par complexes to influence cell fate through processes including EMT, which allows cancer cells to alter their cell morphology and acquire pro-invasive phenotypes that might facilitate their migration to optimally-supportive growth niches (Viloria-Petit et al. 2009; Dow et al. 2008; Ozdamar et al. 2005).

The claudin-low aggressive breast cancer subtype is characterized by near absence of luminal differentiation markers, and increased expression of EMT and cancer stem cell-like markers (Prat et al. 2010). Indeed, gene expression signatures derived from normal human breast cells undergoing EMT in response to snail/slug activation or TGF**β** treatment were recently shown to closely resemble those derived from claudin-low breast cancer tissues (Taube et al. 2010). Poor prognosis claudin-low tumour cells could undergo EMT through changes in several Zeb1 transcription factor–regulated genes. Zeb1 expression, through its repression of junctional proteins, may therefore also have a causal role in cancer types including breast (Aigner et al. 2007) and colorectal (Spaderna et al. 2008). Downregulation of Mir200c in breast cancer cells prevents expression of Zeb1, and reduces cancer cell migration (Cochrane et al. 2010). Furthermore, knockdown of Zeb1 in MDA-MB-231 breast cancer cells promotes EMT reversion whereby induced re-expression of the TJ proteins JAM-A, Occludin, Crumbs and PATJ partially re-establishes cell polarity and epithelial morphology, and significantly decreases cancer cell migration. Encouragingly, Zeb1 knockdown in a mouse model of metastatic colorectal cancer resulted in complete suppression of liver metastasis (Spaderna et al. 2008), suggesting that targeting Zeb1 may be a valuable therapeutic modality.

The TJ peripheral protein Par6 has been demonstrated to be required for TGFβ-induced EMT in breast epithelial cells (Viloria-Petit et al. 2009). Specifically, TGFβ-dependent phosphorylation of Par6 mediated recruitment of Smurf I (an E3 ubiquitin ligase) to promote degradation of RhoA and dissolution of the TJs, a crucial step in EMT (Ozdamar et al. 2005). In addition, TGFβ-Par6 signalling led to a loss of cell polarity and induced local invasion of MMECs in vitro and in vivo (Viloria-Petit et al. 2009).

ZO1 and its associated transcription factor ZONAB have also been implicated in the regulation of epithelial homeostasis and differentiation (Balda et al. 2003; Georgiadis et al. 2010; Sourisseau et al. 2006), through downstream regulation of cell cycle genes such as cyclin D1 and PCNA. Overexpression of ZONAB or knockdown of ZO-1 in mouse epithelial cells resulted in increased proliferation, and induced EMT-like morphological and protein expression changes that disrupted normal epithelial differentiation. This suggests that ZO1 loss, as seen in several cancers (Hoover et al. 1998; Kaihara et al. 2003), may phenocopy ZONAB overexpression in vitro thus altering cell differentiation through the induction of EMT.

Finally, several recent studies have suggested novel roles for TJ-associated proteins in controlling cellular homeostasis through regulation of spindle orientation and cell division. As mentioned above, transplantation of Par3-depleted stem cells into murine mammary fat pads resulted in disrupted ductal morphogenesis

for mechanical movement of cells across biological barriers. One biological process which exerts a key influence on the ability of metastasized tumours (and indeed primary tumours) to survive is the generation of a vascular supply to nourish the growing tumour, termed tumour angiogenesis. Several TJ proteins have been implicated in both physiological and pathological angiogenesis. Junction adhesional molecules, and in particular JAM-A, are known to be important regulators of angiogenesis. JAM-A is expressed in the early vasculature of the developing mouse embryo (Parris et al. 2005), and is a vital component of basic fibroblast growth factor (bFGF)- induced angiogenesis. In the latter context it forms an inhibitory complex with $\alpha v\beta 3$ integrin, which disassembles in response to bFGF signalling. JAM-A then facilitates MAP kinase activation, which in turn induces endothelial tube formation and angiogenesis (Naik et al. 2003; Naik and Naik 2006). Accordingly, transient knockdown of JAM-A has been shown to prevent bFGF-induced endothelial cell migration in an ECM substrate-specific fashion (Naik et al. 2003). Similarly, bFGF cannot induce angiogenesis in JAM-A deficient mice (Cooke et al. 2006), and pancreatic islet cell carcinomas grown in JAM-A null mice have been shown to exhibit a small decrease in angiogenesis compared to JAM-A-expressing mice (Murakami et al. 2010).

Other JAM proteins also appear to regulate angiogenesis in ways that could be relevant to tumour angiogenesis, or the pharmacological antagonism thereof. Soluble JAM-C levels have been shown to be increased in the serum of patients with rheumatoid arthritis, and treatment with exogenous JAM-C has the potential to induce angiogenesis in vitro (Rabquer et al. 2010). Furthermore JAM-C blockade has been shown to reduce angiogenesis by 50% in a mouse model of hypoxia-induced retinopathy (Orlova et al. 2006). Others have reported that functional antagonism of JAM-C with a monoclonal antibody can inhibit angiogenesis both in vitro and in vivo (Rabquer et al. 2010; Lamagna et al. 2005). Conversely, overexpression of JAM-B in a mouse model of Downs syndrome has been shown to inhibit the angiogenic response to vascular endothelial growth factor (VEGF) (Reynolds et al. 2010).

Other TJ proteins such as the claudins also play a complex role in angiogenesis. Claudin-5 has been shown to reduce endothelial cell motility via N-WASP and ROCK signalling cascades (Escudero-Esparza et al. 2011). Claudin-4-expressing ovarian epithelial cells reportedly feature upregulation of several genes encoding pro-angiogenic cytokines, and can induce angiogenesis both in vitro and in in vivo mouse models (Li et al. 2009). Claudins -1, -2 and -5 are expressed in normal murine retinal vessel development; while claudins -2 and -5 are overexpressed in vessels in an oxygen-induced retinopathy model (Luo et al. 2011).

Similarly expression of occludin can be altered by a number of angiogenic factors. Increased occludin expression has been linked with the secretion of angiopoetin-1 from pericytes (Hori et al. 2004), while decreased occludin expression in conjunction with increased paracellular permeability has been noted in retinal endothelial cells treated with vascular endothelial growth factor (VEGF) (Antonetti et al. 1998; Behzadian et al. 2003).

Taken together, the above points illustrate a complex and dynamic relationship between TJ proteins and angiogenic cascades. We believe this shows much potential for interrogation to better understand not only the mechanisms of tumour angiogenesis, but also to drive forward the design of new TJ-based therapeutics aimed at interfering with this process.

1.4 Conclusion

This review has attempted to summarise the molecular aspects of TJs regarding their regulation by normal physiological processes and their contributions to pathophysiological behaviours characteristic of cancers. What has emerged is that TJs are intrinsic downstream components of a number of important cascades regulating physiological processes as diverse as polarity, ion transport and responsiveness to paracrine or endocrine factors. Perhaps more importantly, it has also illustrated that while TJs may act as upstream regulators of functional behaviours intrinsically associated with cancer, there is no universal paradigm whereby simple loss or gain of TJ proteins drives processes like cell proliferation, migration or angiogenesis. Instead this review suggests that complex spatial and temporal regulation of TJ signalling must be elucidated on an individual protein basis, but may bear fruit in the design of future drugs to target tumourigenic behaviour.

References

Abbott NJ, Patabendige AA, Dolman DE, Yusof SR, Begley DJ (2010) Structure and function of the blood-brain barrier. Neurobiol Dis 37:13–25

Adams RB, Planchon SM, Roche JK (1993) IFN-gamma modulation of epithelial barrier function. Time course, reversibility, and site of cytokine binding. J Immunol 150:2356–2363

Afonso PV, Ozden S, Prevost MC, Schmitt C, Seilhean D, Weksler B, Couraud PO, Gessain A, Romero IA, Ceccaldi PE (2007) Human blood-brain barrier disruption by retroviral-infected lymphocytes: role of myosin light chain kinase in endothelial tight-junction disorganization. J Immunol 179:2576–2583

Aigner K, Dampier B, Descovich L, Mikula M, Sultan A, Schreiber M, Mikulits W, Brabletz T, Strand D, Obrist P et al (2007) The transcription factor ZEB1 (deltaEF1) promotes tumour cell dedifferentiation by repressing master regulators of epithelial polarity. Oncogene 26:6979–6988

Albertson R, Chabu C, Sheehan A, Doe CQ (2004) Scribble protein domain mapping reveals a multistep localization mechanism and domains necessary for establishing cortical polarity. J Cell Sci 117:6061–6070

Al-Sadi RM, Ma TY (2007) IL-1beta causes an increase in intestinal epithelial tight junction permeability. J Immunol 178:4641–4649

Al-Sadi R, Ye D, Dokladny K, Ma TY (2008) Mechanism of IL-1beta-induced increase in intestinal epithelial tight junction permeability. J Immunol 180:5653–5661

Al-Sadi R, Ye D, Said HM, Ma TY (2010) IL-1beta-induced increase in intestinal epithelial tight junction permeability is mediated by MEKK-1 activation of canonical NF-kappaB pathway. Am J Pathol 177:2310–2322

Antonetti DA, Barber AJ, Khin S, Lieth E, Tarbell JM, Gardner TW (1998) Vascular permeability in experimental diabetes is associated with reduced endothelial occludin content: vascular endothelial growth factor decreases occludin in retinal endothelial cells. Penn State Retina Research Group. Diabetes 47:1953–1959

Aoyagi-Ikeda K, Maeno T, Matsui H, Ueno M, Hara K, Aoki Y, Aoki F, Shimizu T, Doi H, Kawai-Kowase K et al (2011) Notch induces myofibroblast differentiation of alveolar epithelial cells via transforming growth factor-{beta}-Smad3 pathway. Am J Respir Cell Mol Biol 45:136–144

Aranda V, Haire T, Nolan ME, Calarco JP, Rosenberg AZ, Fawcett JP, Pawson T, Muthuswamy SK (2006) Par6-aPKC uncouples ErbB2 induced disruption of polarized epithelial organization from proliferation control. Nat Cell Biol 8:1235–1245

Arcangeli ML, Frontera V, Bardin F, Obrados E, Adams S, Chabannon C, Schiff C, Mancini SJ, Adams RH, Aurrand-Lions M (2011) JAM-B regulates maintenance of hematopoietic stem cells in the bone marrow. Blood 118:4609–4619

Arshad F, Wang L, Sy C, Avraham S, Avraham HK (2010) Blood-brain barrier integrity and breast cancer metastasis to the brain. Pathol Res Int 2011:920509

Artym VV, Kindzelskii AL, Chen WT, Petty HR (2002) Molecular proximity of seprase and the urokinase-type plasminogen activator receptor on malignant melanoma cell membranes: dependence on beta1 integrins and the cytoskeleton. Carcinogenesis 23:1593–1601

Atisook K, Madara JL (1991) An oligopeptide permeates intestinal tight junctions at glucose-elicited dilatations. Implications for oligopeptide absorption. Gastroenterology 100:719–724

Aveleira CA, Lin CM, Abcouwer SF, Ambrosio AF, Antonetti DA (2010) TNF-alpha signals through PKCzeta/NF-kappaB to alter the tight junction complex and increase retinal endothelial cell permeability. Diabetes 59:2872–2882

Balda MS, Matter K (2000) The tight junction protein ZO-1 and an interacting transcription factor regulate ErbB-2 expression. EMBO J 19:2024–2033

Balda MS, Matter K (2009) Tight junctions and the regulation of gene expression. Biochim Biophys Acta 1788:761–767

Balda MS, Garrett MD, Matter K (2003) The ZO-1-associated Y-box factor ZONAB regulates epithelial cell proliferation and cell density. J Cell Biol 160:423–432

Bamforth SD, Lightman SL, Greenwood J (1997) Ultrastructural analysis of interleukin-1 beta-induced leukocyte recruitment to the rat retina. Invest Ophthalmol Vis Sci 38:25–35

Bazzoni G, Tonetti P, Manzi L, Cera MR, Balconi G, Dejana E (2005) Expression of junctional adhesion molecule-A prevents spontaneous and random motility. J Cell Sci 118:623–632

Behzadian MA, Windsor LJ, Ghaly N, Liou G, Tsai NT, Caldwell RB (2003) VEGF-induced paracellular permeability in cultured endothelial cells involves urokinase and its receptor. FASEB J 17:752–754

Berryman MA, Goldenring JR (2003) CLIC4 is enriched at cell-cell junctions and colocalizes with AKAP350 at the centrosome and midbody of cultured mammalian cells. Cell Motil Cytoskeleton 56:159–172

Boettner B, Van Aelst L (2009) Control of cell adhesion dynamics by Rap1 signaling. Curr Opin Cell Biol 21:684–693

Boivin MA, Roy PK, Bradley A, Kennedy JC, Rihani T, Ma TY (2009) Mechanism of interferon-gamma-induced increase in T84 intestinal epithelial tight junction. J Interferon Cytokine Res 29:45–54

Bolton SJ, Anthony DC, Perry VH (1998) Loss of the tight junction proteins occludin and zonula occludens-1 from cerebral vascular endothelium during neutrophil-induced blood-brain barrier breakdown in vivo. Neuroscience 86:1245–1257

Brandt B, Heyder C, Gloria-Maercker E, Hatzmann W, Rötger A, Kemming D, Zänker KS, Entschladen F, Dittmar T (2005) 3D-extravasation model – selection of highly motile and metastatic cancer cells. Semin Cancer Biol 15:387–395

Brennan K, Offiah G, McSherry EA, Hopkins AM (2010) Tight junctions: a barrier to the initiation and progression of breast cancer? J Biomed Biotechnol 2010:460607

Bruewer M, Luegering A, Kucharzik T, Parkos CA, Madara JL, Hopkins AM, Nusrat A (2003) Proinflammatory cytokines disrupt epithelial barrier function by apoptosis-independent mechanisms. J Immunol 171:6164–6172
Bruewer M, Utech M, Ivanov AI, Hopkins AM, Parkos CA, Nusrat A (2005) Interferon-gamma induces internalization of epithelial tight junction proteins via a macropinocytosis-like process. FASEB J 19:923–933
Calzada MJ, Esteban MA, Feijoo-Cuaresma M, Castellanos MC, Naranjo-Suarez S, Temes E, Mendez F, Yanez-Mo M, Ohh M, Landazuri MO (2006) von Hippel-Lindau tumor suppressor protein regulates the assembly of intercellular junctions in renal cancer cells through hypoxia-inducible factor-independent mechanisms. Cancer Res 66:1553–1560
Capaldo CT, Nusrat A (2009) Cytokine regulation of tight junctions. Biochim Biophys Acta 1788:864–871
Cera MR, Fabbri M, Molendini C, Corada M, Orsenigo F, Rehberg M, Reichel CA, Krombach F, Pardi R, Dejana E (2009) JAM-A promotes neutrophil chemotaxis by controlling integrin internalization and recycling. J Cell Sci 122:268–277
Chen X, Macara IG (2005) Par-3 controls tight junction assembly through the Rac exchange factor Tiam1. Nat Cell Biol 7:262–269
Chen J, Imanaka N, Chen J, Griffin JD (2010) Hypoxia potentiates Notch signaling in breast cancer leading to decreased E-cadherin expression and increased cell migration and invasion. Br J Cancer 102:351–360
Cho SY, Choi HY (1980) Causes of death and metastatic patterns in patients with mammary cancer. Ten-year autopsy study. Am J Clin Pathol 73:232–234
Cochrane DR, Howe EN, Spoelstra NS, Richer JK (2010) Loss of miR-200c: a marker of aggressiveness and chemoresistance in female reproductive cancers. J Oncol 2010:821717
Contreras RG, Shoshani L, Flores-Maldonado C, Lázaro A, Cereijido M (1999) Relationship between Na(+), K(+)-ATPase and cell attachment. J Cell Sci 112:4223–4232
Cooke VG, Naik MU, Naik UP (2006) Fibroblast growth factor-2 failed to induce angiogenesis in junctional adhesion molecule-A-deficient mice. Arterioscler Thromb Vasc Biol 26:2005–2011
Dahan S, Rabinowitz KM, Martin AP, Berin MC, Unkeless JC, Mayer L (2011) Notch-1 signaling regulates intestinal epithelial barrier function, through interaction with CD4+ T cells, in mice and humans. Gastroenterology 140:550–559
Denkins Y, Reiland J, Roy M, Sinnappah-Kang ND, Galjour J, Murry BP, Blust J, Aucoin R, Marchetti D (2004) Brain metastases in melanoma: roles of neurotrophins. Neuro Oncol 6:154–165
Dow LE, Brumby AM, Muratore R, Coombe ML, Sedelies KA, Trapani JA, Russell SM, Richardson HE, Humbert PO (2003) hScrib is a functional homologue of the Drosophila tumour suppressor Scribble. Oncogene 22:9225–9230
Dow LE, Elsum IA, King CL, Kinross KM, Richardson HE, Humbert PO (2008) Loss of human Scribble cooperates with H-Ras to promote cell invasion through deregulation of MAPK signalling. Oncogene 27:5988–6001
Durgan J, Kaji N, Jin D, Hall A (2011) Par6B and atypical PKC regulate mitotic spindle orientation during epithelial morphogenesis. J Biol Chem 286:12461–12474
Ebnet K, Suzuki A, Horikoshi Y, Hirose T, Meyer Zu Brickwedde MK, Ohno S, Vestweber D (2001) The cell polarity protein ASIP/PAR-3 directly associates with junctional adhesion molecule (JAM). EMBO J 20:3738–3748
Escudero-Esparza A, Jiang WG, Martin TA (2011) Claudin-5 participates in the regulation of endothelial cell motility. Mol Cell Biochem 362(1–2):71–85
Etienne-Manneville S (2008) Polarity proteins in migration and invasion. Oncogene 27:6970–6980
Etienne-Manneville S, Hall A (2001) Integrin-mediated activation of Cdc42 controls cell polarity in migrating astrocytes through PKCzeta. Cell 106:489–498
Fazakas C, Wilhelm I, Nagyoszi P, Farkas AE, Haskó J, Molnár J, Bauer H, Bauer HC, Ayaydin F, Dung NT, Siklós L, Krizbai IA (2011) Transmigration of melanoma cells through the

blood-brain barrier: role of endothelial tight junctions and melanoma-released serine proteases. PLoS One 6:e20758
Feng W, Wu H, Chan LN, Zhang M (2008) Par-3-mediated junctional localization of the lipid phosphatase PTEN is required for cell polarity establishment. J Biol Chem 283:23440–23449
Fournier G, Cabaud O, Josselin E, Chaix A, Adelaide J, Isnardon D, Restouin A, Castellano R, Dubreuil P, Chaffanet M et al (2011) Loss of AF6/afadin, a marker of poor outcome in breast cancer, induces cell migration, invasiveness and tumor growth. Oncogene 30:3862–3874
Frederick LA, Matthews JA, Jamieson L, Justilien V, Thompson EA, Radisky DC, Fields AP (2008) Matrix metalloproteinase-10 is a critical effector of protein kinase Ciota-Par6alpha-mediated lung cancer. Oncogene 27:4841–4853
Furuse M, Sasaki H, Fujimoto K, Tsukita S (1998) A single gene product, claudin-1 or -2, reconstitutes tight junction strands and recruits occludin in fibroblasts. J Cell Biol 143:391–401
Fuse C, Ishida Y, Hikita T, Asai T, Oku N (2007) Junctional adhesion molecule-C promotes metastatic potential of HT1080 human fibrosarcoma. J Biol Chem 282:8276–8283
Georgiadis A, Tschernutter M, Bainbridge JW, Balaggan KS, Mowat F, West EL, Munro PM, Thrasher AJ, Matter K, Balda MS et al (2010) The tight junction associated signalling proteins ZO-1 and ZONAB regulate retinal pigment epithelium homeostasis in mice. PloS one 5:e15730
Ghislin S, Obino D, Middendorp S, Boggetto N, Alcaide-Loridan C, Deshayes F (2011) Junctional adhesion molecules are required for melanoma cell lines transendothelial migration in vitro. Pigment Cell Melanoma Res 24:504–511
Glaunsinger BA, Weiss RS, Lee SS, Javier R (2001) Link of the unique oncogenic properties of adenovirus type 9 E4-ORF1 to a select interaction with the candidate tumor suppressor protein ZO-2. EMBO J 20:5578–5586
Gliki G, Ebnet K, Aurrand-Lions M, Imhof BA, Adams RH (2004) Spermatid differentiation requires the assembly of a cell polarity complex downstream of junctional adhesion molecule-C. Nature 431:320–324
Gorodeski GI (2005) Aging and estrogen effects on transcervical-transvaginal epithelial permeability. J Clin Endocrinol Metab 90:345–351
Gorodeski GI (2007) Estrogen decrease in tight junctional resistance involves matrix-metalloproteinase-7-mediated remodeling of occludin. Endocrinology 148:218–231
Gotte M, Mohr C, Koo CY, Stock C, Vaske AK, Viola M, Ibrahim SA, Peddibhotla S, Teng YH, Low JY et al (2010) miR-145-dependent targeting of junctional adhesion molecule A and modulation of fascin expression are associated with reduced breast cancer cell motility and invasiveness. Oncogene 29:6569–6580
Gutwein P, Schramme A, Voss B, Abdel-Bakky MS, Doberstein K, Ludwig A, Altevogt P, Hansmann ML, Moch H, Kristiansen G et al (2009) Downregulation of junctional adhesion molecule-A is involved in the progression of clear cell renal cell carcinoma. Biochem Biophys Res Commun 380:387–391
Hanahan D, Weinberg RA (2011) Hallmarks of cancer: the next generation. Cell 144:646–674
Hao Y, Du Q, Chen X, Zheng Z, Balsbaugh JL, Maitra S, Shabanowitz J, Hunt DF, Macara IG (2010) Par3 controls epithelial spindle orientation by aPKC-mediated phosphorylation of apical pins. Curr Biol 20:1809–1818
Harris TJ, Peifer M (2004) Adherens junction-dependent and -independent steps in the establishment of epithelial cell polarity in Drosophila. J Cell Biol 167:135–147
Harten SK, Shukla D, Barod R, Hergovich A, Balda MS, Matter K, Esteban MA, Maxwell PH (2009) Regulation of renal epithelial tight junctions by the von Hippel-Lindau tumor suppressor gene involves occludin and claudin 1 and is independent of E-cadherin. Mol Biol Cell 20:1089–1101
Heyder C, Gloria-Maercker E, Entschladen F, Hatzmann W, Niggemann B, Zänker KS, Dittmar T (2002) Realtime visualization of tumor cell/endothelial cell interactions during transmigration across the endothelial barrier. J Cancer Res Clin Oncol 128:533–538
Hicks DG, Short SM, Prescott NL, Tarr SM, Coleman KA, Yoder BJ, Crowe JP, Choueiri TK, Dawson AE, Budd GT, Tubbs RR, Casey G, Weil RJ (2006) Breast cancers with brain metas-

tases are more likely to be estrogen receptor negative, express the basal cytokeratin CK5/6, and overexpress HER2 or EGFR. Am J Surg Pathol 30:1097–1104
Hidalgo-Carcedo C, Hooper S, Chaudhry SI, Williamson P, Harrington K, Leitinger B, Sahai E (2011) E Collective cell migration requires suppression of actomyosin at cell-cell contacts mediated by DDR1 and the cell polarity regulators Par3 and Par6. Nat Cell Biol 13:49–58
Hofmann UB, Westphal JR, Zendman AJ, Becker JC, Ruiter DJ, van Muijen GN (2000) Expression and activation of matrix metalloproteinase-2 (MMP-2) and its co-localization with membrane-type 1 matrix metalloproteinase (MT1-MMP) correlate with melanoma progression. J Pathol 191:245–256
Hollander D (1988) Crohn's disease–a permeability disorder of the tight junction? Gut 29: 1621–1624
Hoover KB, Liao SY, Bryant PJ (1998) Loss of the tight junction MAGUK ZO-1 in breast cancer: relationship to glandular differentiation and loss of heterozygosity. Am J Pathol 153:1767–1773
Hori S, Ohtsuki S, Hosoya K, Nakashima E, Terasaki T (2004) A pericyte-derived angiopoietin-1 multimeric complex induces occludin gene expression in brain capillary endothelial cells through Tie-2 activation in vitro. J Neurochem 89:503–513
Huerta M, Munoz R, Tapia R, Soto-Reyes E, Ramirez L, Recillas-Targa F, Gonzalez-Mariscal L, Lopez-Bayghen E (2007) Cyclin D1 is transcriptionally down-regulated by ZO-2 via an E box and the transcription factor c-Myc. Mol Biol Cell 18:4826–4836
Huo L, Wen W, Wang R, Kam C, Xia J, Feng W, Zhang M (2011) Cdc42-dependent formation of the ZO-1/MRCKbeta complex at the leading edge controls cell migration. EMBO J 30:665–678
Hurd TW, Gao L, Roh MH, Macara IG, Margolis B (2003) Direct interaction of two polarity complexes implicated in epithelial tight junction assembly. Nat Cell Biol 5:137–142
Iden S, Collard JG (2008) Crosstalk between small GTPases and polarity proteins in cell polarization. Nat Rev Mol Cell Biol 9:846–859
Itoh M, Sasaki H, Furuse M, Ozaki H, Kita T, Tsukita S (2001) Junctional adhesion molecule (JAM) binds to PAR-3: a possible mechanism for the recruitment of PAR-3 to tight junctions. J Cell Biol 154:491–497
Ivanov AI, Young C, Den Beste K, Capaldo CT, Humbert PO, Brennwald P, Parkos CA, Nusrat A (2010) Tumor suppressor scribble regulates assembly of tight junctions in the intestinal epithelium. Am J Pathol 176:134–145
Javier RT (2008) Cell polarity proteins: common targets for tumorigenic human viruses. Oncogene 27:7031–7046
Kaihara T, Kusaka T, Nishi M, Kawamata H, Imura J, Kitajima K, Itoh-Minami R, Aoyama N, Kasuga M, Oda Y et al (2003) Dedifferentiation and decreased expression of adhesion molecules, E-cadherin and ZO-1, in colorectal cancer are closely related to liver metastasis. J Exp Clin Cancer Res 22:117–123
Kaplan JH (2002) Biochemistry of Na, K-ATPase. Annu Rev Biochem 71:511–535
Kebers F, Lewalle JM, Desreux J, Munaut C, Devy L, Foidart JM, Noël A (1998) Induction of endothelial cell apoptosis by solid tumor cells. Exp Cell Res 240:197–205
Kimura K, Teranishi S, Fukuda K, Kawamoto K, Nishida T (2008) Delayed disruption of barrier function in cultured human corneal epithelial cells induced by tumor necrosis factor-alpha in a manner dependent on NF-kappaB. Invest Ophthalmol Vis Sci 49:565–571
Kimura K, Teranishi S, Nishida T (2009) Interleukin-1beta-induced disruption of barrier function in cultured human corneal epithelial cells. Invest Ophthalmol Vis Sci 50:597–603
Kirschner N, Haftek M, Niessen CM, Behne MJ, Furuse M, Moll I, Brandner JM (2011) CD44 regulates tight-junction assembly and barrier function. J Invest Dermatol 131:932–943
Koch S, Nusrat A (2009) Dynamic regulation of epithelial cell fate and barrier function by intercellular junctions. Ann N Y Acad Sci 1165:220–227
Kurrey NK, Amit K, Bapat SA (2005) Snail and Slug are major determinants of ovarian cancer invasiveness at the transcription level. Gynecol Oncol 97:155–165
Lamagna C, Hodivala-Dilke KM, Imhof BA, Aurrand-Lions M (2005) Antibody against junctional adhesion molecule-C inhibits angiogenesis and tumor growth. Cancer Res 65:5703–5710

Langer HF, Orlova VV, Xie C, Kaul S, Schneider D, Lonsdorf AS, Fahrleitner M, Choi EY, Dutoit V, Pellegrini M et al (2011) A novel function of junctional adhesion molecule-C in mediating melanoma cell metastasis. Cancer Res 71:4096–4105

Larre I, Lazaro A, Contreras RG, Balda MS, Matter K, Flores-Maldonado C, Ponce A, Flores-Benitez D, Rincon-Heredia R, Padilla-Benavides T, Castillo A, Shoshani L, Cereijido M (2010) Ouabain modulates epithelial cell tight junction. Proc Natl Acad Sci U S A 107:11387–11392

Laukoetter MG, Nava P, Lee WY, Severson EA, Capaldo CT, Babbin BA, Williams IR, Koval M, Peatman E, Campbell JA et al (2007) JAM-A regulates permeability and inflammation in the intestine in vivo. J Exp Med 204:3067–3076

Lee BC, Lee TH, Avraham S, Avraham HK (2004) Involvement of the chemokine receptor CXCR4 and its ligand stromal cell-derived factor 1alpha in breast cancer cell migration through human brain microvascular endothelial cells. Mol Cancer Res 2:327–338

LeSimple P, Liao J, Robert R, Gruenert DC, Hanrahan JW (2010) Cystic fibrosis transmembrane conductance regulator trafficking modulates the barrier function of airway epithelial cell monolayers. J Physiol 588:1195–1209

Letessier A, Garrido-Urbani S, Ginestier C, Fournier G, Esterni B, Monville F, Adelaide J, Geneix J, Xerri L, Dubreuil P et al (2007) Correlated break at PARK2/FRA6E and loss of AF-6/Afadin protein expression are associated with poor outcome in breast cancer. Oncogene 26:298–307

Li YM, Pan Y, Wei Y, Cheng X, Zhou BP, Tan M, Zhou X, Xia W, Hortobagyi GN, Yu D, Hung MC (2004) Upregulation of CXCR4 is essential for HER2-mediated tumor metastasis. Cancer Cell 6:459–469

Li Q, Zhang Q, Wang M, Zhao S, Ma J, Luo N, Li N, Li Y, Xu G, Li J (2008) Interferon-gamma and tumor necrosis factor-alpha disrupt epithelial barrier function by altering lipid composition in membrane microdomains of tight junction. Clin Immunol 126:67–80, Orlando, FL

Li J, Chigurupati S, Agarwal R, Mughal MR, Mattson MP, Becker KG, Wood WH 3rd, Zhang Y, Morin PJ (2009) Possible angiogenic roles for claudin-4 in ovarian cancer. Cancer Biol Ther 8:1806–1814

Liao D, Corle C, Seagroves TN, Johnson RS (2007) Hypoxia-inducible factor-1alpha is a key regulator of metastasis in a transgenic model of cancer initiation and progression. Cancer Res 67:563–572

Lie PP, Cheng CY, Mruk DD (2011) Interleukin-1alpha is a regulator of the blood-testis barrier. FASEB J 25:1244–1253

Liedtke M, Ayton PM, Somervaille TC, Smith KS, Cleary ML (2010) Self-association mediated by the Ras association 1 domain of AF6 activates the oncogenic potential of MLL-AF6. Blood 116:63–70

Lingrel JB, Kuntzweiler T (1994) Na+, K(+)-ATPase. J Biol Chem 5:19659–19662

Lioni M, Brafford P, Andl C, Rustgi A, El-Deiry W, Herlyn M, Smalley KS (2007) Dysregulation of claudin-7 leads to loss of E-cadherin expression and the increased invasion of esophageal squamous cell carcinoma cells. Am J Pathol 170:709–721

Liu H, Li M, Wang P, Wang F (2011) Blockade of hypoxia-inducible factor-1alpha by YC-1 attenuates interferon-gamma and tumor necrosis factor-alpha-induced intestinal epithelial barrier dysfunction. Cytokine 56:581–588

Looijer-van Langen M, Hotte N, Dieleman LA, Albert E, Mulder C, Madsen KL (2011) Estrogen receptor-β signaling modulates epithelial barrier function. Am J Physiol Gastrointest Liver Physiol 300:G621–G626

Lu Z, Ding L, Hong H, Hoggard J, Lu Q, Chen YH (2011) Claudin-7 inhibits human lung cancer cell migration and invasion through ERK/MAPK signaling pathway. Exp Cell Res 317:1935–1946

Luo Y, Xiao W, Zhu X, Mao Y, Liu X, Chen X, Huang J, Tang S, Rizzolo LJ (2011) Differential expression of claudins in retinas during normal development and the angiogenesis of oxygen-induced retinopathy. Invest Ophthalmol Vis Sci 52:7556–7564

Madara JL, Pappenheimer JR (1987) Structural basis for physiological regulation of paracellular pathways in intestinal epithelia. J Membr Biol 100:149–164

Madara JL, Stafford J (1989) Interferon-gamma directly affects barrier function of cultured intestinal epithelial monolayers. J Clin Invest 83:724–727
Malik N, Canfield VA, Beckers MC, Gros P, Levenson R (1996) Identification of the mammalian Na, K-ATPase 3 subunit. J Biol Chem 13:22754–22758
Marchiando AM, Graham WV, Turner JR (2010a) Epithelial barriers in homeostasis and disease. Annu Rev Pathol 5:119–144
Marchiando AM, Shen L, Graham WV, Weber CR, Schwarz BT, Austin JR 2nd, Raleigh DR, Guan Y, Watson AJ, Montrose MH et al (2010b) Caveolin-1-dependent occludin endocytosis is required for TNF-induced tight junction regulation in vivo. J Cell Biol 189:111–126
Martin TA, Mason MD, Jiang WG (2011) Tight junctions in cancer metastasis. Front Biosci 16:898–936
Martin-Belmonte F, Gassama A, Datta A, Yu W, Rescher U, Gerke V, Mostov K (2007) PTEN-mediated apical segregation of phosphoinositides controls epithelial morphogenesis through Cdc42. Cell 128:383–397
Martinez-Estrada OM, Culleres A, Soriano FX, Peinado H, Bolos V, Martinez FO, Reina M, Cano A, Fabre M, Vilaro S (2006) The transcription factors Slug and Snail act as repressors of Claudin-1 expression in epithelial cells. Biochem J 394:449–457
Matsuda M, Imaoka T, Vomachka AJ, Gudelsky GA, Hou Z, Mistry M, Bailey JP, Nieport KM, Walther DJ, Bader M, Horseman ND (2004) Serotonin regulates mammary gland development via an autocrine-paracrine loop. Dev Cell 6:193–203
Maxwell PH, Wiesener MS, Chang GW, Clifford SC, Vaux EC, Cockman ME, Wykoff CC, Pugh CW, Maher ER, Ratcliffe PJ (1999) The tumour suppressor protein VHL targets hypoxia-inducible factors for oxygen-dependent proteolysis. Nature 399:271–275
McCaffrey LM, Macara IG (2009) The Par3/aPKC interaction is essential for end bud remodeling and progenitor differentiation during mammary gland morphogenesis. Genes Dev 23:1450–1460
McSherry EA, McGee SF, Jirstrom K, Doyle EM, Brennan DJ, Landberg G, Dervan PA, Hopkins AM, Gallagher WM (2009) JAM-A expression positively correlates with poor prognosis in breast cancer patients. Int J Cancer 125:1343–1351
McSherry EA, Brennan K, Hudson L, Hill AD, Hopkins AM (2011) Breast cancer cell migration is regulated through junctional adhesion molecule-A-mediated activation of Rap1 GTPase. Breast Cancer Res 13:R31
Mendes O, Kim HT, Stoica G (2005) Expression of MMP2, MMP9 and MMP3 in breast cancer brain metastasis in a rat model. Clin Exp Metastasis 22:237–246
Moeser AJ, Nighot PK, Engelke KJ, Ueno R, Blikslager AT (2004) Recovery of mucosal barrier function in ischemic porcine ileum and colon is stimulated by a novel agonist of the ClC-2 chloride channel, lubiprostone. Am J Physiol Gastrointest Liver Physiol 292:G647–G656
Molitoris BA, Nelson WJ (1990) Alterations in the establishment and maintenance of epithelial cell polarity as a basis for disease processes. J Clin Invest 85:3–9
Mori H, Gjorevski N, Inman JL, Bissell MJ, Nelson CM (2009) Self-organization of engineered epithelial tubules by differential cellular motility. Proc Natl Acad Sci USA 106:14890–14895
Murakami M, Murdiastuti K, Hosoi K, Hill AE (2006) AQP and the control of fluid transport in a salivary gland. J Membr Biol 210:91–103
Murakami M, Francavilla C, Torselli I, Corada M, Maddaluno L, Sica A, Matteoli G, Iliev ID, Mantovani A, Rescigno M et al (2010) Inactivation of junctional adhesion molecule-A enhances antitumoral immune response by promoting dendritic cell and T lymphocyte infiltration. Cancer Res 70:1759–1765
Murakami M, Giampietro C, Giannotta M, Corada M, Torselli I, Orsenigo F, Cocito A, D'Ario G, Mazzarol G, Confalonieri S et al (2011) Abrogation of junctional adhesion molecule-A expression induces cell apoptosis and reduces breast cancer progression. PloS one 6:e21242
Naik MU, Naik UP (2006) Junctional adhesion molecule-A-induced endothelial cell migration on vitronectin is integrin alpha v beta 3 specific. J Cell Sci 119:490–499
Naik MU, Vuppalanchi D, Naik UP (2003) Essential role of junctional adhesion molecule-1 in basic fibroblast growth factor-induced endothelial cell migration. Arterioscler Thromb Vasc Biol 23:2165–2171

Nakagawa M, Fukata M, Yamaga M, Itoh N, Kaibuchi K (2001) Recruitment and activation of Rac1 by the formation of E-cadherin-mediated cell-cell adhesion sites. J Cell Sci 114:1829–1838
Naor D, Sionov RV, Ish-Shalom D (1997) CD44: structure, function, and association with the malignant process. Adv Cancer Res 71:241–319
Nava P, Capaldo CT, Koch S, Kolegraff K, Rankin CR, Farkas AE, Feasel ME, Li L, Addis C, Parkos CA et al (2011) JAM-A regulates epithelial proliferation through Akt/beta-catenin signalling. EMBO Rep 12:314–320
Nilsson M, Husmark J, Bjorkman U, Ericson LE (1998) Cytokines and thyroid epithelial integrity: interleukin-1alpha induces dissociation of the junctional complex and paracellular leakage in filter-cultured human thyrocytes. J Clin Endocrinol Metab 83:945–952
Nitta T, Hata M, Gotoh S, Seo Y, Sasaki H, Hashimoto N, Furuse M, Tsukita S (2003) Size-selective loosening of the blood-brain barrier in claudin-5-deficient mice. J Cell Biol 161:653–660
Ogasawara N, Kojima T, Go M, Fuchimoto J, Kamekura R, Koizumi J, Ohkuni T, Masaki T, Murata M, Tanaka S et al (2009) Induction of JAM-A during differentiation of human THP-1 dendritic cells. Biochem Biophys Res Commun 389:543–549
Ohkubo T, Ozawa M (2004) The transcription factor Snail downregulates the tight junction components independently of E-cadherin downregulation. J Cell Sci 117:1675–1685
Orlova VV, Economopoulou M, Lupu F, Santoso S, Chavakis T (2006) Junctional adhesion molecule-C regulates vascular endothelial permeability by modulating VE-cadherin-mediated cell-cell contacts. J Exp Med 203:2703–2714
Osanai M, Murata M, Nishikiori N, Chiba H, Kojima T, Sawada N (2006) Epigenetic silencing of occludin promotes tumorigenic and metastatic properties of cancer cells via modulations of unique sets of apoptosis-associated genes. Cancer Res 66:9125–9133
Osanai M, Murata M, Chiba H, Kojima T, Sawada N (2007) Epigenetic silencing of claudin-6 promotes anchorage-independent growth of breast carcinoma cells. Cancer Sci 98:1557–1562
Oshima T, Laroux FS, Coe LL, Morise Z, Kawachi S, Bauer P, Grisham MB, Specian RD, Carter P, Jennings S et al (2001) Interferon-gamma and interleukin-10 reciprocally regulate endothelial junction integrity and barrier function. Microvasc Res 61:130–143
Oshima T, Kunisaki C, Yoshihara K, Yamada R, Yamamoto N, Sato T, Makino H, Yamagishi S, Nagano Y, Fujii S et al (2008) Reduced expression of the claudin-7 gene correlates with venous invasion and liver metastasis in colorectal cancer. Oncol Rep 19:953–959
Ostermann G, Weber KS, Zernecke A, Schroder A, Weber C (2002) JAM-1 is a ligand of the beta(2) integrin LFA-1 involved in transendothelial migration of leukocytes. Nat Immunol 3:151–158
Ozdamar B, Bose R, Barrios-Rodiles M, Wang HR, Zhang Y, Wrana JL (2005) Regulation of the polarity protein Par6 by TGFbeta receptors controls epithelial cell plasticity. Science 307:1603–1609
Pai VP, Horseman ND (2008) Biphasic regulation of mammary epithelial resistance by serotonin through activation of multiple pathways. J Biol Chem 283:30901–30910
Pai VP, Horseman ND (2011) Multiple cellular responses to serotonin contribute to epithelial homeostasis. PLoS One 6:e17028
Pappenheimer JR, Reiss KZ (1987) Contribution of solvent drag through intercellular junctions to absorption of nutrients by the small intestine of the rat. J Membr Biol 100:123–136
Parris JJ, Cooke VG, Skarnes WC, Duncan MK, Naik UP (2005) JAM-A expression during embryonic development. Dev Dyn 233:1517–1524
Pestalozzi BC, Zahrieh D, Price KN, Holmberg SB, Lindtner J, Collins J, Crivellari D, Fey MF, Murray E, Pagani O, Simoncini E, Castiglione-Gertsch M, Gelber RD, Coates AS, Goldhirsch A, International Breast Cancer Study Group (IBCSG) (2006) Identifying breast cancer patients at risk for Central Nervous System (CNS) metastases in trials of the International Breast Cancer Study Group (IBCSG). Ann Oncol 17:935–944
Piñeiro-Sánchez ML, Goldstein LA, Dodt J, Howard L, Yeh Y, Tran H, Argraves WS, Chen WT (1997) Identification of the 170-kDa melanoma membrane-bound gelatinase (seprase) as a serine integral membrane protease. J Biol Chem 272:7595–7601

Polette M, Gilles C, Nawrocki-Raby B, Lohi J, Hunziker W, Foidart JM, Birembaut P (2005) Membrane-type 1 matrix metalloproteinase expression is regulated by zonula occludens-1 in human breast cancer cells. Cancer Res 65:7691–7698

Polette M, Mestdagt M, Bindels S, Nawrocki-Raby B, Hunziker W, Foidart JM, Birembaut P, Gilles C (2007) Beta-catenin and ZO-1: shuttle molecules involved in tumor invasion-associated epithelial-mesenchymal transition processes. Cells Tissues Organs 185:61–65

Prat A, Parker JS, Karginova O, Fan C, Livasy C, Herschkowitz JI, He X, Perou CM (2010) Phenotypic and molecular characterization of the claudin-low intrinsic subtype of breast cancer. Breast Cancer Res 12:R68

Qi J, Chen N, Wang J, Siu CH (2005) Transendothelial migration of melanoma cells involves N-cadherin-mediated adhesion and activation of the beta-catenin signaling pathway. Mol Biol Cell 16:4386–4397

Rabquer BJ, Amin MA, Teegala N, Shaheen MK, Tsou PS, Ruth JH, Lesch CA, Imhof BA, Koch AE (2010) Junctional adhesion molecule-C is a soluble mediator of angiogenesis. J Immunol 185:1777–1785

Rajasekaran SA, Hu J, Gopal J, Gallemore R, Ryazantsev S, Bok D, Rajasekaran AK (2003) Na, K-ATPase inhibition alters tight junction structure and permeability in human retinal pigment epithelial cells. Am J Physiol Cell Physiol 284:C1497–C1507

Rajasekaran SA, Barwe SP, Gopal J, Ryazantsev S, Schneeberger EE, Rajasekaran AK (2007) Na-K-ATPase regulates tight junction permeability through occludin phosphorylation in pancreatic epithelial cells. Am J Physiol Gastrointest Liver Physiol 292:G124–G133

Rajasekaran SA, Beyenbach KW, Rajasekaran AK (2008) Interactions of tight junctions with membrane channels and transporters. Biochim Biophys Acta 1778:757–769

Reynolds LE, Watson AR, Baker M, Jones TA, D'Amico G, Robinson SD, Joffre C, Garrido-Urbani S, Rodriguez-Manzaneque JC, Martino-Echarri E, Aurrand-Lions M, Sheer D, Dagna-Bricarelli F, Nizetic D, McCabe CJ, Turnell AS, Kermorgant S, Imhof BA, Adams R, Fisher EM, Tybulewicz VL, Hart IR, Hodivala-Dilke KM (2010) Tumour angiogenesis is reduced in the Tc1 mouse model of Down's syndrome. Nature 465:813–817

Rodriguez P, Heyman M, Candalh C, Blaton MA, Bouchaud C (1995) Tumour necrosis factor-alpha induces morphological and functional alterations of intestinal HT29 cl.19A cell monolayers. Cytokine 7:441–448

Schmitz H, Fromm M, Bentzel CJ, Scholz P, Detjen K, Mankertz J, Bode H, Epple HJ, Riecken EO, Schulzke JD (1999) Tumor necrosis factor-alpha (TNFalpha) regulates the epithelial barrier in the human intestinal cell line HT-29/B6. J Cell Sci 112(Pt 1):137–146

Schneeberger EE, Lynch RD (2004) The tight junction: a multifunctional complex. Am J Physiol 286:C1213–C1228

Schoner W, Scheiner-Bobis G (2007) Endogenous and exogenous cardiac glycosides and their mechanisms of action. Am J Cardiovasc Drugs 7:173–189

Schulzke JD, Ploeger S, Amasheh M, Fromm A, Zeissig S, Troeger H, Richter J, Bojarski C, Schumann M, Fromm M (2009) Epithelial tight junctions in intestinal inflammation. Ann N Y Acad Sci 1165:294–300

Severson EA, Lee WY, Capaldo CT, Nusrat A, Parkos CA (2009) Junctional adhesion molecule A interacts with Afadin and PDZ-GEF2 to activate Rap1A, regulate beta1 integrin levels, and enhance cell migration. Mol Biol Cell 20:1916–1925

Simon DB, Lu Y, Choate KA, Velazquez H, Al-Sabban E, Praga M, Casari G, Bettinelli A, Colussi G, Rodriguez-Soriano J et al (1999) Paracellin-1, a renal tight junction protein required for paracellular Mg2+ resorption. Science 285:103–106

Sourisseau T, Georgiadis A, Tsapara A, Ali RR, Pestell R, Matter K, Balda MS (2006) Regulation of PCNA and cyclin D1 expression and epithelial morphogenesis by the ZO-1-regulated transcription factor ZONAB/DbpA. Mol Cell Biol 26:2387–2398

Spaderna S, Schmalhofer O, Wahlbuhl M, Dimmler A, Bauer K, Sultan A, Hlubek F, Jung A, Strand D, Eger A et al (2008) The transcriptional repressor ZEB1 promotes metastasis and loss of cell polarity in cancer. Cancer Res 68:537–544

Stankovich BL, Aguayo E, Barragan F, Sharma A, Pallavicini MG (2011) Differential adhesion molecule expression during murine embryonic stem cell commitment to the hematopoietic and endothelial lineages. PloS one 6:e23810
Stark AM, Anuszkiewicz B, Mentlein R, Yoneda T, Mehdorn HM, Held-Feindt J (2007) Differential expression of matrix metalloproteinases in brain- and bone-seeking clones of metastatic MDA-MB-231 breast cancer cells. J Neurooncol 81:39–48
Stelwagen K, McFadden HA, Demmer J (1999) Prolactin, alone or in combination with glucocorticoids, enhances tight junction formation and expression of the tight junction protein occludin in mammary cells. Mol Cell Endocrinol 156:55–61
Strell C, Lang K, Niggemann B, Zaenker KS, Entschladen F (2007) Surface molecules regulating rolling and adhesion to endothelium of neutrophil granulocytes and MDA-MB-468 breast carcinoma cells and their interaction. Cell Mol Life Sci 64:3306–3316
Sweadner KJ (1989) Isozymes of the Na+/K+-ATPase. Biochim Biophys Acta 988:185–220
Swisshelm K, Machl A, Planitzer S, Robertson R, Kubbies M, Hosier S (1999) SEMP1, a senescence-associated cDNA isolated from human mammary epithelial cells, is a member of an epithelial membrane protein superfamily. Gene 226:285–295
Taube JH, Herschkowitz JI, Komurov K, Zhou AY, Gupta S, Yang J, Hartwell K, Onder TT, Gupta PB, Evans KW et al (2010) Core epithelial-to-mesenchymal transition interactome gene-expression signature is associated with claudin-low and metaplastic breast cancer subtypes. Proc Natl Acad Sci U S A 107:15449–15454
Tester AM, Waltham M, Oh SJ, Bae SN, Bills MM, Walker EC, Kern FG, Stetler-Stevenson WG, Lippman ME, Thompson EW (2004) Pro-matrix metalloproteinase-2 transfection increases orthotopic primary growth and experimental metastasis of MDA-MB-231 human breast cancer cells in nude mice. Cancer Res 64:652–658
Thompson GE (1996) Cortisol and regulation of tight junctions in the mammary gland of the late-pregnant goat. J Dairy Res 63:305–308
Tobioka H, Isomura H, Kokai Y, Tokunaga Y, Yamaguchi J, Sawada N (2004a) Occludin expression decreases with the progression of human endometrial carcinoma. Hum Pathol 35:159–164
Tobioka H, Tokunaga Y, Isomura H, Kokai Y, Yamaguchi J, Sawada N (2004b) Expression of occludin, a tight-junction-associated protein, in human lung carcinomas. Virchows Arch 445:472–476
Tsukada Y, Fouad A, Pickren JW, Lane WW (1983) Central nervous system metastasis from breast carcinoma. Autopsy study. Cancer 52:2359–2354
Tsukita S, Furuse M, Itoh M (2001) Multifunctional strands in tight junctions. Nat Rev Mol Cell Biol 2:285–293
Turner JR, Rill BK, Carlson SL, Carnes D, Kerner R, Mrsny RJ, Madara JL (1997) Physiological regulation of epithelial tight junctions is associated with myosin light-chain phosphorylation. Am J Physiol 273:C1378–C1385
Turner JR, Cohen DE, Mrsny RJ, Madara JL (2000) Noninvasive in vivo analysis of human small intestinal paracellular absorption: regulation by Na+-glucose cotransport. Dig Dis Sci 45:2122–2126
Uchide K, Sakon M, Ariyoshi H, Nakamori S, Tokunaga M, Monden M (2007) Cancer cells cause vascular endothelial cell (vEC) retraction via 12(S)HETE secretion; the possible role of cancer cell derived microparticle. Ann Surg Oncol 14:862–868
Vaira V, Faversani A, Dohi T, Montorsi M, Augello C, Gatti S, Coggi G, Altieri DC, Bosari S (2012) miR-296 regulation of a cell polarity-cell plasticity module controls tumor progression. Oncogene 31:27
Van Itallie CM, Anderson JM (2006) Claudins and epithelial paracellular transport. Annu Rev Physiol 68:403–429
Viloria-Petit AM, David L, Jia JY, Erdemir T, Bane AL, Pinnaduwage D, Roncari L, Narimatsu M, Bose R, Moffat J et al (2009) A role for the TGFbeta-Par6 polarity pathway in breast cancer progression. Proc Natl Acad Sci U S A 106:14028–14033
Wada-Hiraike O, Imamov O, Hiraike H, Hultenby K, Schwend T, Omoto Y, Warner M, Gustafsson JA (2006) Role of estrogen receptor beta in colonic epithelium. Proc Natl Acad Sci USA 103:2959–2964

Wang F, Schwarz BT, Graham WV, Wang Y, Su L, Clayburgh DR, Abraham C, Turner JR (2006) IFN-gamma-induced TNFR2 expression is required for TNF-dependent intestinal epithelial barrier dysfunction. Gastroenterology 131:1153–1163

Wang Z, Wade P, Mandell KJ, Akyildiz A, Parkos CA, Mrsny RJ, Nusrat A (2007) Raf 1 represses expression of the tight junction protein occludin via activation of the zinc-finger transcription factor slug. Oncogene 26:1222–1230

Wiercinska E, Naber HP, Pardali E, van der Pluijm G, van Dam H, ten Dijke P (2011) The TGF-beta/Smad pathway induces breast cancer cell invasion through the up-regulation of matrix metalloproteinase 2 and 9 in a spheroid invasion model system. Breast Cancer Res Treat 128:657–666

Willemsen LE, Hoetjes JP, van Deventer SJ, van Tol EA (2005) Abrogation of IFN-gamma mediated epithelial barrier disruption by serine protease inhibition. Clin Exp Immunol 142:275–284

Wolburg H, Wolburg-Buchholz K, Kraus J, Rascher-Eggstein G, Liebner S, Hamm S, Duffner F, Grote EH, Risau W, Engelhardt B (2003) Localization of claudin-3 in tight junctions of the blood-brain barrier is selectively lost during experimental autoimmune encephalomyelitis and human glioblastoma multiforme. Acta Neuropathol 105:586–592

Yan J, Zhang Z, Shi H (2012) HIF-1 is involved in high glucose-induced paracellular permeability of brain endothelial cells. Cell Mol Life Sci 69:115

Ye D, Guo S, Al-Sadi R, Ma TY (2011) MicroRNA regulation of intestinal epithelial tight junction permeability. Gastroenterology 141:1323–1333

Yeh WL, Lu DY, Lin CJ, Liou HC, Fu WM (2007) Inhibition of hypoxia-induced increase of blood-brain barrier permeability by YC-1 through the antagonism of HIF-1alpha accumulation and VEGF expression. Mol Pharmacol 72:440–449

Yoon CH, Kim MJ, Park MJ, Park IC, Hwang SG, An S, Choi YH, Yoon G, Lee SJ (2010) Claudin-1 acts through c-Abl-protein kinase Cdelta (PKCdelta) signaling and has a causal role in the acquisition of invasive capacity in human liver cells. J Biol Chem 285:226–233

Youakim A, Ahdieh M (1999) Interferon-gamma decreases barrier function in T84 cells by reducing ZO-1 levels and disrupting apical actin. Am J Physiol 276:G1279–G1288

Zettl KS, Sjaastad MD, Riskin PM, Parry G, Machen TE, Firestone GL (1992) Glucocorticoid-induced formation of tight junctions in mouse mammary epithelial cells in vitro. Proc Natl Acad Sci U S A 89:9069–9073

Zhan L, Rosenberg A, Bergami KC, Yu M, Xuan Z, Jaffe AB, Allred C, Muthuswamy SK (2008) Deregulation of scribble promotes mammary tumorigenesis and reveals a role for cell polarity in carcinoma. Cell 135:865–878

Abbreviations

AJ	Adherens junction
BPH	Benign prostatic hyperplasia
BTB	Blood-testis barrier
CAR	Coxsackie adenovirus receptor
CDC42	Cell division cycle 42 (GTP-binding protein 25 kDa)
CD	Coeliac disease
CGN	Cingulin
CLDN10	Claudin 10
CLDN11	Claudin 11 (oligodendrocyte transmembrane protein)
CLDN12	Claudin 12
CLDN14	Claudin 14
CLDN15	Claudin 15
CLDN16	Claudin 16
CLDN17	Claudin 17
CLDN18	Claudin 18
CLDN19	Claudin 19
CLDN1	Claudin 1
CLDN20	Claudin 20
CLDN2	Claudin 2
CLDN3	Claudin 3
CLDN4	Claudin 4
CLDN5	Claudin 5
CLDN6	Claudin 6
CLDN7	Claudin 7
CLDN8	Claudin 8
CLDN9	Claudin 9
CLDN	Claudin
CLDND2	Claudin domain containing 2
CRB3	Crumbs homolog 3 (Drosophila)
CTNNAL1	Catenin (cadherin-associated protein) alpha-like 1
CTNNBIP1	Catenin beta interacting protein 1
ECM	Extracellular matrix
ESCC	Esophageal squamous cell carcinoma
F11R	F11 receptor
FHHNC	Familial hypomagnesemia with hypercalciuria and nephrocalcinosis
FOV	Foveolar epithelium
GLUT1	Glucose transporter1
HB	Hepatoblastoma
HCC	Hepatocellular carcinoma
HCE	Human corneal epithelial cells
HCV	Hepatitis C virus
HPRT1	Hypoxanthine phosphoribosyl-transferase 1
IDC	Invasive ductal carcinoma

IUP	Inverted urothelial papillomas
JAM-A	Junctional adhesion molecule A
JAM-B	Junctional adhesion molecule B
JAM-C	Junctional adhesion molecule C
MAGI1	Membrane-associated guanylate kinase WW and PDZ domain containing 1
MAGIX	MAGI family member X-linked
MAGUK	Membrane-associated guanylate kinase
MARK2	MAP/microtubule affinity-regulating kinase
MLLT4	Myeloid/lymphoid or mixed-lineage leukemia (trithorax homolog Drosophila) translocated to 4
mPCa	Metastatic prostatic adenocarcinoma
MPDZ	Multiple PDZ domain protein
MPP5	Membrane protein palmitoylated 5 (MAGUK p55 subfamily member 5)
NAC	Normal tissue adjacent to carcinoma
OCLN	Occludin
OSE	Ovarian cell surface epithelium
PARD3	par-3 partitioning defective 3 homolog
PARD6A	par-6 partitioning defective 6 homolog alpha
PCa	Prostatic adenocarcinoma
PET	Pancreatic endocrine tumors
PIN	Prostatic intraepithelial neoplasia
PUNLMP	Papillary urothelial neoplasm of low malignant potential
RHOA	Ras homologous gene family member A
RPL13A	Ribosomal protein L13a
SCE	Specialized columnar epithelium
SDHA	Succinate dehydrogenase complex subunit A, flavoprotein (Fp)
SG	Stratum granulosum
SP	Substance P
SqE	Squamous epithelium
SSC	Squamous cell carcinoma
SYMPK	Symplekin
TER	Transepithelial electrical resistance
TGFB1	Transforming growth factor beta 1
TJP3	Tight junction protein 3 (zona occludens 3)
TJ	Tight junction
TRIC	Tricellulin
tTJ	Tricellular tight junctions
UCC	Urothelial cell carcinoma
UP	Urothelial papilloma
ZO1	Tight junction protein 1 (zona occludens 1)
ZO2	Tight junction protein 2 (zona occludens 2)
ZO	Zonula occludens
ZONAB	Zonula occludens 1 (ZO1) –associated nucleic acid binding protein

2.1 Introduction

In epithelial and endothelial tissues cell-cell interaction is mediated by various junctional complexes. Each of these complexes – tight junctions (TJs), adherens junctions (AJs), desmosomes and gap junctions – have typical morphology, composition and function. TJs are the most apical intercellular junctions with diverse functions. It is generally accepted that TJ proteins can be categorized into three groups: *integral membrane proteins* which bridge the intercellular space and create a paracellular seal (occludin, claudins – CLDN, junctional adhesion molecules – JAM, tricellulin and crumbs), *peripherally associated cytoplasmic proteins* which assemble at the cytoplasmic surface of the junctional site (zonula occludens – ZO, partitioning-defective molecules – PAR, MAGUK inverted – MAGI, cingulin, symplectin and others) and *signaling proteins* (protein kinase A, protein kinase C, heterotrimeric G-proteins) (Brennan et al. 2010; Gonzalez-Mariscal et al. 2003, 2007, 2008; Lelievre 2010). So far a high number of proteins locating at tight intercellular contacts have been discovered, and their roles have just partly been unravelled. Recent studies suggest that there are many more transmembrane proteins not yet fully characterized, representing an active area of investigation (Mineta et al. 2011). Intensive research has revealed that the composition of distinct integral membrane proteins determines the specific "tightness" or "leakiness" of an epithelial and endothelial layer (Amasheh et al. 2009; Anderson and Van Itallie 2009). Today TJs are considered as multifunctional complexes which are involved not only in regulation of paracellular diffusion, establishment of polarity, but also in cell signaling and gene expression regulation (Bauer et al. 2011).

2.1.1 Classical Functions of Junctional Proteins: Formation and Regulation of Tissue Barriers

CLDNs have been identified as the major constituents of TJ strands, where they interact with each other in homotypic or heterotypic manner or with other proteins of the TJs such as occludin and tricellulin (TRIC). All current evidence supports a central role for CLDNs in electrical resistance, permselectivity (including size, electrical resistance, ionic charge preference) however, the exact stoichiometry remains unclear and little is known about the molecular mechanisms taking place during assembly and strand formation (Anderson and Van Itallie 2009; Furuse 2006).

Available data support the idea that the first extracellular loop of CLDNs creates an "electrostatic selectivity filter" controlling overall resistance and charge selectivity in different types of tissues (Anderson and Van Itallie 2009). By analysing the transepithelial electrical resistance (TER) epithelial and endothelial tissues are also characterized as "tight" and "leaky" epithelia. The "tight" epithelia are characterized by tight junctions that can maintain the high electrochemical gradient characteristics, for example in the distal nephron, while "leaky" TJs move large amounts of isoosmotic fluids, as in the human gastrointestinal tract (Van Itallie and Anderson 2006). CLDNs 2 and 10 tend to make tight monolayers leakier (Van Itallie and Anderson 2006; Furuse et al. 2001; Colegio et al. 2002; Amasheh et al. 2002), while CLDNs 1, 4, 5, 7, 8, 11, 14, 15, 16, 18, 19 tend to make leaky monolayers tighter (Anderson and Van Itallie 2009; Alexandre et al. 2005; Angelow et al. 2006; Van Itallie et al. 2003; Ben-Yosef et al. 2003; Colegio et al. 2003; Ikari et al. 2008; Jovov et al. 2007).

Interesting data are presented about TRIC, a protein specifically enriched at tricellular tight junctions (tTJs) (Ikenouchi et al. 2005, 2008). TRIC is incorporated into CLDN-based TJs independently of binding ZO1 and the role of TRIC in ion transport and solute is dependent on its localization and the level of expression. Bioinformatic analysis identified an additional protein localizing at TJs, MarvelD3, sharing similar membrane topology with occludin and TRIC (Steed et al. 2009). Similar to occludin and TRIC, MarvelD3 is incorporated into TJ strands, but is unable to form TJ strands by itself. Depletion of MarvelD3 does not disturb TJ formation, but increases the TER (Steed et al. 2009; Raleigh et al. 2010).

2.1.2 Non-classical Functions of Junctional Proteins

2.1.2.1 The Regulation of Gene Expression

Besides their structural function at cell-cell contacts several TJ-associated proteins have been linked to the control of cell proliferation and gene expression. For example ZO1 and ZO2 proteins have been shown to regulate transcription factors and proliferation. In a nicely presented study different transcriptional pathways are described in which several TJ associated proteins are implicated. Reduced expression of the ZO1 protein was found to be associated with increased proliferation of epithelial cells (Balda and Matter 2009). The mechanism by which ZO1 regulates proliferation is not totally elucidated. According to the data presented by Balda et al. (2003), ZO1 and ZONAB regulate G1/S phase transition of the cell cycle in two ways. ZONAB interacts with CDK4, a regulator of G1/S phase transition, which colocalizes with ZO1 at TJs. In another pathway, ZONAB functions as a transcription factor and regulates the expression of the genes involved in cell-cycle regulation (Balda et al. 2000, 2003; Sourisseau et al. 2006).

ZO2 protein has been reported to accumulate transiently in the nucleus of proliferating cells (Traweger et al. 2003; Kausalya et al. 2004) and furthermore, ZO2 binds to DNA scaffolding factor SAF-B (Traweger et al. 2003) and to transcription factors such as Fos, Jun, and c-myc (Betanzos et al. 2004; Huerta et al. 2007). Experimentally induced nuclear overexpression of ZO2 in cerebral endothelial cells led to an increase of pyruvate kinase M2 (M2PK) protein levels and increased proliferation (Traweger et al. 2008). Symplekin can regulate gene expression by its interaction with ZONAB and by regulating RNA processing (Balda and Matter 2009).

Cingulin was demonstrated to regulate gene expression in a RhoA dependent manner and by other unidentified mechanisms (Balda and Matter 2009). It was also demonstrated that the cingulin gene does not affect TJ formation, but may alter gene expression when using mouse embryonic model (Guillemot et al. 2004).

2.1.2.2 Role of Tight Junction Proteins in Epithelial Cell Migration

In 2010, a crucial role for occludin in epithelial cell migration was reported. In their study Du et al. (2010) found that occludin, as a transmembrane TJ protein, is localized at the leading edge of migrating cells and regulates directional cell migration (Du et al. 2010).

The members of the JAM family of immunoglobulin-like TJs localize to sites of intercellular contact in epithelial and endothelial cells. JAMs are capable of mediating homophilic and heterophilic interactions and are known to be involved in the regulation of cell proliferation, migration and invasion (Bazzoni et al. 2005; Mandell and Parkos 2005; Mandell et al. 2005). In 2009 it was shown by Cera MR et al. that JAM-A was indispensable for the internalization of integrins, a pre-requisite for cell movement (Cera et al. 2009). Knockdown of JAM-A was shown to diminish the level of cell-surface-associated β1-integrin, to inhibit cell-ECM interactions and to reduce cell migration (Mandell et al. 2005). In addition, it was presented that the loss of JAM-A in HepG2 cells resulted in the mislocalization of several TJ proteins (i.e. occludin, claudin 1 and ZO1), as well as in the disruption of cell polarity and junction assembly (Konopka et al. 2007).

CAR was initially characterized as a cell surface protein (i.e. receptor) required for the entry of coxsackie B and adenoviruses into cells (Coyne and Bergelson 2005), but was later reported to be a component of the TJ complex and a regulator of TJ assembly (Coyne and Bergelson 2005; Mirza et al. 2005). CAR is also highly homologous to JAM. Loss of CAR expression resulted in weakened cell adhesion, thereby promoting migration of cancer cells (Bruning and Runnebaum 2003, 2004; Matsumoto et al. 2005).

2.2 Tight Junctions in Tissues of Ectodermal Origin

2.2.1 Tight Junctions in the Epidermis

A variety of TJ proteins have been identified in mammalian epidermis, however our knowledge and our understanding of the role and function of TJ proteins in epidermis are still limited. Even though TJ proteins have been intensively investigated in simple epithelia and endothelia for many years, the first description of TJ proteins in the human epidermis was in 2001/2002 (Kirschner et al. 2010; Brandner et al. 2002, 2006; Pummi et al. 2001) and nowadays a large variety of TJ associated proteins, like occludin, several CLDNs (1, 3, 4, 5, 7), tricellulin, JAM-A, ZO1, ZO2, PAR3, PAR6 have been identified in mammalian epidermis (Brandner 2009; Kubo et al. 2009; Niessen 2007). At RNA level, additional TJ molecules have been identified in human keratinocytes, like CLDNs 8, 17 (Brandner 2009).

The localization patterns of TJ proteins in the epidermis and in skin appendages are very complex. While occludin, cingulin, PAR6 are restricted to stratum granulosum (SG), some proteins like ZO1 and CLDN4 are found in several suprabasal layers and other proteins such as CLDN1 are localized in all epidermal layers (Brandner 2009).

The localization of some of the TJ proteins in different layers of the epidermis and in various skin diseases is nicely presented in a recent study by Kirschner et al. (2010) and in a study by Brandner (2009), but the question of the distinct functions of TJ proteins in the skin is controversial. One of the most prominent functions of the skin is its barrier function. TJs play an important role in inside-out barrier of the skin. An important finding demonstrating the involvement of TJs in mammalian barrier function was made with CLDN1 deficient mice. These mice die within one day of birth due to dehydration (Furuse et al. 2002). Interestingly the outside-in barrier of these mice was not investigated or described (Furuse et al. 2002; Tunggal et al. 2005).

The role of occludin and ZO1 in the barrier function of the epidermis was also studied. Occludin was found to be dispensable for TJ barrier formation and occludin deficient mice did not show obvious defect in skin barrier (Furuse and Moriwaki 2009). According to the data of Smalley et al. (2005) ZO1 seems to be involved in non-barrier related functions, too (Smalley et al. 2005).

Studies on the role of TJ in epidermal polarity are only at the beginning. In CLDN1 deficient mice an alteration of stratum corneum structure was observed, even though no functional alterations have been described up to now (Furuse et al. 2002). CLDN6 overexpressing mice show alterations in stratum corneum composition and barrier function (Troy et al. 2005; Turksen and Troy 2002).

TJ proteins are also considered to regulate epithelial proliferation and differentiation and are engaged in delivering signals to the cell interior (Balda and Matter 2009; Matter and Balda 2003). ZO1 influences gene expression and cell-cycle progression in a cell density-dependent manner, ZO2 was shown to take part in the regulation of gene expression. In the skin CLDN6 overexpression was associated with increased proliferation of keratinocytes and dysregulated epidermal and hair follicle differentiation (Troy et al. 2005; Turksen and Troy 2002).

In various human skin diseases altered expression of TJ proteins has been observed. Several data suggest for example the altered expression of TJs in psoriasis (Kirschner et al. 2010; Yoshida et al. 2001). Psoriasis vulgaris, ichthyosis vulgaris are characterized by a broadened localization of TJ proteins, i.e. CLDN4, occludin, ZO1 that are normally restricted to the upper layers of the epidermis (Brandner et al. 2006; Pummi et al. 2001; Yoshida et al. 2001). The data of Kirschner et al. (2010) also demonstrated that alteration of TJ proteins is an early event in psoriasis vulgaris (Kirschner et al. 2010). It was interesting to find that CLDN1, which is normally expressed in all layers of the epidermis, is down-regulated in psoriatic epidermis (Brandner et al. 2006; Watson et al. 2007). In infections with Staphylococcus aureus a down-regulation of TJ proteins in the upper and also the lower layers of the epidermis was observed (Ohnemus et al. 2008), suggesting again the barrier function of TJs for pathogen invasion (Brandner 2009).

Alterations in expression of TJ proteins are prominent features of human carcinomas. In oral squamous cell carcinoma (SSC) overexpression of CLDN1 is associated with increased invasiveness (Dos Reis et al. 2008), while decreased levels of CLDN1 were correlated with recurrence in esophageal SSC (Miyamoto et al. 2008). In cutaneous SSC our knowledge of TJ is more restricted. According to data presented by Morita et al. (2004), by investigating cases of cutaneous SSC they found that in unkeratinized tumor cells CLDN1 presented a heterogenous expression, while the expression of ZO1 was weak and occludin and CLDN4 were absent (Morita et al. 2004). The numerous data, publications presented above strongly suggest the important role played by TJ proteins in dermatological diseases.

2.2.2 *Tight Junctions in the Mammary Glands*

According to the classical function of TJs in the mammary gland, TJs have barrier and fence functions, regulate polarity and differentiation as well as adhesion and migration (Brennan et al. 2010; Tsukita et al. 2001). Fascinating data are presented about TJ proteins and genes involved in the function of the powerful glandular activity of the mammary epithelium and about the role of TJs in breast carcinomas (Brennan et al. 2010; Martin et al. 2004; Szasz et al. 2011a, b). TJ structures are considered to be highly dynamic in the breast epithelium and are under the control of several factors. In non-lactating breast, fewer interconnections are seen in electron microscopy compared with the very tight organization necessary to prevent leakiness in the lactation period (Nguyen and Neville 1998; Pitelka et al. 1973). TJ permeability increases with milk statis suggesting that environmental factors such as pressure may affect apical organization. The effect of hormones and growth factors on TJ permeability in the breast was also described (Nguyen and Neville 1998). Epithelial barrier function of TJs relies heavily on the CLDN family of TJ proteins (Furuse 2006; Van Itallie and Anderson 2006), while polarity is partly under the control of the assembly of three main polarity complexes, namely CRB3, PAR complex, and Scrib complex (Brennan et al. 2010). The CRB complex is the most apically located polarity complex in epithelial cells, acting as an anchor for cytoplasmic proteins (Benton and St Johnston 2003; Hurd et al. 2003). The loss of apical polarity proteins from the cell membrane appears to be a key aspect of breast cancer cell behavior like proliferation and invasive potential (Brennan et al. 2010). The same group has found that the loss of apical polarity is a paramount for very early stages of breast tumor development. The expression and function of several TJ proteins in breast carcinoma, however, are not known and some of the published data are contradictory. The relevant publications in this field, without claim of completeness, are presented in Table 2.1.

In our previous studies we found significant loss of CLDN1 protein in breast cancer cells compared with normal breast tissue, with downregulation of CLDN4 noted in ductal carcinoma grade 1, in special types of breast carcinoma (mucinous, papillary, tubular) and in areas of apocrine metaplasia (Tokes et al. 2005). Other studies have

Table 2.1 Relevant literary data on TJ genes and protein expression in breast carcinomas

Nr.	Gene symbol	Gene name	Expression in breast carcinomas	Reference
1	CDC42-Hs00741586_mH	Cell division cycle 42 (GTP binding protein, 25 kDa)	–	Nolan et al. (2008)
2	CGN-Hs00430426_m1	Cingulin	–	Citi et al. (2009)
3	CLDN10-Hs00199599_m1	Claudin 10	–	Hewitt et al. (2006)
4	CLDN12-Hs00273258_s1	Claudin 12	–	Hewitt et al. (2006)
5	CLDN14-Hs00273267_s1	Claudin 14	–	Hewitt et al. (2006)
6	CLDN15-Hs00204982_m1	Claudin 15	–	Hewitt et al. (2006)
7	CLDN15-Hs00370756_m1	Claudin 15	–	Hewitt et al. (2006)
8	CLDN16-9Hs00198134_m1	Claudin 16	Down	Martin et al. (2008)
9	CLDN1-Hs00221623_m1	Claudin 1	Down	Hoevel et al. (2002)
				Tokes et al. (2005)
10	CLDN2-Hs00252666_s1	Claudin 2	Down	Kim et al. (2008)
11	CLDN3-Hs00265816_s1	Claudin 3	Up	Kominsky et al. (2004)
12	CLDN4-Hs00533616_s1	Claudin 4	Down	Kominsky et al. (2004)
			Up	Kulka et al. (2009)
13	CLDN5-Hs00533949_s1	Claudin 5 (transmembrane protein deleted in velocardiofacial syndrome)	–	Soini (2004)
14	CLDN6-Hs00607528_s1	Claudin 6	–	Osanai et al. (2007)
15	CLDN7-Hs00600772_m1	Claudin 7	Down	Kominsky et al. (2003)
				Tokes et al. (2005)
16	CLDN8-HS00273282-s1	Claudin 8	–	Hewitt et al. (2006)
17	CLDN9-Hs00253134_s1	Claudin 9	–	Hewitt et al. (2006)
18	CRB3-Hs00373616_m1	Crumbs homolog 3 (Drosophila)	–	Fogg et al. (2005)

(continued)

Table 2.1 (continued)

Nr.	Gene symbol	Gene name	Expression in breast carcinomas	Reference
19	F11R-Hs00170991_m1	F11 receptor	Down	Naik et al. (2008), Kominsky et al. (2003)
20	OCLN-Hs00170162_m1	Occludin	Down	Polette et al. (2005)
21	PARD6A-Hs00180947_m1	Partitioning defective 6 homolog alpha (C elegans)	Up	Nolan et al. (2008)
22	RHOA-Hs00357608_m1	Ras homolog gene family, member A	Up	Bellizzi et al. (2008)
23	TJP1-Hs00268480_m1	Tight junction protein 1 (zona occludens 1)	Down	Hoover et al. (1998)
24	TJP2-Hs00178081_m1	Tight junction protein 2 (zona occludens 2)	Down	Chlenski et al. (2000)
25	TJP3-Hs00274276_m1	Tight junction protein 3 (zona occludens 3)	Down	Martin et al. (2004)

also reported dowregulation of occludin, CLDN1, 4 in breast cancer (Osanai et al. 2007; Hoevel et al. 2004; Michl et al. 2003). Contrary to the above presented, we have found that in the basal-like breast carcinomas compared with the non-basal-like grade 3 breast carcinomas the CLDN4 expression was significantly higher ($p=0.017$) (Kulka et al. 2009). In some studies expressions of CLDN3, 4 and 7 have been found to be increased in breast carcinoma (Hewitt et al. 2006; Tokes et al. 2005; Kominsky et al. 2003; Lanigan et al. 2009). Recent studies showed that CLDN16 expression was also reduced in human breast cancer, particularly in patients developing aggressive tumors with high mortality rate (Martin et al. 2008). Furthermore, JAM-A is robustly expressed in normal human mammary epithelium, and its expression is downregulated in metastatic breast cancer (Naik et al. 2008; Naik and Naik 2008). Representative images about the expression of different CLDNs in human breast are presented in Fig. 2.1a–j.

In a recent study by us, ten TJ associated genes were found to be significantly downregulated in tumors compared with normal breast tissues (CLDNs 5, 10, 16, 18, 19, CTNNAL1, JAM-B, ZO1, ZO2 and PARD3), whereas one gene (CLDN17) was significantly up-regulated in tumors when compared with normal breast. At protein level CLDNs 5, 10, 16, 18, ZO1 and ZO2 were downregulated in tumors compared with normal breast tissue. CLDN17 showed variable expression in tumor tissues compared with the normal breast (Tőkés et al. 2012).

The expression of ZO1 protein has been widely studied, but there are only few data about the expression of ZO2 in breast carcinomas. In a review published by Brennan et al. (2010) it is concluded that ZO proteins play important roles in migratory events associated with breast cancer progression (Brennan et al. 2010). We found that ZO1 and ZO2 are significantly downregulated at mRNA and protein levels in tumors compared with normal breast epithelium. The alterations mentioned above suggest a relationship between TJ alterations and the malignant potential of breast carcinomas.

It is to be mentioned that in the last years, based on DNA microarray analysis a new breast carcinoma subtype was identified and defined as a claudin-low subtype. These tumors exhibit low expression of many of the claudin genes, including 3, 4, and 7 and lack cell–cell junction proteins, including E-cadherin. The claudin-low tumors are also triple negative and are somewhat similar to basal-like cancers. Other important features of claudin-low tumors are that they have an intense immune cell infiltrate and stem cell features. In the absence of therapy, patients with claudin-low tumors are considered to have poor prognosis, similar to the prognosis of patients with basal-like tumors and HER-2–enriched subtypes (Prat and Perou 2011; Prat et al. 2010; Perou 2010).

2.2.3 Tight Junctions in the Sensory Epithelium of Ears, Nose and Eyes

Little is known about the role of TJ expression in the sensory epithelium of ears, nose and eyes. As it was presented earlier, TJs form the principal barrier to passive movement of fluids, macromolecules, electrolytes.

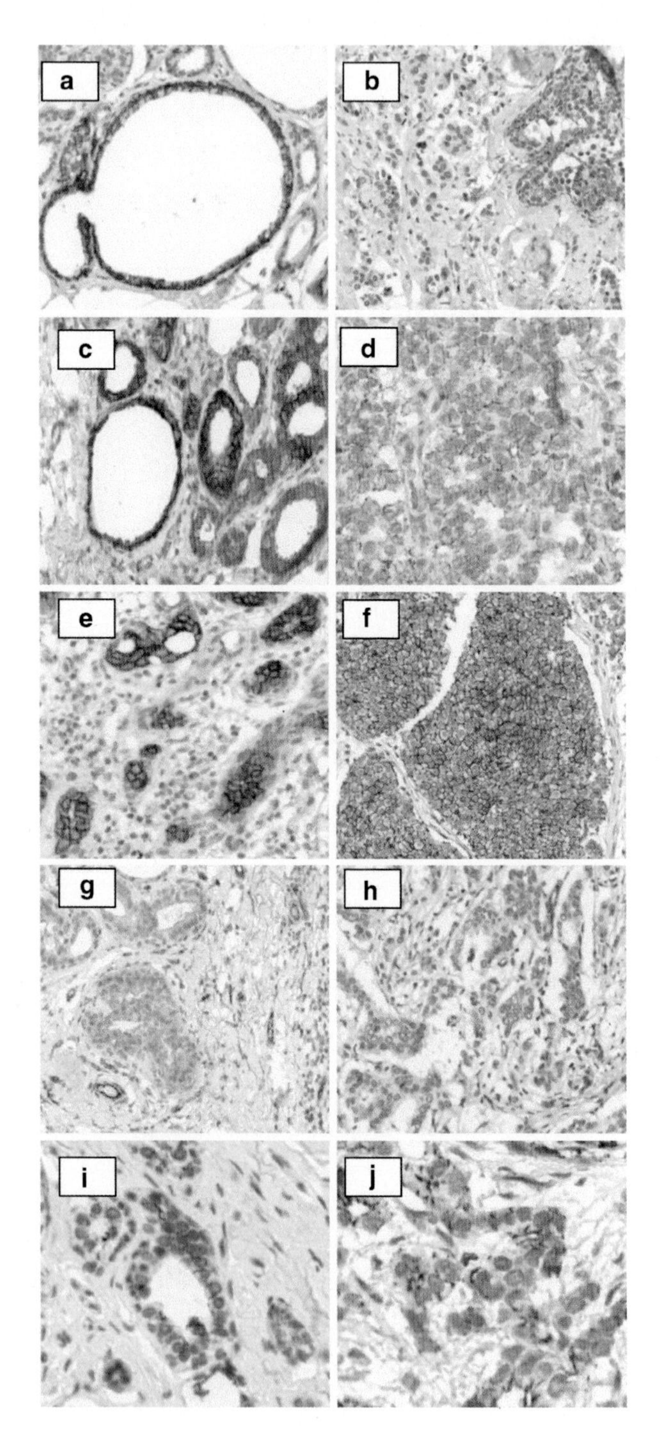
a
b
c
d
e
f
g
h
i
j

2.2.3.1 Tight Junctions in the Acoustic Organs

Earlier it was presented by Ben-Yosef et al. (2003) that CLDN14 deficient mice are deaf due to the degeneration of cochlear hair cells. CLDN14 is expressed in TJs lining the intracochlear space, which contains fluid high in K+ (Ben-Yosef et al. 2003). According to data published in 2010, mutations in CLDN14 cause profound deafness in mice and humans (Bashir et al. 2010). CLDN9 was shown to be essential for the preservation of sensory cells in the hearing organ by protecting the basolateral side of hair cells from the K3 endolymph (Nakano et al. 2009).

Four different recessive mutations of TRIC have been shown to cause non-syndromic deafness. In the inner ear TRIC is concentrated at the tricellular TJs in cochlear and vestibular epithelia, including the structurally complex and extensive junctions between supporting and hair cells (Riazuddin et al. 2006).

2.2.3.2 Tight Junctions in the Corneal Epithelial Cells

Among the various protein components of TJs ZO1 is expressed in the superficial cells of the corneal epithelium (Sugrue and Zieske 1997; Ko et al. 2009a). In a study by Ko et al. (2009a) analysis of the effect of substance P (SP) as neurotransmitter on the expression of ZO1, occludin, CLDNs 1, 2 and 7 found that incubation of the human corneal epithelial cells (HCE) with SP resulted in significant increase in ZO1 expression, but did not affect the expression of occludin and CLDNs 1, 2 or 7, suggesting that ZO1 is an important component of TJs at least in HCE cells (Ko et al. 2009a, b, c).

2.2.3.3 Tight Junctions in the Nasal Olfactory Mucosa

In the olfactory epithelium TJs provide barrier and adhesion properties between neurons, epithelial and glial cells. Studies of Tserentsoodol et al. (1998, 1999) showed that occludin and GLUT1 were specifically expressed in the cells of the

Fig. 2.1 Claudin expression in breast tissue. (**a**) Epithelial cells exhibit high CLDN1 positivity in the cell membranes in benign breast lesion. (**b**) Note the absence of membrane staining in tumor cells compared to epithelial cells observed in benign breast epithelium. (**c**) CLDN3 in benign breast tissues. Continuous membrane staining characterizes most of the epithelial cells. (**d**) Decreased CLDN3 positivity is apparent in the membranes of the majority of carcinoma cells. (**e**) Intense CLDN4 positivity is seen in benign luminal breast epithelium. (**f**) CLDN4 is highly expressed in this case of invasive ductal breast carcinoma of grade 3. Positive reaction is evident in the membranes of the tumor cells. (**g**) CLDN5 in benign breast. CLDN5 is expressed in endothelial cells. (**h**) CLDN5 is expressed in endothelial cells and scattered CLDN5 expression was also seen in the membrane of some of the tumor cells. (**i**) CLDN7 expression in benign breast epithelium. Membranous CLDN7 positivity is seen in some of the luminal epithelial cells. (**j**) CLDN7 expression in invasive breast carcinoma

2.3.3 Tight Junction Expression in Epithelial Lining of the Guts

Several recently published data have reviewed the structure and function of TJs as having principal role in regulating paracellular transport across the intestinal epithelium (Anderson and Van Itallie 2009; de Kort et al. 2011).

TJs are considered to be dynamic structures. TJs are open and closed in response to various stimuli, such as viral or bacterial agents, dietary products, inflammatory mediators (de Kort et al. 2011). Zonulin is considered a physiological modulator of intercellular tight junctions in intestinal mucosa, as being involved in trafficking of macromolecules and, therefore, in tolerance/immune response balance. When the finely tuned zonulin pathway is deregulated in genetically susceptible individuals, both intestinal and extraintestinal autoimmune, inflammatory, and neoplastic disorders may occur (Fasano 2011). Zonulin binds to the zonulin receptor on intestinal epithelial cells inducing in this way rearrangement of the cytoskeleton, downregulation of ZO1, occludin and finally causing disruption of TJ complex integrity (Groschwitz and Hogan 2009). It is intriguing to find how the intestinal barrier function is altered in different diseases where intestinal permeability is affected. Ultrastructural examination of the duodenum from diabetic patients revealed altered TJ structure and an increase in the paracellular space between epithelial cells as compared with healthy control subjects (Secondulfo et al. 2004). In a study performed at our institution, increased CLDN2 and 3 expressions were observed in the proximal and distal part of the duodenum in patients with Coeliac disease (CD) compared with controls. The authors observed a significant difference concerning CLDN2 in case of mild villous atrophy (Group I) in the distal region and in pronounced atrophy (Group II) in the bulb and the distal region as well in CD patients, when compared with controls. CLDN2 expression was significantly increased in the proximal and distal part of CD patients with severe atrophy (Group II) in comparison to the mild atrophy (Group I). Similar changes were observed with CLDN3, too. CLDN3 expression was significantly increased in the proximal and distal part in CD patients with pronounced atrophy (Group II) in comparison to the mild atrophy (Group I). Expression of CLDN4 was similar in all groups studied (Szakal et al. 2010).

2.3.4 Tight Junctions in the Parenchyma of Liver

TJs in hepatocytes play crucial role as barriers to keep bile in bile canaliculi away from the blood circulation. TJ proteins of hepatocytes are regulated by several factors, like cytokines and different growth factors (Kojima et al. 2011). Several evidences of changes in the molecular composition of TJs are noticeable in a number of pathological conditions, especially during tumorigenesis. Since the discovery that receptor CLDN1 plays an important role in the entry of Hepatitis C virus (HCV) after viral binding to CD81, more and more papers present results on TJ expression in hepatocytes (Helle and Dubuisson 2008; Evans et al. 2007). Representative images presenting the expression of different CLDNs in human normal and neoplastic liver are presented in Fig. 2.2a–h. In 2009 Meertens L et al. found that CLDNs 1, 6, and 9 are entry co-factors for hepatitis C virus (Meertens et al. 2008). A recent result of Zadori et al. (2011) by means of analysing CLDN1 expression on 24 hepatic biopsies from liver transplant patients found that CLDN1 immunostaining localized to both the basolateral and the apical surfaces of the hepatocytes, but the immunoreaction was stronger at the apical surface. Greater fibrosis showed higher CLDN1 expression in these localizations. The authors also raised the possibility that low CLDN1 expression at the time of HCV recurrence correlates with a better chance of the patient to achieve an end-of-therapy virologic response and a lower fibrosis score (Zadori et al. 2011). Mensa et al. (2011) found that hepatitis C recurrence after liver transplantation was associated with increased levels of CLDN1 and occludin in the hepatocyte cell membrane (Mensa et al. 2011).

Different types of CLDNs were also associated with diverse forms of carcinomas, including hepatocellular carcinoma (HCC). In 2007 Higashi Y et al. found that loss of CLDN1 expression correlates with aggressive behaviour of HCC (Higashi et al. 2007). A recent study by Huang et al. (2011) concluded that CLDN10 protein is highly expressed in HCC tissue and HCC patients with high CLDN10 expression had significantly shorter overall survival (Huang et al. 2011).

In 2006 Halasz J et al. aimed to explain the molecular mechanism underlying the main epithelial components of hepatoblastomas (HBs) based on the composition of TJs. They analysed fourteen formalin-fixed, paraffin-embedded surgical resection specimens of HB by immunohistochemistry for CLDNs 1, 2, 3, 4 and 7. They found significantly increased protein and messenger RNA expression of CLDNs 1 and 2 in the fetal as compared with the embryonal component. Both cell types displayed negative or weak immunostainings for CLDNs 3, 4, and 7. The authors concluded that increased expressions of CLDNs 1 and 2 characterize the more differentiated fetal component in HBs and are reliable markers for differentiating fetal and embryonal cell types in HBs (Halasz et al. 2006).

2.3.5 *Tight Junctions in the Biliary Tract*

TJs play an essential role in maintaining cell polarity and determining paracellular permeability in organs of epithelial origin. Cell adhesion, polarity and intercellular communication are especially important in the trabecular structure of the liver and in the formation of the bile duct system, where tight junctions separate bile flow from plasma (Lodi et al. 2006). Interesting data were presented by Lódi et al. (2006) by analysing CLDN4 expression in biliary tract cancers, in hepatocellular carcinomas, in normal liver and normal extrahepatic biliary duct samples. They found intense membranous immunolabeling for CLDN4 in all biliary tract cancers unrelated to the primary site of origin, namely intrahepatic, extrahepatic or gallbladder cancers. According to their results normal biliary epithelium showed weak positivity for CLDN4 (Lodi et al. 2006). The studies of Nemeth et al. (2009a, b) analysed CLDNs 1, 2, 3, 4, 7, 8, 10, ZO1 and occludin in normal and neoplastic biliary tract

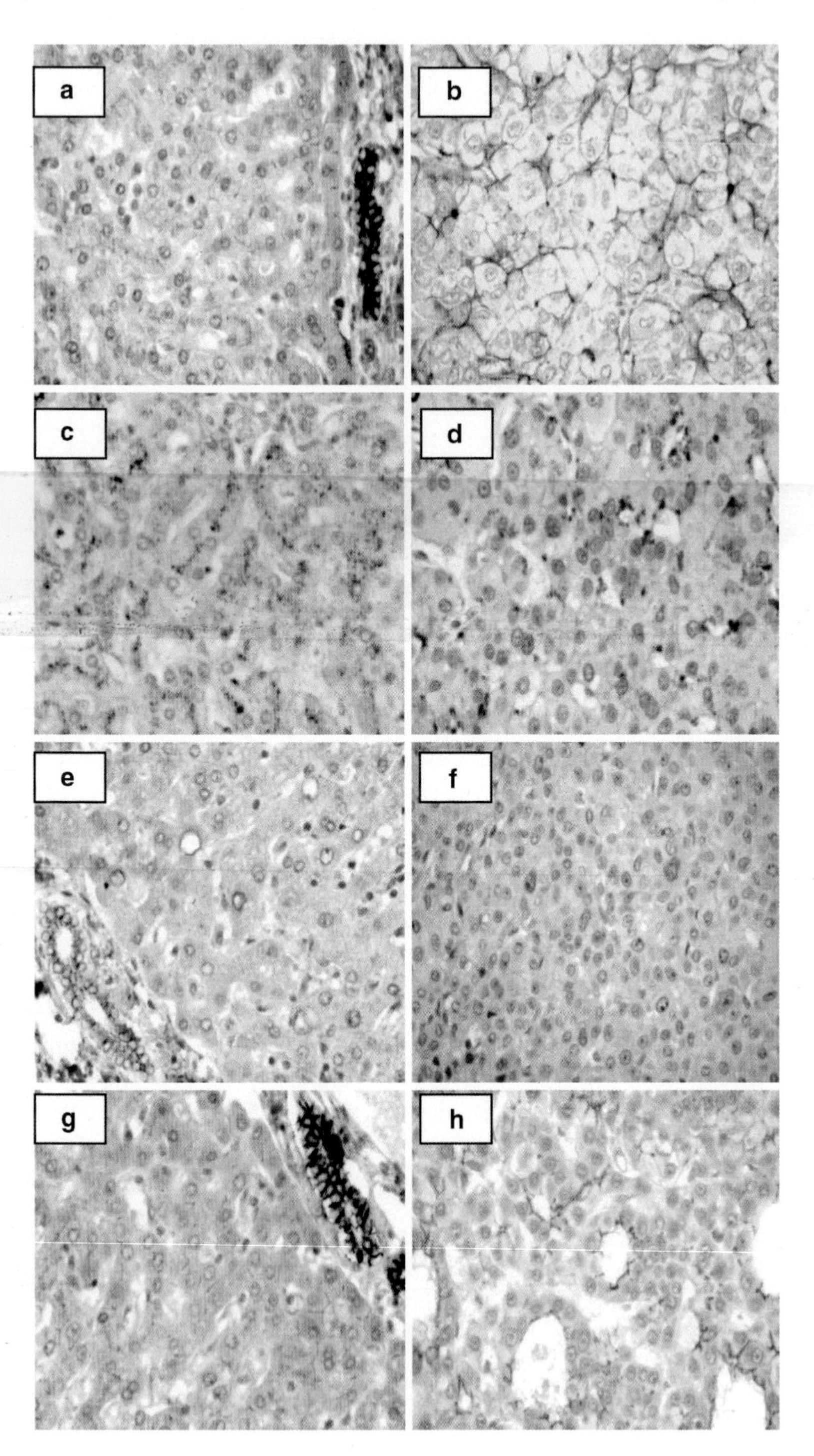
a
b
c
d
e
f
g
h

samples and revealed that CLDN expressions differed in the normal tissue samples: the normal gallbladder strongly expressed CLDNs 2, 3, 4, and 10, but only weak reaction was observed in normal intrahepatic bile ducts. Although each cancer type expressed several CLDNs with various intensities, only CLDN4 presented especially strong immunoreactions in extrahepatic bile duct cancers and gallbladder carcinomas, whereas CLDNs 1 and 10 were present in intrahepatic bile duct cancers. When they compared the normal and carcinoma groups, the most significant decrease was detected in the expression of CLDN10. The authors concluded that the expression pattern of CLDNs is different in the various parts of the normal and neoplastic biliary tract and an unequivocal decrease characterizes the carcinomas compared with their corresponding normal samples. They also found that ZO1 and occludin are downregulated as well in carcinomas arising from various compartments of the biliary tract (normal intrahepatic and extrahepatic bile ducts, gallbladder) as compared with their normal sites of origin (Nemeth et al. 2009a, b).

2.3.6 Tight Junctions in the Parenchyma of Pancreas

Relatively little is known about TJs not only in normal human pancreatic cells but also in pancreatic duct carcinomas. Figure 2.3a–h shows some representative images about the expression of different TJ proteins in the parenchyma of pancreas.

According to the data presented by Borka et al. (2007) who analysed CLDNs 1, 2, 3, 4 and 7 in normal pancreatic tissue, CLDN immunoreaction was predominantly seen in cell membranes as linear staining except for CLDN2, which showed granular cytoplasmic reaction. They reported CLDN1 protein positivity in normal acini and pancreatic ducts; however, the endocrine islands were negative. CLDN2 reaction was expressed scattered in the ducts whereas the authors reported that acini and Langerhans islands were negative. They detected CLDN3 protein in the exocrine glands, ducts, and endocrine cells as well. CLDN4 protein was expressed in pancreatic ducts as well as in the acini; however, endocrine islands were negative. CLDN7 positivity was described in the pancreatic ducts, the acini, and the Langerhans islands, too (Borka et al. 2007).

Recently it was presented that CLDN18 is highly expressed in pancreatic intraepithelial neoplasia, including precursor PanIN lesion and pancreatic duct carcinoma, this protein may therefore serve as a diagnostic marker (Karanjawala et al. 2008). However, little is known about how CLDN18 is regulated, not only in pancreatic

Fig. 2.2 Claudin expression in the liver. (**a**) Note the faint membranous expression of CLDN1 in normal liver cells and high CLDN1 expression in the biliary duct epithelium. (**b**) High CLDN1 expression in hepatocellular carcinoma compared to normal liver. (**c**) CLDN2 revealed perimembranous and cytoplasmic granular reaction in normal liver. (**d**) Reduced granular CLDN2 reaction in hepatocellular carcinoma. (**e** and **f**) CLDN4 is observed neither in normal liver cells nor in hepatocellular carcinoma cells. In contrast CLDN4 is detected in the biliary duct epithelium. (**g**) Moderate CLDN7 is detected in normal liver. High CLDN7 expression is present in biliary duct epithelium. (**h**) CLDN7 is expressed by hepatocellular carcinoma cells

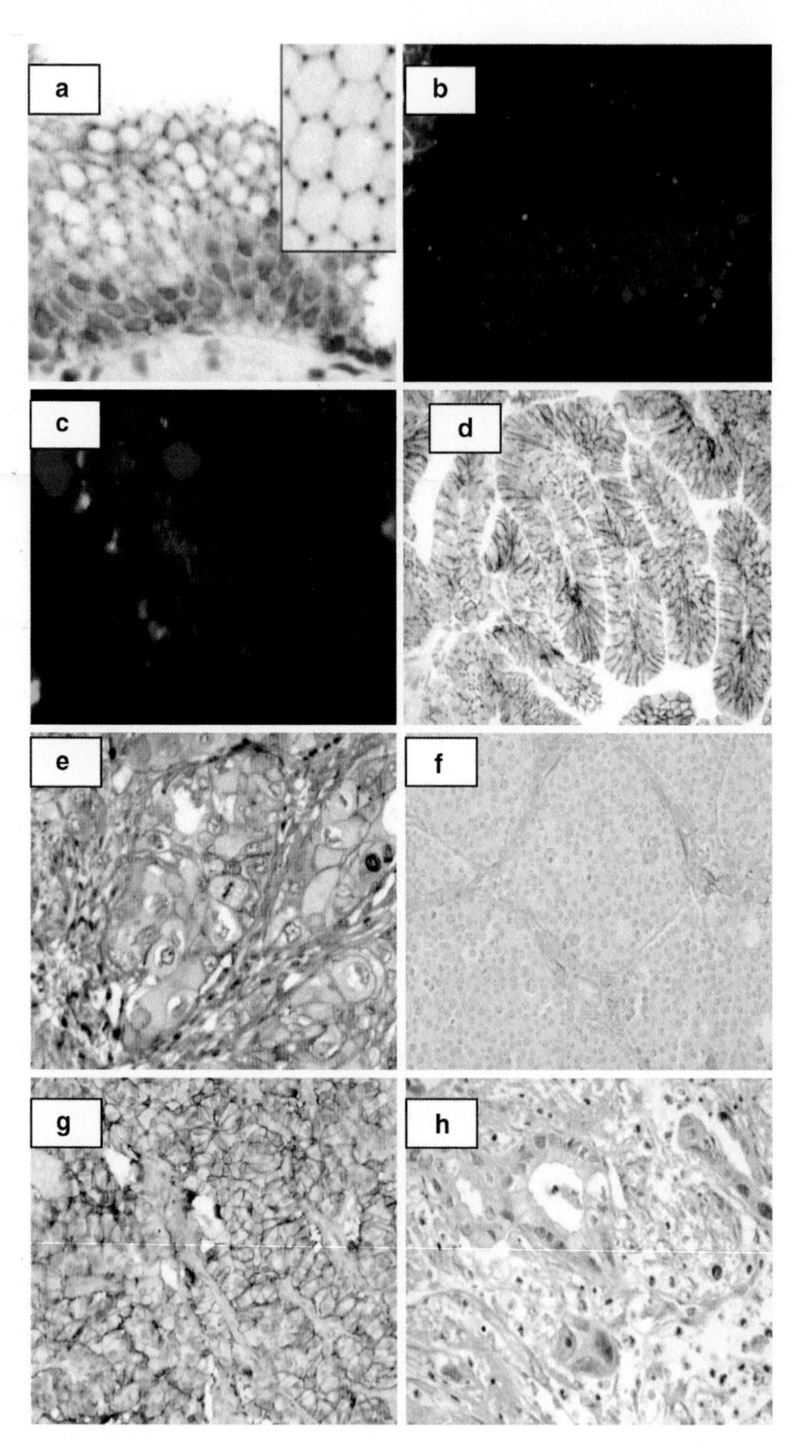
a
b
c
d
e
f
g
h

duct carcinoma, but also in normal human pancreatic duct epithelial cells. In a very recent study by Ito T et al. it was suggested that in human pancreatic cancer cells, CLDN18 is primarily regulated at transcriptional level via specific PKC signaling pathways and modified by DNA methylation (Ito et al. 2011).

As observed in other tumor types, different TJ signatures were observed in pancreatic tumors of different origin and biological behavior. Data on the expression of different CLDNs in pancreatic endocrine tumors (PET) were first presented by Borka et al. (2007) who found that CLDN2 was expressed in half of ductal adenocarcinomas, while the vast majority of endocrine tumors were negative. CLDN1, 4, and 7 immunohistochemistry was positive in all adenocarcinomas, whereas endocrine tumors were completely negative for CLDNs 1 and 4. CLDN3 and 7 proteins were detected in all endocrine tumors, while CLDN3 was not detected in ductal adenocarcinomas. The authors suggested that PET and ductal carcinomas are specifically characterized by different expression patterns of CLDNs (Borka et al. 2007) and Fig. 2.3e–h.

Comper et al. (2009) analysed the expression patterns of CLDNs 1, 2, 3, 4, 5 and 7 in different types of pancreatic tumors with the finding that all solid-pseudopapillary tumors of the pancreas showed intense membrane CLDN5 and cytoplasmic CLDN2 staining, lack of CLDNs 3 and 4 and positive cytoplasmic CLDN1 and 7 stainings in a few cases. Conversely, pancreatic endocrine tumors, pancreatic acinar cell carcinomas and pancreatoblastomas showed strong membrane expression of CLDN 7 and lack of CLDN5, whereas CLDNs 1, 2, 3, and 4 showed variable expression among the samples (Comper et al. 2009).

2.3.7 Tight Junctions in the Epithelial Lining of Urinary Bladder and Urethra

Epithelial cells that line the interior surface of the urinary bladder form an important barrier against noxious substances in urine. This barrier results from the formation of TJs as well as adherens junctions (Keay et al. 2011). The importance of tight junction morphology in the maintenance of integrity of uroepithelium and further, in the prevention of recurrence of urothelial cell carcinoma (UCC) has been recognized for a long time, but the molecular composition of TJs, including the expression of CLDNs has been the focus of studies since

Fig. 2.3 Tight junctions in the parenchyma of pancreas. (**a** and **b**) Tricellulin positivity seen at tricellular contacts in normal pancreatic duct detected by immunohistochemistry (**a**) and by using immunofluorescent techniques (**b**). (**c** and **d**) Tricellulin positivity detected in ductal adenocarcinoma of the pancreas. (**c**) immunofluorescent techniques and (**d**) immunohistochemistry. (**e**) High CLDN4 detection in ductal pancreatic adenocarcinoma. (**f**) Note the absence of CLDN4 expression in pancreatic endocrine tumor. (**g**) Positive linear membrane reaction for claudin 7 in an insulinoma. (**h**) Negative CLDN7 reaction in a ductal adenocarcinoma of the pancreas

recent years (Stravoravdi et al. 1996). In a nicely presented study of Nakanishi et al. (2008) it was described that among the analysed TJ proteins in normal urothelial cells, occludin was expressed in the apicolateral and basolateral surfaces of cells in the umbrella cell layer, but was not found in the intermediate layer or the basal layer (Nakanishi et al. 2008). CLDN1 was mainly expressed in the plasma membranes of the basal and intermediate layers, but was sometimes detected in all the layers. CLDN3 was expressed in the apicolateral surface of the umbrella cell layer, but was not found in the intermediate or basal layer. CLDN4 expressed in the basolateral surface of the umbrella cell layer and sometimes in the apicolateral and basolateral surfaces of the umbrella cell layer and the lateral surface of the intermediate layer, whereas CLDN7 was expressed in the plasma membrane of all three layers. Törzsök et al. (2011), in order to reveal potential prognostic and differential diagnostic values of certain CLDNs (1, 2, 4 and 7), investigated the expression of the mentioned CLDNs in normal bladder mucose, in inverted urothelial papillomas (IUPs), urothelial papillomas (UPs), papillary urothelial neoplasms of low malignant potential (PUNLMPs), and intraepithelial (Ta), low-grade urothelial cell carcinomas (LG-UCCs). They found that the distribution of CLDNs showed urothelium-specific topographical distribution. CLDN1 showed membranous reaction in the basal layers, mainly at the basal surface of cells having connection with the connective tissue, while upper layers showed no staining. CLDN2 revealed perimembranous and cytoplasmic granular reaction, with the reaction being stronger in the basal/parabasal layers. In some cases, the umbrella cells also showed positivity. CLDN3 presented only weak scattered expression, mainly at the membrane of the umbrella cells. CLDN4 positivity was detected in the upper layers, diminishing towards the basal layers. CLDN7 positivity was weak and membranous, detectable in a similar localization as CLDN4. According to the data presented by Törzsök et al. LG-UCCs showed significantly decreased CLDN1 expression in comparison to IUPs. By semiquantitative evaluation, LG-UCCs expressing CLDN4 above the median value were associated with significantly shorter recurrence-free survival. PUNLMPs expressing CLDN1 above the median revealed significantly longer recurrence-free survival (Torzsok et al. 2011) and Fig. 2.4a–h.

Fig. 2.4 CLDN expression in normal urothel and in noninvasive urothelial neoplasms. (**a**) CLDN1 positivity in normal urothel. CLDN1 showed memranous reaction in the basal layers. (**b**) High membranous CLDN1 expression in inverted urothelial papilloma. (**c**) CLDN2 positivity in normal urothel. CLDN2 revealed perimembranous and cytoplasmic granular reaction with the reaction being stronger in the basal/parabasal layers. (**d**) Perimembranous and cytoplasmic granular reaction of CLDN2 in an urothelial papilloma. (**e**) CLDN4 expression in normal urothel. CLDN4 was detected mainly in the upper layers diminishing towards the basal layers. (**f**) High CLDN4 expression in inverted urothelial papilloma. (**g**) CLDN7 expression in normal urothel. CLDN7 positivity was weak and membranous, detectable mainly in umbrella cells. (**h**) CLDN7 positivity in urothelial papilloma. Shown is the weak CLDN7 expression mainly in the uppermost umbrella cell layer

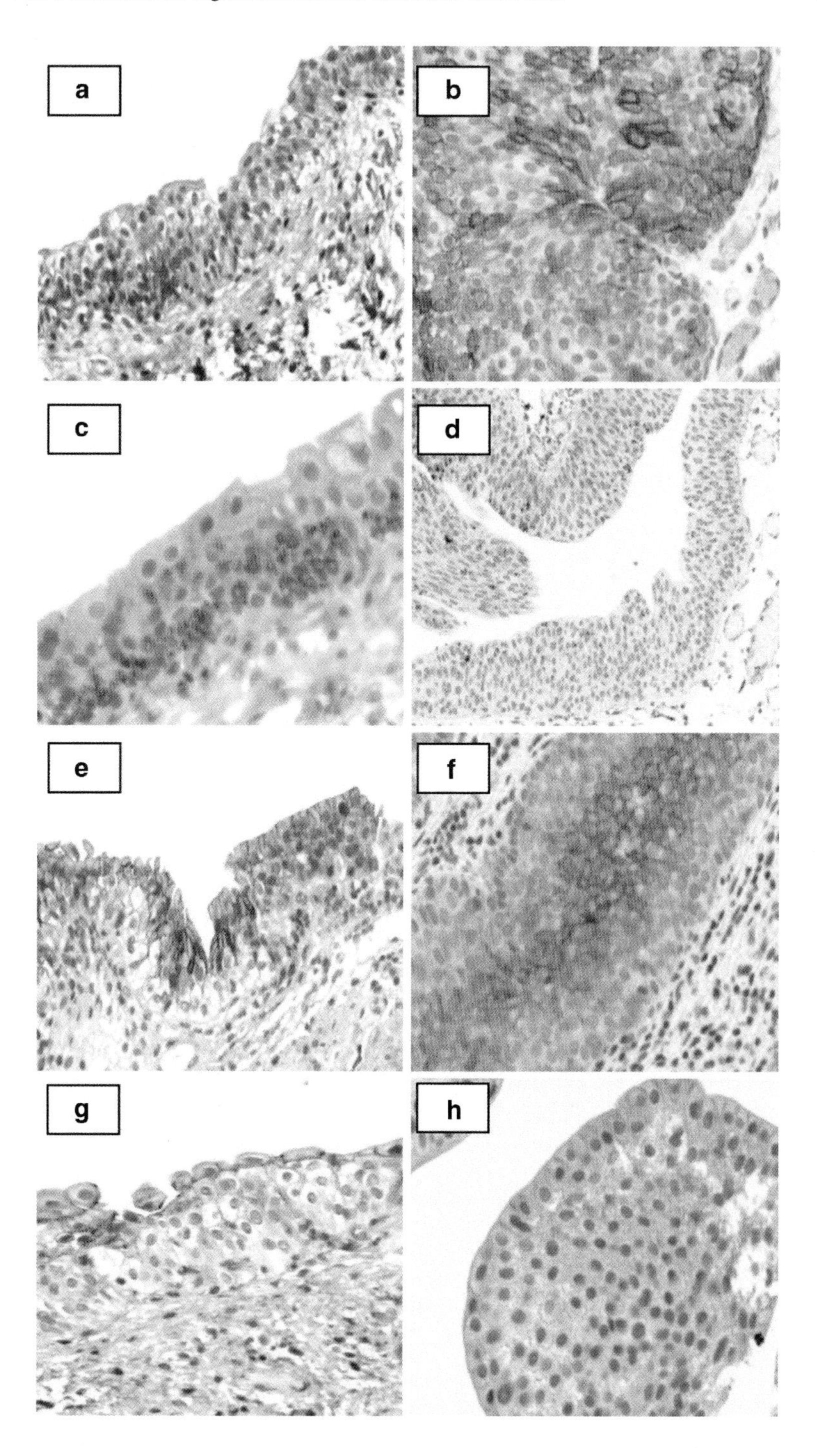
a
b
c
d
e
f
g
h

2.3.8 Tight Junctions in the Parenchyma of Thyroid Glands

Follicular cells of the thyroid gland are arranged in a single polarized layer and function as a barrier between the lumen of the follicle, where thyroglobulin and thyroid hormones are stored, and the extrafollicular space. Epithelial cell polarity and follicular space entrenchment are due to the presence of firm tight junctions (Tzelepi et al. 2008). There are only few publications dealing with the composition of TJs in thyroid glands. In a study by Hewitt et al. (2006), gene expression of several CLDNs (i. e. CLDNs 3, 4, 7) was found and reported in normal and malignant thyroid tissues (Hewitt et al. 2006).

The protein expressions of CLDN1, 4, 7 and occludin in thyroid neoplastic samples were investigated by Tzelepi VN et al. in 2008 with the finding that CLDN1, 4 and 7 expressions were frequently seen in undifferentiated thyroid carcinomas (Tzelepi et al. 2008). They also found that in non-neoplastic thyroid tissues CLDNs and occludin were focally expressed, mainly in hyperplastic follicular cells (Tzelepi et al. 2008). Nemeth et al. (2009), by analysing CLDN1 protein expression in papillary thyroid carcinoma and regional lymph node metastases, found that CLDN1 immunostaining was detected in the majority of primary-papillary thyroid carcinomas and papillary microcarcinomas and in the corresponding lymph node metastases, respectively. They only found weak or no CLDN1 expression in the follicular adenomas and peritumoral non-malignant thyroid tissues. The authors suggest that high CLDN1 protein expression is specific for papillary thyroid carcinoma and its regional lymph node metastases and as a consequence, CLDN1 may be a useful tumor marker for papillary thyroid carcinoma (Nemeth et al. 2010).

2.4 Tight Junctions in Tissues of Mesodermal Origin

2.4.1 Tight Junctions in the Kidney

The role of TJs in transporting epithelia, such as renal tubules, is highly important and has been extensively studied. The net transport of water and solutes across the renal tubular epithelia occurs via two pathways: transcellular and paracellular transport (Muto et al. 2010, 2011). The paracellular transport is governed partly by the TJ complex.

In general, the leakiness of the paracellular pathway largely depends on its ionic conductance between cells and across the TJs and decreases along the nephron from the proximal tubule (the leakiest) to the collecting ducts (Muto et al. 2011). However, knowledge about the detailed profiles is incomplete and controversy remains as concerns the distribution of the different TJ proteins. Several studies analysed the expression of different type of CLDNs, but precisely how the CLDN subtypes are involved in paracellular electrical resistance and charge selectivity remains unclear. Immunolocalization studies have shown that multiple CLDNs are expressed at TJs of individual nephron segments in a nephron segment-specific manner. For example, the glomerulus expresses CLDN1 (Kiuchi-Saishin et al. 2002), the proximal tubule and the thin descending limb express CLDNs 2, 10 (Kiuchi-Saishin et al. 2002; Enck et al. 2001; Van Itallie et al. 2006), the thick ascending limb expresses CLDNs 3, 16 (Kiuchi-Saishin et al. 2002; Haisch et al. 2011) and 19 (Angelow et al. 2007; Konrad et al. 2006), the distal convoluted tubule expresses CLDNs 7 (Li et al. 2004), 8 (Li et al. 2004), 16 (Haisch et al. 2011; Konrad et al. 2006), and 19 (Konrad et al. 2006), whereas the collecting duct expresses CLDNs 4 (Kiuchi-Saishin et al. 2002), 7 (Li et al. 2004) and 8 (Li et al. 2004). In the kidney tubules, CLDNs 16 and 19 share similar expression pattern. Mutation in the CLDN16 gene causes a selective disturbance in renal Mg2+ and Ca2+ reabsorbtion in the thick ascending limb (Blanchard et al. 2001). The importance of CLDNs 16 and 19 was emphasized by the discovery of mutations in these two members of the CLDN family. Familial hypomagnesemia with hypercalciuria and nephrocalcinosis (FHHNC) is an autosomal recessive renal tubular disorder that typically presents with disturbances in Mg and Ca homeostasis, recurrent urinary tract infections, as well as consecutive polyuria and/or polydipsia. Multiple distinct mutations in the CLDN16 gene have been found responsible for this disorder. In a subset of patients carrying this disease mutations in CLDN19 were also identified (Haisch et al. 2011).

2.4.2 Tight Junctions in the Uterine Cervix

Cervical epithelia have numerous functions that include proliferation, differentiation, maintenance of fluid balance, protection from environmental hazards, and paracellular transport of solutes via TJs (Timmons et al. 2007). Several molecular changes have been described during the progression of cervical cancer, even from the early stage (Szabo et al. 2009). Alterations of CLDN expression have been observed in cervical and endometrial cancers as well as premalignant lesion (Sobel et al. 2005, 2006). By using in situ hybridization for the detection of CLDN1 expression Chen et al. (2003) detected low levels of CLDN1 in normal basal epithelial cells (Chen et al. 2003). In a study by Sobel et al. (2005), by analysing samples including cervical intraepithelial neoplasias (CINI-II-III), in situ carcinomas (CIS) and normal cervical samples, it was demonstrated that occludin and CLDN2 colocalized in the normal cervical squamous epithelium. CLDNs 1, 4 and 7 were found coexpressed in the parabasal and intermedier layers in normal epithelia, whereas intensity of occludin staining was decreased in CIN/CIS lesions. The authors detected CLDNs 1, 2, 4 and 7 in the entire epithelium in CIN cases and denoted decreased expression of these proteins in CIS cases. They suggested that significant changes occur in the composition of TJ complexes even in early stages of cervical carcinogenesis (Sobel et al. 2005). According to the data presented by Lee et al. (2005), gradually increased expression of CLDNs 1 and 7 in accordance with progression from low grade intraepithelial lesion to high grade squamous intraepithelial lesion was observed (Lee et al. 2005). Contrary to the data presented by Sobel et al. (2005) the group of Lee could not detect CLDNs 1 and 7 in normal cervical epithelia (Lee et al. 2005).

2.4.3 Tight Junctions in the Endometrium

In a relatively recent study by Gaetje et al. (2008) it is reported that the downregulation of various members of the CLDN family may contribute to endometrial cell detachment and may increase the number of cells colonizing in pelvic organs (Gaetje et al. 2008). In their study Gaetje R et al. analyzed the expression of 13 members of the CLDN family in the endometrium and peritoneum by microarray analysis and found diminished expression of CLDN3, 4 and 7 genes in ectopic endometrium. Altered expression of CLDNs 3 and 4 was also detected in ectopic endometrium in a study performed by Pan et al. (2008). They found significantly lower CLDN3 and 4 expressions in ectopic endometrium than in healthy controls both at RNA and protein level. In 2007, the same group found that CLDNs 3 and 4 were upregulated in endometrial atypical hyperplasia and endometroid adenocarcinoma, as compared with normal endometrium (Pan et al. 2007). Sobel et al. (2006) differentiated endometrial carcinoma on the basis of CLDN expression. Type I endometrial carcinoma and endometrial glandular hyperplasia expressed low levels of CLDN1 and high CLDN2 protein and mRNA expression. Type II (seropapillary, non-endometrioid) endometrial carcinoma showed high CLDN1 and low CLDN2 expressions. The same group described significantly higher CLDN3 expression in both types of carcinomas compared with the normal proliferative phase, as well as higher CLDN4 expression in Type I carcinomas compared with the proliferative phase. They also suggested that CLDN1 may serve as a marker to differentiate Type I and Type II endometrial cancer (Sobel et al. 2006).

2.4.4 Tight Junction Expression in Ovarian Cell Surface and Ovarian Cancer

Human ovarian surface epithelium is considered as a not fully developed epithelium made up of a single layer of mesothelial-type epithelial cells (Zhu et al. 2004). This single layer of cells is considered to be the origin of approximately 90% of all ovarian cancers. Incomplete TJ structure has earlier been demonstrated in the normal ovarian cell surface epithelium (OSE). The results of Zhu et al. (2004) showed that normal human OSE expresses ZO1, occludin, and CLDN1 localized to OSE cell borders both in ovarian biopsies and in cultured OSE. Later the same group, by investigating the distribution of CLDN1, 2, 3, 4 and 5 proteins in cultured OSE, normal ovarian, benign, borderline and ovarian cancer tissues found weak or absent expression of CLDNs 3 and 4 on the surface of OSE. They also described that CLDN3 was significantly increased in ovarian adenocarcinomas compared with benign and borderline-type tumors, whereas CLDN4 was significantly increased in both borderline-type and ovarian adenocarcinomas compared with benign tumors. They found no changes for CLDNs1 or 5. The authors concluded that CLDNs 3 and 4 might be used as novel markers for ovarian tumors (Zhu et al. 2004, 2006).

2.4.5 *Tight Junctions in the Prostate*

Although it has been reported that several CLDN proteins are expressed in the prostate, little is known concerning the regulation of prostatic tight junctions and their potential role(s) in association with prostatic inflammation and other pathological conditions (Sakai et al. 2007).

In a report by Krajewska et al. (2007), who determined the pattern of CLDN1 protein expression in normal prostate, preneoplastic prostatic tissue, and prostate adenocarcinomas (PCa) by using immunohistochemistry, it was found that in benign prostatic epithelium, pronounced CLDN1 expression was observed in the basal cell layer, showing cytosolic and membranous intracellular localization. They observed no staining in the luminal cells. Benign prostatic hyperplasia showed normal CLDN1 staining pattern. The authors found that the majority (98%) of prostate cancers were negative for CLDN1 expression, concluding therefore that CLDN1 expression is uniformly lost in prostate cancers (Krajewska et al. 2007). Concerning the regulation of TJ functions in the prostate, Meng et al. (2011) hypothesized that testosterone regulates components of prostate tight junctions. In their study the authors found that low serum testosterone is associated with reduced transcript and protein levels of CLDNs 4 and 8, resulting in defective tight junction ultrastructure in benign prostate glands, whereas testosterone supplementation in castrated mice resulted in re-expression of tight junction components in prostate epithelium (Meng et al. 2011). In another study by Zheng et al. (2003) it is demonstrated that CLDN7 has both structural and regulatory functions in the prostate. The group of Zheng described two forms of CLDN7: a full-length form with 211 amino-acid residues and a C-terminal truncated form with 158 amino-acid residues. The authors found that both forms of CLDN7 are expressed in human prostate, kidney and lung samples, however in some prostate samples from healthy individuals, the truncated form of CLDN7 was predominant. By analysing LNCaP prostate cell line, the authors found the followings: the two forms of CLDN7 are able to regulate the expression of the prostate-specific antigen (PSA) and the expression of CLDN7 is responsive to androgen stimulation in the LNCaP cell line, suggesting that this protein is involved in the regulatory mechanism of androgen (Zheng et al. 2003). As described earlier in other cancers, TJ components varied according to tumor differentiation and tumor type. In a study by Bartholow et al. (2011) analysis of CLDN3 protein expression in benign prostatic hyperplasia (BPH), prostatic intraepithelial neoplasia (PIN), normal tissue adjacent to prostatic adenocarcinoma (NAC), primary prostatic adenocarcinoma (PCa), and metastatic prostatic adenocarcinoma (mPCa) revealed that PCa and mPCa presented higher CLDN3 expression. Both had significantly higher intensity staining than BPH and NAC. The authors found that PIN had a lower, non-significant staining score than PCa and mPCa, but a statistically higher score than both BPH and NAC (Bartholow et al. 2011). In a very recent study by Szasz et al. (2010) the authors evaluated the expression of CLDNs 1, 2, 3, 4, 5, 7, 8 and 10 on samples of patients who underwent radical prostatectomy for organ confined cancer (pT2N0M0), on samples of clinically advanced cancer cases, and in a control group with benign prostatic hyperplasia. They found that claudin-1 expression could be a novel prognostic marker to distinguish benign and malignant prostatic lesions and CLDN4 seems to be important in cellular differentiation with possible use as a marker of progression of prostatic adenocarcinomas in a clinical setting (Szasz et al. 2010).

2.4.6 *Tight Junctions in the Testis*

In the testis, TJs are found between adjacent Sertoli cells at the level of the blood-testis barrier (BTB). The BTB physically divides the seminiferous epithelium into a basal (where spermatogonia and early spermatocytes are found) and an adluminal compartment (where more developed germ cells are sequestered from the systemic circulation) (Dym and Fawcett 1970; Mruk and Cheng 2011).

BTB is constituted by several different types of coexisting junctions: tight junctions (TJs), basal ectoplasmic specializations (ES) and desmosome-gap junctions (Vogl et al. 2008). For example, JAM-A and -B were found in Sertoli cells, localizing specifically at the BTB (Gliki et al. 2004), whereas JAM-A, -B and -C were present at the site of the apical basal ectoplasmic specializations (ES) (Gliki et al. 2004; Shao et al. 2008). Of these molecules, JAM-A was also localized to the head and flagellum of the sperm (Shao et al. 2008), whereas JAM-C was essential for the polarization of round spermatids during spermiogenesis and for fertility (Gliki et al. 2004). CAR was also found to be expressed by Sertoli cells, localizing to the BTB and apical ES (Wang et al. 2007). While the exact role of these molecules is not fully elucidated, interesting studies are presented about the eventual role of TJ proteins in the function of BTB.

Spermatogenesis takes place in the seminiferous tubules in the adult testes in which developing germ cells must traverse the seminiferous epithelium. It is highly accepted that this complex function involves extensive junction restructuring particularly at the BTB. A cross-talk between TJs and anchoring junctions at the BTB was discussed in the studies of (Yan et al. 2008a, b, c).

The data of Cyr et al. (2011) revealed that the epididymis is altered in infertile patients and mRNA levels of over 400 genes including CLDNs 1, 10 and ZO1 are also altered in infertile men. The same group demonstrated that downregulation of a single CLDN could alter the formation of TJs and is sufficient to compromise the blood-epididymis barrier (Cyr et al. 2011). Nah et al. (2011) found that the expression of CLDN11, as a component of BTB, was increased in impaired spermatogenesis, including hypospermatogenesis and maturation arrest. They also described increased CLDN11 immunoreactivity at the inter-Sertoli tight junctions in maturation arrest (Nah et al. 2011).

References

Alexandre MD, Lu Q, Chen YH (2005) Overexpression of claudin-7 decreases the paracellular Cl- conductance and increases the paracellular Na+conductance in LLC-PK1 cells. J Cell Sci 118:2683–2693

Amasheh S, Meiri N, Gitter AH et al (2002) Claudin-2 expression induces cation-selective channels in tight junctions of epithelial cells. J Cell Sci 115:4969–4976

Amasheh S, Milatz S, Krug SM et al (2009) Tight junction proteins as channel formers and barrier builders. Ann NY Acad Sci 1165:211–219

Anderson JM, Van Itallie CM (2009) Physiology and function of the tight junction. Cold Spring Harb Perspect Biol 1:a002584

Angelow S, Kim KJ, Yu AS (2006) Claudin-8 modulates paracellular permeability to acidic and basic ions in MDCK II cells. J Physiol 571:15–26

Angelow S, El-Husseini R, Kanzawa SA et al (2007) Renal localization and function of the tight junction protein, claudin-19. Am J Physiol Renal Physiol 293:F166–F177

Balda MS, Matter K (2009) Tight junctions and the regulation of gene expression. Biochim Biophys Acta 1788:761–767

Balda MS, Flores-Maldonado C, Cereijido M et al (2000) Multiple domains of occludin are involved in the regulation of paracellular permeability. J Cell Biochem 78:85–96

Balda MS, Garrett MD, Matter K (2003) The ZO-1-associated Y-box factor ZONAB regulates epithelial cell proliferation and cell density. J Cell Biol 160:423–432

Bartholow TL, Chandran UR, Becich MJ et al (2011) Immunohistochemical profiles of claudin-3 in primary and metastatic prostatic adenocarcinoma. Diagn Pathol 6:12

Bashir R, Fatima A, Naz S (2010) Mutations in CLDN14 are associated with different hearing thresholds. J Hum Genet 55:767–770

Bauer HC, Traweger A, Zweimueller-Mayer J et al (2011) New aspects of the molecular constituents of tissue barriers. J Neural Transm 118:7–21

Bazzoni G, Tonetti P, Manzi L et al (2005) Expression of junctional adhesion molecule-A prevents spontaneous and random motility. J Cell Sci 118:623–632

Bellizzi A, Mangia A, Chiriatti A et al (2008) RhoA protein expression in primary breast cancers and matched lymphocytes is associated with progression of the disease. Int J Mol Med 22:25–31

Benton R, St Johnston D (2003) A conserved oligomerization domain in drosophila Bazooka/PAR-3 is important for apical localization and epithelial polarity. Curr Biol 13:1330–1334

Ben-Yosef T, Belyantseva IA, Saunders TL et al (2003) Claudin 14 knockout mice, a model for autosomal recessive deafness DFNB29, are deaf due to cochlear hair cell degeneration. Hum Mol Genet 12:2049–2061

Betanzos A, Huerta M, Lopez-Bayghen E et al (2004) The tight junction protein ZO-2 associates with Jun, Fos and C/EBP transcription factors in epithelial cells. Exp Cell Res 292:51–66

Blanchard A, Jeunemaitre X, Coudol P et al (2001) Paracellin-1 is critical for magnesium and calcium reabsorption in the human thick ascending limb of Henle. Kidney Int 59:2206–2215

Borka K, Kaliszky P, Szabo E et al (2007) Claudin expression in pancreatic endocrine tumors as compared with ductal adenocarcinomas. Virchows Arch 450:549–557

Brandner JM (2009) Tight junctions and tight junction proteins in mammalian epidermis. Eur J Pharm Biopharm 72:289–294

Brandner JM, Kief S, Grund C et al (2002) Organization and formation of the tight junction system in human epidermis and cultured keratinocytes. Eur J Cell Biol 81:253–263

Brandner JM, Kief S, Wladykowski E et al (2006) Tight junction proteins in the skin. Skin Pharmacol Physiol 19:71–77

Brennan K, Offiah G, McSherry EA et al (2010) Tight junctions: a barrier to the initiation and progression of breast cancer? J Biomed Biotechnol 2010:460607

Bruning A, Runnebaum IB (2003) CAR is a cell-cell adhesion protein in human cancer cells and is expressionally modulated by dexamethasone, TNFalpha, and TGFbeta. Gene Ther 10:198–205

Kausalya PJ, Phua DC, Hunziker W (2004) Association of ARVCF with zonula occludens (ZO)-1 and ZO-2: binding to PDZ-domain proteins and cell-cell adhesion regulate plasma membrane and nuclear localization of ARVCF. Mol Biol Cell 15:5503–5515

Keay S, Kaczmarek P, Zhang CO et al (2011) Normalization of proliferation and tight junction formation in bladder epithelial cells from patients with interstitial cystitis/painful bladder syndrome by d-proline and d-pipecolic acid derivatives of antiproliferative factor. Chem Biol Drug Des 77:421–430

Kim KJ, Malik AB (2003) Protein transport across the lung epithelial barrier. Am J Physiol Lung Cell Mol Physiol 284:L247–L259

Kim TH, Huh JH, Lee S et al (2008) Down-regulation of claudin-2 in breast carcinomas is associated with advanced disease. Histopathology 53:48–55

Kirschner N, Bohner C, Rachow S et al (2010) Tight junctions: is there a role in dermatology? Arch Dermatol Res 302:483–493

Kiuchi-Saishin Y, Gotoh S, Furuse M et al (2002) Differential expression patterns of claudins, tight junction membrane proteins, in mouse nephron segments. J Am Soc Nephrol 13:875–886

Ko JA, Yanai R, Nishida T (2009a) Up-regulation of ZO-1 expression and barrier function in cultured human corneal epithelial cells by substance P. FEBS Lett 583:2148–2153

Ko JA, Murata S, Nishida T (2009b) Up-regulation of the tight-junction protein ZO-1 by substance P and IGF-1 in A431 cells. Cell Biochem Funct 27:388–394

Ko JA, Yanai R, Morishige N et al (2009c) Upregulation of connexin43 expression in corneal fibroblasts by corneal epithelial cells. Invest Ophthalmol Vis Sci 50:2054–2060

Kojima T, Takasawa A, Kyuno D et al (2011) Downregulation of tight junction-associated MARVEL protein marvelD3 during epithelial-mesenchymal transition in human pancreatic cancer cells. Exp Cell Res 317:2288–2298

Kominsky SL, Argani P, Korz D et al (2003) Loss of the tight junction protein claudin-7 correlates with histological grade in both ductal carcinoma in situ and invasive ductal carcinoma of the breast. Oncogene 22:2021–2033

Kominsky SL, Vali M, Korz D et al (2004) Clostridium perfringens enterotoxin elicits rapid and specific cytolysis of breast carcinoma cells mediated through tight junction proteins claudin 3 and 4. Am J Pathol 164:1627–1633

Konopka G, Tekiela J, Iverson M et al (2007) Junctional adhesion molecule-A is critical for the formation of pseudocanaliculi and modulates E-cadherin expression in hepatic cells. J Biol Chem 282:28137–28148

Konrad M, Schaller A, Seelow D et al (2006) Mutations in the tight-junction gene claudin 19 (CLDN19) are associated with renal magnesium wasting, renal failure, and severe ocular involvement. Am J Hum Genet 79:949–957

Krajewska M, Olson AH, Mercola D et al (2007) Claudin-1 immunohistochemistry for distinguishing malignant from benign epithelial lesions of prostate. Prostate 67:907–910

Kubo A, Nagao K, Yokouchi M et al (2009) External antigen uptake by Langerhans cells with reorganization of epidermal tight junction barriers. J Exp Med 206:2937–2946

Kulka J, Szasz AM, Nemeth Z et al (2009) Expression of tight junction protein claudin-4 in basal-like breast carcinomas. Pathol Oncol Res 15:59–64

LaFemina MJ, Rokkam D, Chandrasena A et al (2010) Keratinocyte growth factor enhances barrier function without altering claudin expression in primary alveolar epithelial cells. Am J Physiol Lung Cell Mol Physiol 299:L724–L734

Lanigan F, McKiernan E, Brennan DJ et al (2009) Increased claudin-4 expression is associated with poor prognosis and high tumour grade in breast cancer. Int J Cancer 124:2088–2097

Lee JW, Lee SJ, Seo J et al (2005) Increased expressions of claudin-1 and claudin-7 during the progression of cervical neoplasia. Gynecol Oncol 97:53–59

Lelievre SA (2010) Tissue polarity-dependent control of mammary epithelial homeostasis and cancer development: an epigenetic perspective. J Mammary Gland Biol Neoplasia 15:49–63

Li WY, Huey CL, Yu AS (2004) Expression of claudin-7 and -8 along the mouse nephron. Am J Physiol Renal Physiol 286:F1063–F1071

Lodi C, Szabo E, Holczbauer A et al (2006) Claudin-4 differentiates biliary tract cancers from hepatocellular carcinomas. Mod Pathol 19:460–469

Mandell KJ, Parkos CA (2005) The JAM family of proteins. Adv Drug Deliv Rev 57:857–867
Mandell KJ, Babbin BA, Nusrat A et al (2005) Junctional adhesion molecule 1 regulates epithelial cell morphology through effects on beta1 integrins and Rap1 activity. J Biol Chem 280:11665–11674
Martin TA, Watkins G, Mansel RE et al (2004) Loss of tight junction plaque molecules in breast cancer tissues is associated with a poor prognosis in patients with breast cancer. Eur J Cancer 40:2717–2725
Martin TA, Harrison GM, Watkins G et al (2008) Claudin-16 reduces the aggressive behavior of human breast cancer cells. J Cell Biochem 105:41–52
Matsumoto K, Shariat SF, Ayala GE et al (2005) Loss of coxsackie and adenovirus receptor expression is associated with features of aggressive bladder cancer. Urology 66:441–446
Matter K, Balda MS (2003) Functional analysis of tight junctions. Methods 30:228–234
Mazzon E, Cuzzocrea S (2007) Role of TNF-alpha in lung tight junction alteration in mouse model of acute lung inflammation. Respir Res 8:75
Meertens L, Bertaux C, Cukierman L et al (2008) The tight junction proteins claudin-1, -6, and -9 are entry cofactors for hepatitis C virus. J Virol 82:3555–3560
Mehta D (2004) p120: the guardian of endothelial junctional integrity. Am J Physiol Lung Cell Mol Physiol 286:L1140–L1142
Meng J, Mostaghel EA, Vakar-Lopez F et al (2011) Testosterone regulates tight junction proteins and influences prostatic autoimmune responses. Horm Cancer 2:145–156
Mensa L, Crespo G, Gastinger MJ et al (2011) Hepatitis C virus receptors claudin-1 and occludin after liver transplantation and influence on early viral kinetics. Hepatology 53:1436–1445
Merikallio H, Kaarteenaho R, Paakko P et al (2011) Impact of smoking on the expression of claudins in lung carcinoma. Eur J Cancer 47:620–630
Michl P, Barth C, Buchholz M et al (2003) Claudin-4 expression decreases invasiveness and metastatic potential of pancreatic cancer. Cancer Res 63:6265–6271
Mineta K, Yamamoto Y, Yamazaki Y et al (2011) Predicted expansion of the claudin multigene family. FEBS Lett 585:606–612
Mirza M, Raschperger E, Philipson L et al (2005) The cell surface protein coxsackie- and adenovirus receptor (CAR) directly associates with the Ligand-of-Numb Protein-X2 (LNX2). Exp Cell Res 309:110–120
Mitchell LA, Overgaard CE, Ward C et al (2011) Differential effects of claudin-3 and claudin-4 on alveolar epithelial barrier function. Am J Physiol Lung Cell Mol Physiol 301:L40–L49
Miyamoto K, Kusumi T, Sato F et al (2008) Decreased expression of claudin-1 is correlated with recurrence status in esophageal squamous cell carcinoma. Biomed Res 29:71–76
Moldvay J, Jackel M, Paska C et al (2007) Distinct claudin expression profile in histologic subtypes of lung cancer. Lung Cancer 57:159–167
Morita K, Tsukita S, Miyachi Y (2004) Tight junction-associated proteins (occludin, ZO-1, claudin-1, claudin-4) in squamous cell carcinoma and Bowen's disease. Br J Dermatol 151:328–334
Mruk DD, Cheng CY (2011) An in vitro system to study Sertoli cell blood-testis barrier dynamics. Methods Mol Biol 763:237–252
Muto S, Hata M, Taniguchi J et al (2010) Claudin-2-deficient mice are defective in the leaky and cation-selective paracellular permeability properties of renal proximal tubules. Proc Natl Acad Sci USA 107:8011–8016
Muto S, Furuse M, Kusano E (2012) Claudins and renal salt transport. Clin Exp Nephrol 16(1):61–67
Nah WH, Lee JE, Park HJ et al (2011) Claudin-11 expression increased in spermatogenic defect in human testes. Fertil Steril 95:385–388
Naik UP, Naik MU (2008) Putting the brakes on cancer cell migration: JAM-A restrains integrin activation. Cell Adhes Migr 2:249–251
Naik MU, Naik TU, Suckow AT et al (2008) Attenuation of junctional adhesion molecule-A is a contributing factor for breast cancer cell invasion. Cancer Res 68:2194–2203
Nakanishi K, Ogata S, Hiroi S et al (2008) Expression of occludin and claudins 1, 3, 4, and 7 in urothelial carcinoma of the upper urinary tract. Am J Clin Pathol 130:43–49

Nakano Y, Kim SH, Kim HM et al (2009) A claudin-9-based ion permeability barrier is essential for hearing. PLoS Genet 5:e1000610
Nemeth Z, Szasz AM, Somoracz A et al (2009a) Zonula occludens-1, occludin, and E-cadherin protein expression in biliary tract cancers. Pathol Oncol Res 15:533–539
Nemeth Z, Szasz AM, Tatrai P et al (2009b) Claudin-1, -2, -3, -4, -7, -8, and -10 protein expression in biliary tract cancers. J Histochem Cytochem 57:113–121
Nemeth J, Nemeth Z, Tatrai P et al (2010) High expression of claudin-1 protein in papillary thyroid tumor and its regional lymph node metastasis. Pathol Oncol Res 16:19–27
Nguyen DA, Neville MC (1998) Tight junction regulation in the mammary gland. J Mammary Gland Biol Neoplasia 3:233–246
Niessen CM (2007) Tight junctions/adherens junctions: basic structure and function. J Invest Dermatol 127:2525–2532
Nolan ME, Aranda V, Lee S et al (2008) The polarity protein Par6 induces cell proliferation and is overexpressed in breast cancer. Cancer Res 68:8201–8209
Ohnemus U, Kohrmeyer K, Houdek P et al (2008) Regulation of epidermal tight-junctions (TJ) during infection with exfoliative toxin-negative Staphylococcus strains. J Invest Dermatol 128:906–916
Okuyama M, Fujiwara Y, Tanigawa T et al (2007) Roles of ZO-1 and epidermal growth factor in esophageal epithelial defense against acid. Digestion 75:135–141
Osanai M, Murata M, Nishikiori N et al (2007) Occludin-mediated premature senescence is a fail-safe mechanism against tumorigenesis in breast carcinoma cells. Cancer Sci 98:1027–1034
Pan XY, Wang B, Che YC et al (2007) Expression of claudin-3 and claudin-4 in normal, hyperplastic, and malignant endometrial tissue. Int J Gynecol Cancer 17:233–241
Pan XY, Weng ZP, Wang B (2008) Expression of claudin-4 in eutopic and ectopic endometrium of women with endometriosis. Zhonghua Fu Chan Ke Za Zhi 43:418–421
Perou CM (2010) Molecular stratification of triple-negative breast cancers. Oncologist 15(Suppl 5):39–48
Piontek J, Winkler L, Wolburg H et al (2008) Formation of tight junction: determinants of homophilic interaction between classic claudins. FASEB J 22:146–158
Pitelka DR, Hamamoto ST, Duafala JG et al (1973) Cell contacts in the mouse mammary gland. I. Normal gland in postnatal development and the secretory cycle. J Cell Biol 56:797–818
Polette M, Gilles C, Nawrocki-Raby B et al (2005) Membrane-type 1 matrix metalloproteinase expression is regulated by zonula occludens-1 in human breast cancer cells. Cancer Res 65:7691–7698
Prat A, Perou CM (2011) Deconstructing the molecular portraits of breast cancer. Mol Oncol 5:5–23
Prat A, Parker JS, Karginova O et al (2010) Phenotypic and molecular characterization of the claudin-low intrinsic subtype of breast cancer. Breast Cancer Res 12:R68
Pummi K, Malminen M, Aho H et al (2001) Epidermal tight junctions: ZO-1 and occludin are expressed in mature, developing, and affected skin and in vitro differentiating keratinocytes. J Invest Dermatol 117:1050–1058
Raleigh DR, Marchiando AM, Zhang Y et al (2010) Tight junction-associated MARVEL proteins marveld3, tricellulin, and occludin have distinct but overlapping functions. Mol Biol Cell 21:1200–1213
Riazuddin S, Ahmed ZM, Fanning AS et al (2006) Tricellulin is a tight-junction protein necessary for hearing. Am J Hum Genet 79:1040–1051
Sakai N, Chiba H, Fujita H et al (2007) Expression patterns of claudin family of tight-junction proteins in the mouse prostate. Histochem Cell Biol 127:457–462
Secondulfo M, Iafusco D, Carratu R et al (2004) Ultrastructural mucosal alterations and increased intestinal permeability in non-celiac, type I diabetic patients. Dig Liver Dis 36:35–45
Shao M, Ghosh A, Cooke VG et al (2008) JAM-A is present in mammalian spermatozoa where it is essential for normal motility. Dev Biol 313:246–255
Smalley KS, Brafford P, Haass NK et al (2005) Up-regulated expression of zonula occludens protein-1 in human melanoma associates with N-cadherin and contributes to invasion and adhesion. Am J Pathol 166:1541–1554

Sobel G, Paska C, Szabo I et al (2005) Increased expression of claudins in cervical squamous intraepithelial neoplasia and invasive carcinoma. Hum Pathol 36:162–169
Sobel G, Nemeth J, Kiss A et al (2006) Claudin 1 differentiates endometrioid and serous papillary endometrial adenocarcinoma. Gynecol Oncol 103:591–598
Soini Y (2004) Claudins 2, 3, 4, and 5 in Paget's disease and breast carcinoma. Hum Pathol 35:1531–1536
Sourisseau T, Georgiadis A, Tsapara A et al (2006) Regulation of PCNA and cyclin D1 expression and epithelial morphogenesis by the ZO-1-regulated transcription factor ZONAB/DbpA. Mol Cell Biol 26:2387–2398
Steed E, Rodrigues NT, Balda MS et al (2009) Identification of MarvelD3 as a tight junction-associated transmembrane protein of the occludin family. BMC Cell Biol 10:95
Stravoravdi P, Natsis K, Kirtsis P et al (1996) Ultrastructural study of the urothelial lysosomal system in patients with transitional cell carcinoma after transurethral resection and interferon therapy. J Exp Ther Oncol 1:222–225
Sugrue SP, Zieske JD (1997) ZO1 in corneal epithelium: association to the zonula occludens and adherens junctions. Exp Eye Res 64:11–20
Sung CO, Han SY, Kim SH (2011) Low expression of claudin-4 is associated with poor prognosis in esophageal squamous cell carcinoma. Ann Surg Oncol 18:273–281
Szabo I, Kiss A, Schaff Z et al (2009) Claudins as diagnostic and prognostic markers in gynecological cancer. Histol Histopathol 24:1607–1615
Szakal DN, Gyorffy H, Arato A et al (2010) Mucosal expression of claudins 2, 3 and 4 in proximal and distal part of duodenum in children with coeliac disease. Virchows Arch 456:245–250
Szasz AM, Nyirady P, Majoros A et al (2010) beta-catenin expression and claudin expression pattern as prognostic factors of prostatic cancer progression. BJU Int 105:716–722
Szasz AM, Nemeth Z, Gyorffy B et al (2011a) Identification of a claudin-4 and E-cadherin score to predict prognosis in breast cancer. Cancer Sci 102:2248–2254
Szasz AM, Tokes AM, Micsinai M et al (2011b) Prognostic significance of claudin expression changes in breast cancer with regional lymph node metastasis. Clin Exp Metastasis 28:55–63
Timmons BC, Mitchell SM, Gilpin C et al (2007) Dynamic changes in the cervical epithelial tight junction complex and differentiation occur during cervical ripening and parturition. Endocrinology 148:1278–1287
Tokes AM, Kulka J, Paku S et al (2005) Claudin-1, -3 and -4 proteins and mRNA expression in benign and malignant breast lesions: a research study. Breast Cancer Res 7:R296–R305
Tőkés AM, Szász AM, Zuhást E (2012) Expression of tight junction molecules in breast carcinomas analysed by array PCR and immunohisto chemistry. Path Omc Res 18:593–606
Torzsok P, Riesz P, Kenessey I et al (2011) Claudins and ki-67: potential markers to differentiate low- and high-grade transitional cell carcinomas of the urinary bladder. J Histochem Cytochem 59:1022–1030
Traweger A, Fuchs R, Krizbai IA et al (2003) The tight junction protein ZO-2 localizes to the nucleus and interacts with the heterogeneous nuclear ribonucleoprotein scaffold attachment factor-B. J Biol Chem 278:2692–2700
Traweger A, Lehner C, Farkas A et al (2008) Nuclear Zonula occludens-2 alters gene expression and junctional stability in epithelial and endothelial cells. Differentiation 76:99–106
Troy TC, Rahbar R, Arabzadeh A et al (2005) Delayed epidermal permeability barrier formation and hair follicle aberrations in Inv-Cldn6 mice. Mech Dev 122:805–819
Tserentsoodol N, Shin BC, Suzuki T et al (1998) Colocalization of tight junction proteins, occludin and ZO-1, and glucose transporter GLUT1 in cells of the blood-ocular barrier in the mouse eye. Histochem Cell Biol 110:543–551
Tserentsoodol N, Shin BC, Koyama H et al (1999) Immunolocalization of tight junction proteins, occludin and ZO-1, and glucose transporter GLUT1 in the cells of the blood-nerve barrier. Arch Histol Cytol 62:459–469
Tsukita S, Furuse M, Itoh M (2001) Multifunctional strands in tight junctions. Nat Rev Mol Cell Biol 2:285–293
Tunggal JA, Helfrich I, Schmitz A et al (2005) E-cadherin is essential for in vivo epidermal barrier function by regulating tight junctions. EMBO J 24:1146–1156

Turksen K, Troy TC (2002) Permeability barrier dysfunction in transgenic mice overexpressing claudin 6. Development 129:1775–1784
Tzelepi VN, Tsamandas AC, Vlotinou HD et al (2008) Tight junctions in thyroid carcinogenesis: diverse expression of claudin-1, claudin-4, claudin-7 and occludin in thyroid neoplasms. Mod Pathol 21:22–30
Van Itallie CM, Anderson JM (2006) Claudins and epithelial paracellular transport. Annu Rev Physiol 68:403–429
Van Itallie CM, Fanning AS, Anderson JM (2003) Reversal of charge selectivity in cation or anion-selective epithelial lines by expression of different claudins. Am J Physiol Renal Physiol 285:F1078–F1084
Van Itallie CM, Rogan S, Yu A et al (2006) Two splice variants of claudin-10 in the kidney create paracellular pores with different ion selectivities. Am J Physiol Renal Physiol 291:F1288–F1299
Vogl AW, Vaid KS, Guttman JA (2008) The Sertoli cell cytoskeleton. Adv Exp Med Biol 636:186–211
Wang CQ, Mruk DD, Lee WM et al (2007) Coxsackie and adenovirus receptor (CAR) is a product of Sertoli and germ cells in rat testes which is localized at the Sertoli-Sertoli and Sertoli-germ cell interface. Exp Cell Res 313:1373–1392
Watson RE, Poddar R, Walker JM et al (2007) Altered claudin expression is a feature of chronic plaque psoriasis. J Pathol 212:450–458
Yan HH, Mruk DD, Cheng CY (2008a) Junction restructuring and spermatogenesis: the biology, regulation, and implication in male contraceptive development. Curr Top Dev Biol 80:57–92
Yan HH, Mruk DD, Lee WM et al (2008b) Cross-talk between tight and anchoring junctions-lesson from the testis. Adv Exp Med Biol 636:234–254
Yan HH, Mruk DD, Lee WM et al (2008c) Blood-testis barrier dynamics are regulated by testosterone and cytokines via their differential effects on the kinetics of protein endocytosis and recycling in Sertoli cells. FASEB J 22:1945–1959
Yeo NK, Jang YJ (2010) Rhinovirus infection-induced alteration of tight junction and adherens junction components in human nasal epithelial cells. Laryngoscope 120:346–352
Yoshida Y, Morita K, Mizoguchi A et al (2001) Altered expression of occludin and tight junction formation in psoriasis. Arch Dermatol Res 293:239–244
Zadori G, Gelley F, Torzsok P et al (2011) Examination of claudin-1 expression in patients undergoing liver transplantation owing to hepatitis C virus cirrhosis. Transplant Proc 43:1267–1271
Zheng JY, Yu D, Foroohar M et al (2003) Regulation of the expression of the prostate-specific antigen by claudin-7. J Membr Biol 194:187–197
Zhu Y, Maric J, Nilsson M et al (2004) Formation and barrier function of tight junctions in human ovarian surface epithelium. Biol Reprod 71:53–59
Zhu Y, Brannstrom M, Janson PO et al (2006) Differences in expression patterns of the tight junction proteins, claudin 1, 3, 4 and 5, in human ovarian surface epithelium as compared to epithelia in inclusion cysts and epithelial ovarian tumours. Int J Cancer 118:1884–1891

Chapter 3
Methods to Study Tight Junctions

María Isabel Larre, Catalina Flores-Maldonado, and Marcelino Cereijido

Abstract A cell in the ocean exchanges with a constant reservoir, that is not exhausted of nutrients consumed by the cell nor polluted by the wastes it excretes. On the contrary, when a cell belongs to a metazoan, the situation is completely different, as the ocean is now replaced by an extracellular milieu less than one micron thick, that would be quickly exhausted and spoiled, were it not by a circulatory apparatus that continuously carries nutrients and wastes to and from to enormous areas of epithelia, where the exchange with the extracellular environment actually takes place. Thanks to this continuous purification and stability of the internal milieu performed mainly by "*transporting epithelia*", metazoan cells can enormously simplify their housekeeping efforts, and engage instead in differentiation and multiple forms of organization (tissues, organs, systems) that enable them produce an astonishing diversity of higher organisms. Metazoan exist thanks to transporting epithelia.

This chapter summarizes the main methods to study the structure and function of the tight junctions, with natural epithelia, as well as monolayers of cell lines grown *in vitro* on permeable supports.

Keywords Tight junctions • Epithelia • Transporting epithelial phenotype • Polarity • Claudins • Freeze-fracture • TER (transepithelial electrical potential) • Calcium switch

M.I. Larre • C. Flores-Maldonado • M. Cereijido (✉)
Center For Research and Advanced Studies of Mexico (Cinvestav),
Avenida Instituto Politécnico Nacional 2508, Del. Gustavo A. Madero,
México, D.F 07360, Mexico
e-mail: cereijido@fisio.cinvestav.mx

T.A. Martin and W.G. Jiang (eds.), *Tight Junctions in Cancer Metastasis*,
Cancer Metastasis - Biology and Treatment 19, DOI 10.1007/978-94-007-6028-8_3,

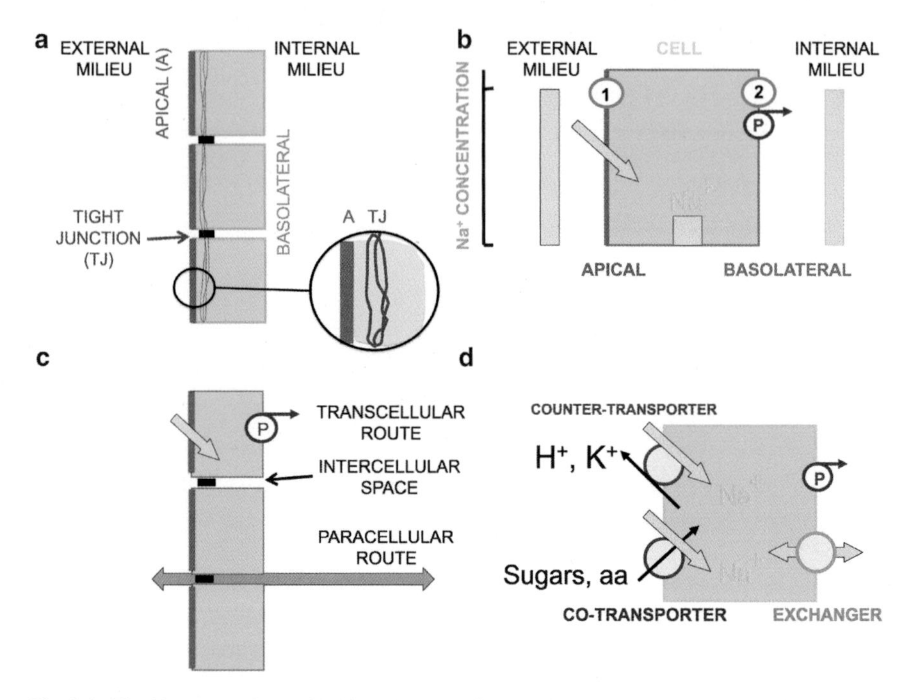

Fig. 3.1 **The "transporting epithelial phenotype"**. (**a**) Epithelial cells form layers with an apical (*magenta*) and a basolateral side (*gray*). The intercellular space is closed on its outermost end by tight junctions (TJs) observed in freeze fracture replicas as a continuous belt of anastomosing intramembrane fibrils (*red*), and in transmission electron microscopy as a fusion of the outer leaflet of the plasma membrane, that together with the submembrane cytoskeleton and specific molecules form an obscure, osmium stained spot (*black*) (see also Fig. 3.2a, b). (**b**) The model of Koeffoed-Johnson and Ussing (KJU) proposes a first step "1" of penetration of Na^+ (*yellow*), down its gradient from the outer milieu into the cytoplasm, followed by its extrusion "2" towards the internal milieu. This step is operated by the Na^+,K^+-ATPase (P *red*). (**c**) Permeating substances traverse the epithelium through a transcellular route, as proposed by the KJU model, plus a paracellular route that crosses the TJ and proceeds through the intercellular space. (**d**) As long as the pump keeps the concentration of Na at a low level, this ion is continuously entering the cytoplasm through the apical membrane. Nature uses this Na-gradient to drive counter-transporters and co-transporters

3.1 Introduction

Transporting epithelia are vast areas of cells (e.g. 90 m^2 of lung, and 270 m^2 of intestinal mucosa in humans) that have two basic differentiated features: (1) ***Tight junctions*** (TJs), that form a continuous belt of intramembrane fibers that seals the outermost end of the lateral intercellular space, completely surrounds the cells, and transforms the layer of epithelial cells into an effective permeability barrier (Fig. 3.1a, *red*). And (2) ***Polarity*** of the cells, consisting in having an apical domain (Fig. 3.1a, *magenta*) in contact with the external milieu, and a basolateral domain (Fig. 3.1a, *gray*) in contact with blood through the interstitial fluid. The apical and basolateral domains have drastically different anatomical, molecular and physiological properties, that constitute the structural basis that allow cells to transport substances vectorially, i.e. in a net amount towards the outer or the internal milieu (Cereijido et al. 1988).

Given that 90% of the cancers in mature humans start or compromise an epithelium, we expect a strong correlation between cancers and altered TJs (Chen et al. 2006; Martin et al. 2011; Cereijido et al. 2000, 2007; Escudero-Esparza et al. 2011). Several claudins, proteins of the tight junction have been evaluated in primary human tumors to examine their expression levels and cancer progression. For example, claudin-1 exhibits a consistent elevation in colon carcinoma (Miwa et al. 2001; Dhawan et al. 2005). Yet claudin-3 and -4 are frequently overexpressed in ovarian, breast, pancreatic, and prostate cancers (Morin 2005; Hewitt et al. 2006). Down regulation of claudins could contribute to epithelial transformation by increasing the paracellular permeability of nutrients and growth factors to cancerous cells. For example, claudin-11 decreases the invasiveness of bladder cancer cell (Awsare et al. 2011), Claudin-7 decreases or disappears in breast cancer, and head and neck squamous cell carcinoma (Kominsky et al. 2003; Al Moustafa et al. 2002). On the other hand, the barrier function of neoplastic cells could also be altered. At this moment, it is not easy to predict whether this correlation will involve causality, i.e. a cancer would interfere with the normal expression of a TJ or, conversely, alterations of the TJ would cause or favor the development of a cancer, or interfere with pharmacologic treatment of a cancer. This may be the case of a cancer of the nervous system, when it cannot be reached by pharmaceutical agents because of the blood/brain barrier. It is in this case that one would wish that knowledge of the modulation of TJs would afford a way of circumvent the hermeticity of this structure.

In this chapter we will focus on TJs, and the main technical procedures and protocols to study them.

3.2 The Tight Junction (TJ) in Retrospect

It takes a good microscope to –scarcely- see a TJs, that is why this structure only started to be mentioned in biomedical publications in the second half of the nineteenth century, under a wide variety of names, all of them pejorative, as cell contacts were regarded as little more than neutral bolts and fasteners that secured the framework of the tissue, lest it would disintegrate on deformation. Even a century later the TJ continued to be disregarded, and was not even represented in the seminal model put forward by Koeffoed-Johnson and Ussing (1958) that served as fertile blueprint for all transporting epithelia (Fig. 3.1b) (Koefoed-Johnsen and Ussing 1958). This disregard was due to the fact that in those days it was taken for granted that substances only cross an epithelium through the so called "transcellular route", in two successive steps: a first one to penetrate from the external milieu into the cytoplasm of epithelial cells, and a second one to exit from the cytoplasm towards the internal milieu.

Later on, several circumstances brought TJs to the forefront: (1) epithelia with an exclusive transcellular route ("tight epithelia") are an extreme case, as most natural epithelia have also a "paracellular route" that does not cross through the cytoplasm of the cells (Fig. 3.1c, *lilac*). (2) the structure that determines whether the paracellular route is important or negligible is, precisely, the TJ. (3) the overall permeability (transcellular *plus* paracellular) is reflected by the transepithelial electrical resistance (TER) of the epithelium, so this parameter is widely used, as it is also an easy to measure one (*see below*). Epithelial TER ranges from a mere 8–10 $\Omega.cm^2$ as in the proximal tube of the nephron, to hundreds of thousands as in epithelia like the urinary bladder. (4) A given TJ does not have a constant structure nor TER, as these are adjusted to physiological conditions, hormones and pharmacological challenge. When required, the TJ can relax its hermeticity enough to allow the passage of an entire cell, as it is the case of macrophages en route to a spot invaded by microorganism. (5) The TJ is formed by more than 50 different species of proteins, most of them so specific that are used as markers of this cell contact. Some of them are membrane proteins exposed to the intercellular space and contact with analogous proteins belonging to a neighboring cell (e.g. occludin, claudins, JAMs). There is a group of TJ-proteins that form the submembrane scaffold (e.g. ZO-1, ZO-2, ZO-3, ZONAB). There is another a group of proteins that relate the TJ to the cytoskeleton (e.g. cingulin); (6) some of these peptides exist in several chemical states, because they undergo a variety of phosphorylations, and are governed by several routes of intracellular signaling, a scenario that seems far too complex to fulfill the humble role of neutral bolt that secures the epithelial framework (Turksen and Troy 2004). (7) There is an ever growing body of evidence that the TJ is involved in some grave human diseases; in particular those associated to auto-immunity. Thus one century ago, sages of the stature of Paul Ehrlich regarded the intestinal flora as a nuisance that intoxicated the organism, to the point that some enthusiastic followers resorted to abdominal surgery to remove large segments of gross intestine, hoping to favor a longer and healthier life. Today the intestinal fauna is considered instead a quasi-organ, whose cells can even transiently penetrate into the epithelial mucosa to participate in a now re-assessed collaboration. (8) The intestinal fauna exchanges substances and signals with the rest of the organism. (9) Yet if for some abnormal reason (e.g. a molecular defect in a molecule constituting the TJ) some of the peptides produced by the flora gain access to the blood, the immune system may generate antibodies that also attack normal proteins in the thyroid gland, the brain, pancreas, etc. thereby triggering terrible diseases such as Hashimoto's thyroiditis, multiple sclerosis, diabetes, Crohn disease, etc. (Cereijido et al. 2007).

3.3 The Ability of the TJ to Adjust Its Hermeticity to Physiological Needs

The fluid just filtered in the renal glomerulus has virtually the same composition of water, amino acids, glucose, urea, and ions than plasma; therefore the chemical gradient across the walls of the proximal tubule is not sharp, and the hermeticity of the epithelial wall is indeed very low (some 8–10 $\Omega.cm^2$). Yet as the fluid in the lumen flows towards more distal parts of the nephron, the active removal of chemical components (e.g. sugars, vitamins, amino acids) and the addition of some others (e.g. urea, urobilin, certain ions such as potassium, hydrogen ions), make the tubular fluid more and more different from plasma, requiring tighter TJs to withstand the now sharp gradients between the lumen and the interstitial fluid. Accordingly, the value of TER across the wall of the urinary tract increases, reaching 1,000 $\Omega.cm^2$ in the distal nephron, 2,000 in the collecting tube, and 60,000–100,000 $\Omega.cm^2$ in the urinary bladder.

While this is teleologically sound, we still ignore the mechanisms that sense the asymmetry of composition and adjust TER accordingly. This information would be very valuable, as it may help to develop molecular tools to correct the leakiness of the TJs (Flores-Benitez et al. 2009; Contreras et al. 2006) that, as mentioned above, is deemed responsible for many autoimmune diseases.

3.3.1 Specific Techniques an Approaches to Study TJs

This description will be based on the rational as well as use of a given technique, not in the detailed protocol to really use it in a given experiment. These are provided in the Methods section of the references to a publication that uses it.

3.3.1.1 Transmission Electron Microscopy (TEM)

The religious tradition of body/soul, as well as the nineteenth century idea that life consists in structures that function, are misleading and unsuited to understand life at the atomic, molecular and organelle levels, as function comports a change in structure, and *vice versa*: changes of structure have functional consequences. Therefore, in order to "see"a TJ its function must be suddenly stopped with extremely low temperature and quickly penetrating fixatives that crosslink its proteins, so that in spite of being now dead, its structure resembles one of the configurations it had when it was alive. In spite of the fact that water constitutes up to 70–90% of the substance of a cell, it must be totally removed while preserving the micro-anatomy of the epithelium. Then another problem is in line: an epithelium is too thick to permit the passage of the electron beam of the ultramicroscope. Therefore the preparation has to go to several steps of dehydration, while it is infiltrated by embedding resins, again, without distorting the ultramicroanatomy of the TJ, so that the epithelium becomes rock-hard and can be cut with a diamond knife in ultrathin layers that will allow the passage of the electron beam. A third difficulty arises from the fact that the plasma membrane offers almost no resistance to the passage of the electron beam, is invisible, and therefore one cannot see how does it form a TJ. Nevertheless a treatment with OsO_4 binds Os to the two lipid leaflets of the plasma membrane, that now can be seen as a double line resembling the track of a railroad (Fig. 3.2b). Therefore when the plasma membrane of neighboring cells approach each other,

3.3.1.3 Flux of Dextran

A flux is the amount of a given substance that crosses an epithelium per unit time and unit area, and is usually expressed as $\mu mole.h^{-1}.cm^{-2}$. It can be measured under a mindboggling variety of conditions: e.g. as a function of the concentration of the substance, in the presence of competing analog molecules, in epithelia treated with all sort of inhibitors, etc. One usually starts with the measurement of a small inert molecule that would serve as comparison for the flux of other substances, or the flux of the same substance but in a variety of epithelia (e.g. ileum mucosa, gallbladder epithelium), or the same epithelium but from different animal species. Although we will illustrate the measurement with Dextran, it can be substituted by any other substance that would not damage the cells.

Dextran is a complex, branched polysaccharide made of many glucose molecules, that are available in chains of the desired length and molecular weight. The permeability of electrically neutral molecules across the TJ is measured with dextran of some 3 kDa, that it is too big to cross through the transcellular route, a circumstance that makes it ideal to study the paracellular one. To make its flux easy to measure, Dextran is commonly bound to fluorescent tags, e.g. FITC-Dextran. A freshly prepared solution with 10 μg/ml of FITC-Dextran is dissolved in P-buffer, and placed in the chamber in contact with the apical compartment. The one in contact with thc basolatcral sidc only contains P-buffer with or without the drug whose effect on TJs is investigated (e.g. 10 nM ouabain). After 1 h incubation at 37°C, the basal medium is collected, and the fluorescence of the transported FITC-Dextran is measured with a fluorescence spectrometer at 492 nm (excitation) and 520 nm (emission). The quantity of FITC is calculated by comparing with a standard curve. The unidirectional flux of Dextran from the apical to the basolateral direction (J_{DEX} in $ng.cm^{-2}.h^{-1}$) is calculated by dividing the fluorescence intensity of a given sample of the bottom solution by the corresponding value of the upper solution conveniently diluted.

3.3.1.4 Transepithelial Electrical Resistance (TER)

The degree of sealing of TJs can be assessed by measuring the TER of the monolayer as depicted in Fig. 3.3. A small current ($20\ \mu A.cm^{-2}$) causes a voltage deflection which is measured with a voltohmeter (EVOM) and an EndOhm-6 systems (World Precision Instruments, Sarasota, FL). Ohm's Law is then used to calculate an overall TER, and the component due to the resistance of monolayer itself can be obtained by subtracting the resistance of the bathing solution and the empty support. Results are expressed as $\Omega\ cm^2$. The specific conductance of the TJ to a given ion can be studied by substituting control ions (usually Na^+, K^+, or Cl^- by other univalent cations and anions (Larre et al. 2010, 2011; Cereijido et al. 1978)).

Frömter and coworkers have used a more refined procedure based on impedance measurement (see (Kottra et al. 1989, 1996)).

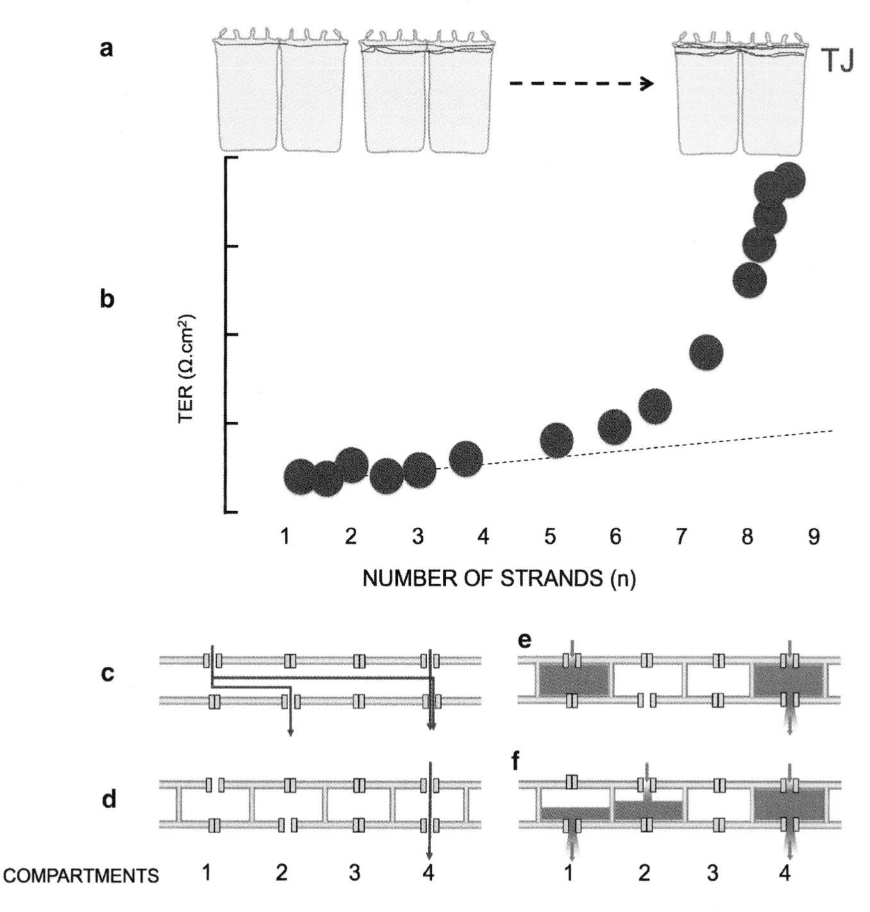

Fig. 3.3 The relationship between junctional structure, TER and permeability of the TJ. (**a**) TJs (*red*) vary in the number of strands (**n**). (**b**) If strands were simple ohmic resistors, then TER would increase with **n** in a linear manner (*dashed black line*). Yet experimental measurements in a wide variety of epithelia, show that TER increases exponentially with **n** (*red circles*). (**c**) Flickering channels (*1*, *2*, *3* and *4*) that alternatively switch from an open to a closed state would not explain the discrepancy between TER and permeation studies though, because ions may cross a given strand (*red lines*) by any channel that happens to be open at the time. (**d**) Yet strands have frequent anastomoses and trabeculae that compartmentalizes the TJ, so that for electric current to flow across the whole TJ it is necessary that channels in the upper and lower strands be open simultaneously. Only compartment 4 would be conducting in this example. (**e**) Permeation would instead obey a different set of rules. Thus a diffusing substance like mannitol or dextran (*pink*) may flow into compartments 1 and 2 in spite of the lower channel being open (1) or closed (4). (**f**) When channels in the lower strand open, the solute would leave the compartment regardless of whether the channel in the above strand remains open or closed. Therefore, in a segment of TJ with compartments imposed by the presence of trabeculae, only those compartments with channels in the open state in the upper ***and*** the inner strand will let the electric current through (**d** # 4, and **f** # 4). On the contrary, in a study of permeation to a given substance, this substance will enter the compartment whenever the channel in the upper strand is open, and will proceed its movement towards the basal side of the epithelium whenever the channel in the lower strand is in the open state. In summary, the difference between electrical measurements and solute permeability studies is that, while the first requires that all channels in the series should be simultaneously in the open state, solute permeation does not

3.3.1.5 Electric Current Paths Through an Epithelium

When an electric current is applied through an epithelium it can follow several paths. If the epithelium is a very tight one, current will only successively flow through the apical and the basolateral domain of the plasma membrane of the cells. But it can in principle flow through spurious paths. The most common one would be ***edge damage***, i.e. the points where the pressure of the rim of the Lucite chamber would destroy cells at the perimeter of the exposed area (Fig. 3.3f). Yet today these nuisances are avoided by a suitable design (e.g. rubber O-rings), greasing the halves of the chamber, etc. The other potential source of artifacts would be damaged or dying cells. Yet the adherence of cells to a solid support entails a very competitive process; as cells continuously strive to adhere to a larger area of support, including the formation of an actin ring that effectively deprives the unhealthy cell of attachment to the support, as well as interrupts the access of nutrients coming from the organism through the basolateral side (Peralta Soler et al. 1996). Eventually, unhealthy cells are forced to abandon the monolayer and are discarded towards the lumen. In a natural tissue, this is the usual destiny of aging cells.

Apart from these spurious paths due to technical inefficiencies, in epithelia or monolayers with low TER (usually below 2,000 $\Omega.cm^2$) more than 98% of the applied current follows the paracellular route, that is one of the reasons that TER is routinely used to gauge the hermeticity of the TJ (Cereijido et al. 1984).

Frömter and Diamond (1972; Frömter 1972), and Steffani and Cereijido (1983) have develop *ad hoc* procedures in which a glass microelectrode is used to scan the surface of an epithelium, to detect the points where current flows through the epithelium of the gall bladder. Invariably these coincide with the intercellular space.

3.3.1.6 "Discrepancies" Between Permeability and TER

The electrical resistance of an epithelium measures how easily an ion can cross it. Yet at times resistance measurements indicate that an ion crossing the epithelium meets a relatively high degree of difficulty, while its permeability, measured through the flux of a tracer (e.g. a radioisotope) is nevertheless high. This puzzling result entails no discrepancy and, in fact, is quite common. Imagine this analogy: five persons open the front door, enter into the living room, close the door behind them, and proceed likewise to the dining room, the kitchen, the back room: they would "flow" through the entire house, i.e. the house is "permeable" to people. This compares to the flux of a substance that penetrates a route of compartments in series, each one limited by flickering channels (Fig. 3.4e, f). On the contrary, the measurement of the electrical conductance/resistance requires instead that doors be simultaneously open. Solving this "discrepancies" between conductance (or resistance) and flux measurements may afford information on the TJ.

Since each TJ strand is a resistor, one would expect that TER would increase linearly with the number of strands (Fig. 3.4a, b). Yet in natural epithelia TER increases exponentially with the number of strands, as indicated in Fig. 3.4a, b. This situation is also explained by the existence of trabeculae that compartmentalize the TJ.

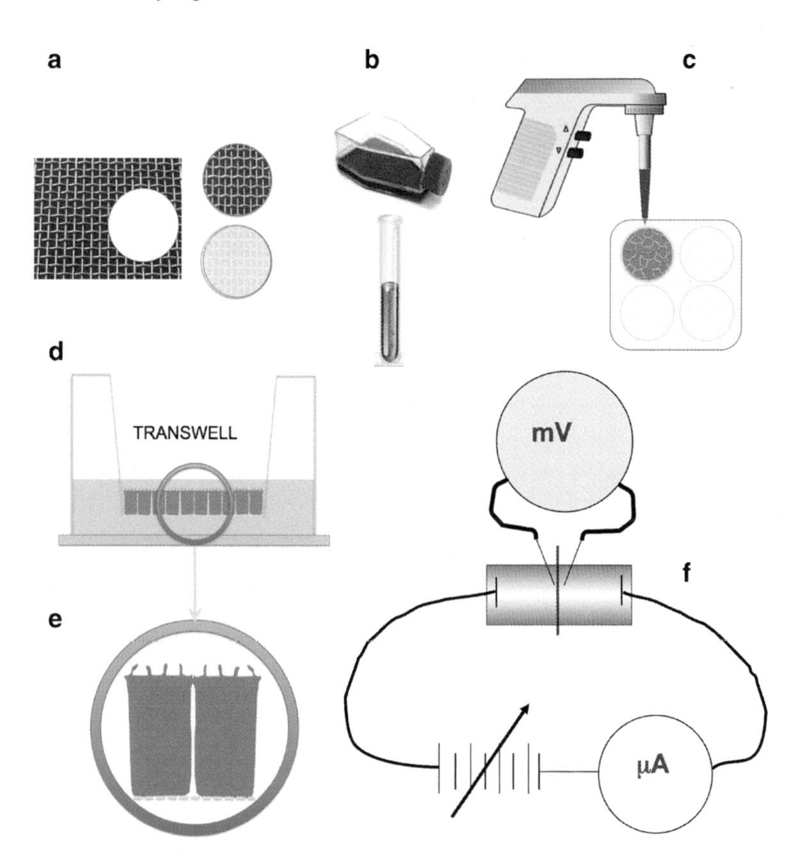

Fig. 3.4 An epithelial model system: Epithelial cells can be cultured as monolayers that resemble natural epithelia and offer a series of experimental advantages. (**a**) Disks are cut out of nylon cloth with square windows of 100 µm on the side, which is coated with collagen and sterilized. (**b**) MDCK cells (epithelial from dog kidney) are mass-cultured in plastic bottles, harvested with trypsin-EDTA, and (**c**) plated at saturating density on the collagen coated disc. (**d, e**) Alternatively, monolayers can be prepared in a Transwell assembly. (**f**) The disk with the monolayer is mounted as a flat sheet between two Lucite chambers with Ringer solution, a current of 20 $A.cm^{-2}$ is passed and the voltage deflection measured. These parameters are used to calculate TER with Ohm's Law

3.3.1.7 Monolayers of Epithelial Cell Lines Cultured on a Permeable Support

By the end of the decade of 1960 TJs and polarity were sufficiently characterized, and their basic properties as well as physiological role were reasonably understood, and the next step was to study when, how and why these structures are synthesized, assembled and become ready to function (Cereijido et al. 2008). It was not possible to learn about the mechanisms involved using natural epithelia as model systems, because these have TJs and polarity already established. In this respect, a great breakthrough was made by using epithelial cell lines cultured on permeable supports (Cereijido et al. 1978, 2004). These cell lines can be mass cultured in bottles, harvested with trypsin-EDTA and plated at confluence on a Nitex cloth coated with collagen, or on Millipore filters (Fig. 3.3d). This harvesting is so harsh that makes cell lose their TJs and polarity. Yet upon re-seeding, cells re-establish these features under culturing conditions amenable for experimental control, e.g. in the presence of inhibitors of the synthesis of proteins, DNA, RNA, glycosylation, and assembly of microtubules and microfilaments. There are finer procedures to block protein synthesis more specifically, resorting to RNA interference (RNAi), or force the cell to express an engineered version of a given molecule, block phosphorylation, or mutate a given amino acid suspected to be crucial in the making of a TJ or during the process of polarization. Monolayers prepared in this way can be then inspected by a variety of electron microscope techniques, epifluorescence with tagged molecules, etc.

Cells can be plated at low density, followed by waiting until they proliferate and achieve confluence. This protocol has the disadvantage of mixing cells that are still engaged in proliferation, with those that in the meanwhile have recovered from trypsin, contacted several neighboring cells and are ready to synthesize and assemble TJs. Most often instead, cells are plated at saturating density and allowed to attach for 20–30 min, followed by a change to a bathing medium without cells. This protocol has the advantage of removing cells that were unable to find room in the support, and those that did attach would not waste time in proliferation.

3.3.1.8 The Calcium Switch

The procedure described in the previous paragraph can be perfected in still another way, based on the observation that removal and restoration of Ca^{2+} opens and reseal TJs (Cereijido et al. 1978). Thus Fig. 3.5 (*left*) shows two sort of monolayers; the first is basically the one described in the previous paragraph (*open circles*), but in the second the assembly of TJs as well as polarization are arrested by the removal of Ca^{2+} 30 min after cell attachment to the support (*filled circles*). If this ion is re-admitted at the 20th hour (*blue arrow*), TER is observed to increase with a much faster kinetics, because cells are already recovered from trypsination and had adjusted their borders to each other. This assembly and sealing can be so abrupt, that a fraction of Na^+,K^+-ATPase, that in mature monolayers of MDCK cells only occupy the basolateral side can be trapped on the apical (wrong) side as well. Even the observation of the fate of these misplaced subpopulation of the enzyme provides valuable knowledge on the removal and relocation of membrane molecules during polarization. (Contreras et al. 1989; Shoshani et al. 2005)

Besides of being a useful technical procedure to observe the kinetics of TJs assembly and polarization, this technique provided a way to study some intricacies of the mechanism triggered by Ca^{2+}, that results in the development of the two basic characteristics of the "epithelial transporting phenotype". Cells incubated overnight in the absence of Ca^{2+} lose this ion, that achieves a very low concentration in the cytoplasm, and markedly increase their specific Ca^{2+}-permeability. Hence 20 h later, when this ion is restored to the bathing fluid, it rapidly penetrates into the cells and increases its

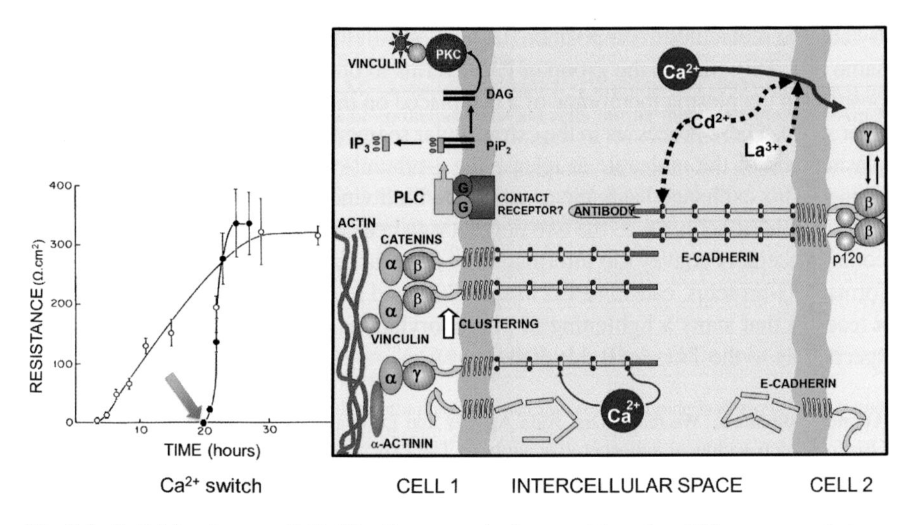

Fig. 3.5 (*left*) Monolayers of MDCK cells progressively assemble and seal TJs, a process that can be followed by the value of TER (open circles). Monolayers of cells plated in the presence of Ca^{2+}, are allowed to attach for 20–40 min, switched to a medium without this ion and incubated overnight, have a negligible TER at thc 20th hour. At this time the restoration of Ca^{2+} triggers a fast assembly and sealing of TJs (*filled circles*). This maneuver can be performed under a variety of experimental conditions, such as the presence of inhibitors of RNA or protein synthesis, glycosylation, the presence of competing cations such as Cd^{2+} and La^{3+}. (*right*) Two neighboring epithelial cells showing that Ca^{2+} acts primarily on the extracellular segment of E-cadherin, making this segment straight so that it can interact with other E-cadherin molecules located on the same cell membrane. Once in this position, extracellular segments can interact through van der Waals' forces, that decay with the 5th power of distance. Since these events are occurring in the neighboring cell as well, straightened molecules of E-cadherin can reach those on the other side of the intercellular space and establish a firm cell-cell adhesion (see text)

concentration in the cytoplasm. For a while this penetration mislead research workers that investigated the role of *intracellular* Ca^{2+}, and resorted to prevent the increase concentration in the cytoplasm by, for instance, using Ca-buffers. On the contrary, Contreras et al. (1992) and González-Mariscal et al. (1990) demonstrated that it is instead the *extracellular* Ca^{2+} that triggers junction formation and polarization. The demonstration is as follows: (1) the concentration of Ca^{2+} necessary to promote TJ formation and polarization is so low, that it may trigger these processes without modifying the concentration of calcium in the cytoplasm (Nigam et al. 1992). (2) The penetration of Ca^{2+} can be blocked by Cd^{2+} or La^{3+} that are nevertheless unable to trigger differentiation. Put in other words, Ca^{2+} triggers the making of TJs and polarization from outside of thc cytoplasm. (3) In spite of their affinity for the mechanisms used by Ca^{2+} to penetrate, Cd^{2+} and La^{3+} cannot substitute calcium in triggering TJ formation and polarization. (4) Later on it was found that extracellular Ca^{2+} acts on the extracellular domain of E-cadherin, at the joins between the five repeats in the extracellular segment of this molecule (Fig. 3.5 *right*) (Ringwald et al. 1987). This causes this molecular segment to straighten, and attach to similar segments in neighbor

macromolecular drug delivery. In contrast to macromolecular drugs, when [^{14}C] methylglucose, a representative low-molecular-weight drug mimic, was studied under hypertensive condition, accumulation of this agent in tumor was much less than that of polymeric drugs and lasted no longer than 10 min (Li et al. 1993). Such low-molecular weight drugs seem to be washed out rapidly into the general circulation and are excreted via the urine.

In a converse approach, vasodilators, such as the NO-releasing agent isosorbide dinitrate (ISDN; Nitrol), were utilized to enhance the EPR effect via widening the tumor-feeding artery. This result was accomplished by local arterial infusion of ISDN by catheter (Greish et al. 2003; Greish 2007).

Another vascular mediator that greatly influences the EPR effect is bradykinin, which induces intense pain as well as increases vascular permeability. The increase of vascular permeability is associated with a down-regulation of the expression of the tight junction proteins, claudin-5, ZO-1 and occludin and a rearrangement of F-actin in a blood tumor barrier model (Liu et al. 2008). Furthermore, bradykinin cross-talks with prostaglandin and NO, resulting in greater vasodilatation as well as angiogenesis. A significant activation of the bradykinin generating cascade in the tumor compartment was reported, as well as [^{3}hydroxyprolyl] bradykinin. Further, BK was found to be involved in the accumulation of malignant ascetic and pleural fluid (Matsumura et al. 1988, 1991; Maeda et al. 1988). Angiotensin-converting enzyme (ACE) inhibitors such as enalapril and other similar agents can inhibit degradation of bradykinin in vivo and lead to higher bradykinin concentrations at sites of tumor and infection, because of an amino acid sequence homology to that of bradykinin near the C-termini. Consequently, ACE inhibitors did enhance the EPR effect (Matsumura et al. 1988, 1991; Hori et al. 2000) mediated by either bradykinin or NO. Therefore, increasing the local concentration of bradykinin by means of ACE inhibitors, and thereby improving tumor-selective delivery of macromolecular drugs would be possible. Interestingly, ACE inhibitors were found to beneficial on treatment of hepatocellular carcinoma (Noguchi et al. 2003) and prostate cancer (Uemura et al. 2005).

Another important mediators for EPR effect are prostaglandins (PGs) particularly PG_{E2}, generated via cyclooxygenase isozymes (COX 2), which is markedly elevated in inflammatory and cancer tissues. These increased levels of PGs can also enhance vascular permeability in solid tumor, as evidenced by significant suppression of vascular permeability in sarcoma 180 and other solid tumor models by the COX inhibitors indomethacin and salicylic acid (Wu et al. 1998; Maeda et al. 1996). It was found that a PG_{I2} analogue (beraprost sodium) with a much longer in vivo half-life (about 30 min vs. 3 s for PG_{I2}) was useful for the delivery of macromolecules (Tanaka et al. 2003); although a therapeutic advantage of beraprost sodium needs to be demonstrated. Prostaglandin E2 reverses the effect of the epidermal growth factor in epithelial cells and increases the permeability (Flores-Benitez et al. 2009).

Another potential modulator of the EPR effect is hydrogen peroxide. The role of hydrogen peroxide in regulating vascular permeability is currently attracting the interest of many researchers. Several studies have shown that H_2O_2 is involved in the increase of the vascular permeability in various types of cells (Meyer et al. 2001; Jepson 2003). Hydrogen peroxide increases paracellular permeability by affecting the expression and localization of occludin and ZO-1 (Lee et al. 2004). Drummond et al. (2000) have discovered a role of hydrogen peroxide in transcriptional and posttranscriptional regulation of endothelial NO synthases expression by endothelial cells. Direct addition of 100 and 150 mmol/l H_2O_2 caused increases in bovine aortic endothelial cell eNOS mRNA that were time and concentration dependent (i.e. 3.1- and 5.2-fold increases), and elevated eNOS protein expression and enzyme activity, accordingly. In other studies, it had been found that elevated levels of H_2O_2 cause calcium dependent release of NO from the endothelium and potassium channel-dependent relaxation of vascular smooth muscles (Weir and Archer 1995; Yang et al. 1999). In addition, H_2O_2 was reported to stimulate multiple forms of vascular phospholipases and directly modify lipids to species that are vasoactive (Rao et al. 1995; Natarajan et al. 1998). Cseko et al. (2004) proposed that H_2O_2 in a concentration dependent manner activates several endothelial and smooth muscle pathways, resulting in biphasic changes on the diameter and myogenic tone of isolated skeletal muscle arterioles. The constrictions induced by H_2O_2 are mediated by endothelial PGH2/TxA2, whereas the dilations are caused primarily by the activation of both endothelial NO synthase and various Kþ channels in vascular smooth muscle cells. It seems plausible that exogenous administration of H_2O_2 upstream into a tumor feeding artery could enhance the anticancer drug delivery, similar to the effect produced by ISDN, however this needs to be verified. Maeda's group have demonstrated the role of H_2O_2 in enhancing the EPR effect utilizing polyethylene glycol conjugated D-amino acid oxidase, which can selectively produce H_2O_2 in tumor tissues upon injection of D-proline (Fang et al. 2002).

Photodynamic therapy, in which a photosensitizer is administrated systemically or locally and subsequently activated by illumination with visible light, leading to the generation of cytotoxic reactive oxygen species in the presence of oxygen, has been identified to have an active role in enhancing tumor vascular permeability (Fingar 1996; Dougherty et al. 1998). Chen et al. (2006) found that the concentration of 2,000-kDa FITC-dextran were fivefold higher in orthotropic MatLyLu rat prostate tumors treated with vascular-targeting photodynamic therapy verteporfin, at 15 min following light irradiation, compared to non-irradiated control group. When they studied the effects of verteporfin photosensitization on endothelial cell morphology, and cytoskeleton, they found that photosensitization causes endothelial cell microtubule depolymerization and induces the formation of actin stress fibers. Thus, endothelial cells were found to retract, disrupting the tight junctions and leading to the formation of intercellular gaps, which result in enhanced vascular permeability. In addition, endothelial cell damage leads to the establishment of thrombogenic sites within the vessel lumen and this initiates a physiological cascade of responses including platelet aggregation, the release of vasoactive molecules, leukocyte adhesion and increases in vascular permeability (Fingar 1996).

Table 4.2 (continued)

Type of Nanomedicine	Name	Therapeutic agent	Status	Cancer type	Reference-(Clinical trial number)[a]
	Hepacid/ADI-PEG	Arginine deaminase	Phase II	Hepatocellular carcinoma	(NCT00056992)[a] Shen and Shen (2006)
			Phase I	Melanoma	(NCT00029900)[a] Shen and Shen (2006)
	PK2, FCE28069	Doxorubicin	Phase I	Hepatocellular carcinoma	Seymour et al. (2002)
	PNU166945	Paclitaxel	Phase I	Breast cancer	Meerum Terwogt et al. (2001)
	MAG-CPT	Campothecin	Phase I	Solid tumor	Wachters et al. (2004), Bissett et al. (2004)
	AP5280	Platinum	Phase I	Solid tumor	Rademaker-Lakhai et al. (2004)
	AP5346	Platinum	Phase I	Solid tumor	Campone et al. (2007)
	DOX-OXD/AD-70	Doxorubicin	Phase I	Various cancers	Danhauser-Riedl et al. (1993)
	DE-310	Topoisomerase-I-inhibitor, exatecan mesylate	Phase I	Solid tumors	Soepenberg et al. (2005)
	BIND-014	Doxetaxel	Phase I	Solid tumors	(NCT01300533)[a]
Albumin based Nanomedicine	Abraxane (ABI-007)	Paclitaxel	Approved	Metastatic breast cancer	Gradishar (2006)
			Phase II	Non-small-cell lung cancer	Reynolds et al. (2009)
			Phase II	Melanoma	Kottschade et al. (2011), Hersh et al. (2010)
			Phase II	Ovarian cancer	Coleman et al. (2011), Teneriello et al. (2009)
			Phase I	Pancreatic cancer	Stinchcombe et al. (2007)
			Phase I	Bladder cancer	McKiernan et al. (2011)
	MTX-HSA	Methotrexate	Phase II	Kidney carcinoma	Vis et al. (2002)
	ABI-008	Docetaxel	Phase I/II	Metastatic breast cancer	(NCT00531271)[a]
			Phase I/II	Prostate cancer	(NCT00477529)[a]
	ABI-009	Rapamycin	Phase I	Solid tumors	(NCT00635284)[a]
	ABI-010	Tanespimycin (17-AAG)	Phase I	Solid tumors	(NCT00820768)[a]
	ABI-011	Microtubule and topoisomerase inhibitor	Phase I	Advanced solid tumors and lymphomas	(NCT01163071)[a]

[a] ClinicalTrials.gov Identifier number

external pH (Riggio et al. 2011). Genexol-PM is a polymeric micelle of paclitaxel approved for the treatment of metastatic breast cancer in South Korea (Oerlemans et al. 2010).

The dendrimers are a new class of synthetic macromolecules with a well-defined tree-like structure organized in a series of radially homocentric layers (Cheng et al. 2011). Physiochemical properties of these dendrimers make them suitable for anticancer drug delivery (Cheng et al. 2011). Encapsulation of cisplatin within the dendrimers demonstrated a higher accumulation within the solid tumor compare to the free drug (Malik et al. 1999). Other Nanoparticle drug delivery platforms that have been extensively include metal Nanoparticle and molecular targeted Nanoparticles. Their efficacy of drug delivery systems to enhance the pharmacokinetic properties of drugs has been confirmed in preclinical trials (Rippel and Seifalian 2011; Wang et al. 2008a).

4.8 Challenges to the Enhanced Permeability Strategy to Target Nanosize Drug Carriers

Recognition of the EPR effect among researchers in drug delivery field resulted in a considerable momentum to the emerging field of Nanomedicine. The expectation for realization of a selective anticancer drug as a result of adapting the EPR effect in Nanotechnology was high, however as shown in Table 4.2, only few Nanomedicine have found its way to the clinic (Duncan 2003). Following we discuss some of the factors that might interfere with the full exploitation of EPR effect for anticancer specific targeting.

4.8.1 EPR Effect in Animal Models

The EPR effect has been repeatedly proved in animal models through the use of Evans blue dye (EBD). EBD binds instantly to plasma albumin which results in large molecular weight complex about 7 nm in diameter that can simulate the effect of a Nanomedicine. After 6 h, usually there is a profound accumulation in tumor lesion compared to surrounding tissues. Similarly, many Nanomedicines have been proven to accumulate in tumor tissues compare to other organs, such as 11-fold higher concentration of SMA-doxorubicin micelle compare to free drug (Greish et al. 2004). The question whether the results of EPR-based drug targeting results in animal models can be translated clinically remained unanswered. A great difference between the tumor models in animal and human is the progression rate; animals usually develop a large clinically relevant tumor (>5 mm) 1 week after tumor cell inoculation subcutaneously, while such a tumor volume can take few years to develop in a human. This fast progression rate carries many confounding factors towards more effective EPR effect in animal models. Animal tumors developing

this problem with different level of success such as devising specific bonds between the Nanocarrier and the drug that will only cleave upon exposure to specific enzymes in tumor cells or merely the high acidic condition in tumor tissue (Greish et al. 2003; Greish 2007).

4.8.4 Biocompatibility

After intravenous administration of a specific dose of a drug carried by Nanocarrier, a relatively high concentration is targeted to the tumor by the EPR effect. However this portion of the drug is usually less than 10% of the total administered dose. The remaining 90% still find its way to different organs and tissue (Bae and Park 2011). Unless the Nanocarrier is biodegradable this amount of Nanocarrier can remain in the body after releasing its anticancer drug cargo. The Nanosize particles used to carry the drug load are frequently recognized and dealt with as a foreign body. The innate elements of the immune system are non-specifically stimulated by many Nanocarrier through toll like receptor (TLR's), i.e. TLR-4 (Kedmi et al. 2010). Following activation by Nanocarrier, immune cells can produce cytokines which trigger inflammation and the Nanocarrier is phagocytized by monocyte/macrophage. At that point the phagocytic cells will try to degrade the Nanocarrier in the lysosomal compartment through lysosomal enzymes. Failure of this process can lead to (frustrated macrophage) or the formation of a giant foreign body cells that closely resemble the formation of granuloma. This will result in a pathological capsule with dense fibrous capsule replacing the original functional tissue, which can compromise organ function, especially in the liver (Kao and Lee 2001). Another concern of the Nanocarrier not being biodegradable is malignancy induction, where the prolonged unresolved inflammation may result in malignant transformation. A successful Nanocarrier thus must be biodegradable.

4.8.5 Intracellular Internalization

While the EPR effect will result in a high drug concentration in the tumor tissues in a specific subset of tumors, it cannot guarantee the internalization of these drugs through the tumor cell membrane into the cytoplasm or the nucleus. Usually Nanosize anticancer drugs are internalized into tumor cells through endocytosis with the final localization in the endosomes, then the lysosomal compartment (Zaki and Tirelli 2010). A major limitation in tumor cells is their phagocytic capability. It was found that macrophages are much more susceptible to toxicity associated with Nanosized silica Nanoparticles compared to different tumor cells in vitro, and this deferential cytotoxicity activity in macrophage-derived cell lines was clearly correlated with the higher intercellular uptake by the professional phagocytic macrophage (Yu et al. 2011). Based on this observation, it is clear that the relatively large carriers (macromolecules and Nanocarriers) that adhere to cell surfaces without intracellular trans-localization may not give any additional benefit by the *retention* effect. Many Nanosize particles may locate juxtaposition to the leaky sites due to limited permeability or mobility in the extracellular space and non-specific interactions with the extracellular matrix. Thus, cell internalization is essential for Nanocarriers for effective drug delivery besides the enhanced permeation and retention (EPR) effect.

4.9 Conclusion

The EPR effect can be considered a hallmark concept that exploits the anatomical and pathophysiological defects in the tumor vasculature. It plays a critical role in selective delivery of Nanomedicine-based-anticancer agents to tumor tissues. EPR effect outcome can be influenced by variables such as tumor diversity, animal models, biodistribution, intracellular interaction, and release rate of active cytotoxic cargo from its Nanosize carrier. Understanding and manipulating the different variables contributing to the EPR effect, can further improve the selective targeting of high-molecular-weight biocompatible or anticancer Nanomedicine to tumor, thus ensuring bright future for EPR based anticancer Nanomedicine.

Acknowledgment The author gratefully acknowledges the support of Professor Hiroshi Maeda. The EPR effect was first described and extensively studied by Professor Maeda's group in the department of Microbiology, Kumamoto University, Japan. This work has been supported by Departmental fund No.; (PL. 108403.01.S. LM) to KG from the department of pharmacology and toxicology, Otago University. KG thanks Ms Rebecca Cookson for proof reading the article.

References

Aird WC (2007) Phenotypic heterogeneity of the endothelium: I. Structure, function, and mechanisms. Circ Res 100(2):158–173

Allen T (2002) Ligand-targeted therapeutics in anticancer therapy. Nat Rev Cancer 2(10):750–763

Ando K, Mori K, Corradini N, Redini F, Heymann D (2011) Mifamurtide for the treatment of nonmetastatic osteosarcoma. Expert Opin Pharmacother 12(2):285–292

Angeletti CA, Lucchi M, Fontanini G, Mussi A, Chella A, Ribechini A, Vignati S, Bevilacqua G (1996) Prognostic significance of tumoral angiogenesis in completely resected late stage lung carcinoma (stage IIIA-N2). Impact of adjuvant therapies in a subset of patients at high risk of recurrence. Cancer 78(3):409–415

Apostolidou E, Swords R, Alvarado Y, Giles FJ (2007) Treatment of acute lymphoblastic leukaemia: a new era. Drugs 67(15):2153–2171

Bachem M, Schunemann M, Ramadani M, Siech M, Beger H, Buck A, Zhou S, Schmid-Kotsas A, Adler G (2005) Pancreatic carcinoma cells induce fibrosis by stimulating proliferation and matrix synthesis of stellate cells. Gastroenterology 128(4):907–921

Bae Y, Park K (2011) Targeted drug delivery to tumors: myths, reality and possibility. J Control Release 153(3):198–205

Seymour LW, Miyamoto Y, Maeda H, Brereton M, Strohalm J, Ulbrich K, Duncan R (1995) Influence of molecular weight on passive tumour accumulation of a soluble macromolecular drug carrier. Eur J Cancer 31A(5):766–770

Seymour LW, Ferry DR, Anderson D, Hesslewood S, Julyan PJ, Poyner R, Doran J, Young AM, Burtles S, Kerr DJ, Committee ftCRCPIICT (2002) Hepatic drug targeting: phase I evaluation of polymer-bound doxorubicin. J Clin Oncol 20(6):1668–1676

Seymour LW, Ferry DR, Kerr DJ, Rea D, Whitlock M, Poyner R, Boivin C, Hesslewood S, Twelves C, Blackie R, Schatzlein A, Jodrell D, Bissett D, Calvert H, Lind M, Robbins A, Burtles S, Duncan R, Cassidy J (2009) Phase II studies of polymer-doxorubicin (PK1, FCE28068) in the treatment of breast, lung and colorectal cancer. Int J Oncol 34(6):1629–1636

Shen LJ, Shen WC (2006) Drug evaluation: ADI-PEG-20 – A PEGylated arginine deiminase for arginine-auxotrophic cancers. Curr Opin Mol Ther 8(3):240–248

Shi J, Xiao Z, Kamaly N, Farokhzad O (2011) Self-assembled targeted Nanoparticles: evolution of technologies and bench to bedside translation. Acc Chem Res 44(10):1123–1157

Soepenberg O, de Jonge MJ, Sparreboom A, de Bruin P, Eskens FA, de Heus G, Wanders J, Cheverton P, Ducharme MP, Verweij J (2005) Phase I and pharmacokinetic study of DE-310 in patients with advanced solid tumors. Clin Cancer Res 11(2 Pt 1):703–711

Soker S, Takashima S, Miao HQ, Neufeld G, Klagsbrun M (1998) Neuropilin-1 is expressed by endothelial and tumor cells as an isoform-specific receptor for vascular endothelial growth factor. Cell 92(6):735–745

Stinchcombe TE, Socinski MA, Walko CM, O'Neil BH, Collichio FA, Ivanova A, Mu H, Hawkins MJ, Goldberg RM, Lindley C, Claire Dees E (2007) Phase I and pharmacokinetic trial of carboplatin and albumin-bound paclitaxel, ABI-007 (Abraxane) on three treatment schedules in patients with solid tumors. Cancer Chemother Pharmacol 60(5):759–766

Stopeck AT, Jones A, Hersh EM, Thompson JA, Finucane DM, Gutheil JC, Gonzalez R (2001) Phase II study of direct intralesional gene transfer of allovectin-7, an HLA-B7/beta2-microglobulin DNA-liposome complex, in patients with metastatic melanoma. Clin Cancer Res 7(8):2285–2291

Suzuki M, Hori K, Abe I, Saito S, Sato H (1981) A new approach to cancer chemotherapy: selective enhancement of tumor blood flow with angiotensin II. J Natl Cancer Inst 67(3):663–669

Swenson CE, Bolcsak LE, Batist G, Guthrie TH Jr, Tkaczuk KH, Boxenbaum H, Welles L, Chow SC, Bhamra R, Chaikin P (2003) Pharmacokinetics of doxorubicin administered i.v. as Myocet (TLC D-99; liposome-encapsulated doxorubicin citrate) compared with conventional doxorubicin when given in combination with cyclophosphamide in patients with metastatic breast cancer. Anticancer Drugs 14(3):239–246

Tanaka S, Akaike T, Wu J, Fang J, Sawa T, Ogawa M, Beppu T, Maeda H (2003) Modulation of tumor-selective vascular blood flow and extravasation by the stable prostaglandin 12 analogue beraprost sodium. J Drug Target 11(1):45–52. doi:10.1080/1061186031000086072

Teneriello MG, Tseng PC, Crozier M, Encarnacion C, Hancock K, Messing MJ, Boehm KA, Williams A, Asmar L (2009) Phase II evaluation of Nanoparticle albumin-bound paclitaxel in platinum-sensitive patients with recurrent ovarian, peritoneal, or fallopian tube cancer. J Clin Oncol 27(9):1426–1431

Torchilin V (2005) Recent advances with liposomes as pharmaceutical carriers. Nat Rev Drug Discov 4(2):145–205

Tsukita S, Furuse M, Itoh M (1996) Molecular dissection of tight junctions. Cell Struct Funct 21(5):381–385

Tsukita S, Furuse M, Itoh M (2001) Multifunctional strands in tight junctions. Nat Rev Mol Cell Biol 2(4):285–293

Tsukita S, Yamazaki Y, Katsuno T, Tamura A (2008) Tight junction-based epithelial microenvironment and cell proliferation. Oncogene 27(55):6930–6938

Uchino H, Matsumura Y, Negishi T, Koizumi F, Hayashi T, Honda T, Nishiyama N, Kataoka K, Naito S, Kakizoe T (2005) Cisplatin-incorporating polymeric micelles (NC-6004) can reduce nephrotoxicity and neurotoxicity of cisplatin in rats. Br J Cancer 93(6):678–687

Uemura H, Hasumi H, Kawahara T, Sugiura S, Miyoshi Y, Nakaigawa N, Teranishi J, Noguchi K, Ishiguro H, Kubota Y (2005) Pilot study of angiotensin II receptor blocker in advanced hormone-refractory prostate cancer. Int J Clin Oncol 10(6):405–410. doi:10.1007/s10147-005-0520-y

Unemori EN, Ferrara N, Bauer EA, Amento EP (1992) Vascular endothelial growth factor induces interstitial collagenase expression in human endothelial cells. J Cell Physiol 153(3):557–562

Valle JW, Armstrong A, Newman C, Alakhov V, Pietrzynski G, Brewer J, Campbell S, Corrie P, Rowinsky EK, Ranson M (2011) A phase 2 study of SP1049C, doxorubicin in P-glycoprotein-targeting pluronics, in patients with advanced adenocarcinoma of the esophagus and gastroesophageal junction. Invest New Drugs 29(5):1029–1037

Van Itallie CM, Anderson JM (2006) Claudins and epithelial paracellular transport. Annu Rev Physiol 68:403–429

Van Itallie CM, Fanning AS, Anderson JM (2003) Reversal of charge selectivity in cation or anion-selective epithelial lines by expression of different claudins. Am J Physiol Renal Physiol 285(6):F1078–F1084. doi:10.1152/ajprenal.00116.2003

Van Itallie CM, Holmes J, Bridges A, Gookin JL, Coccaro MR, Proctor W, Colegio OR, Anderson JM (2008) The density of small tight junction pores varies among cell types and is increased by expression of claudin-2. J Cell Sci 121(Pt 3):298–305

van Moorselaar RJ, Voest EE (2002) Angiogenesis in prostate cancer: its role in disease progression and possible therapeutic approaches. Mol Cell Endocrinol 197(1–2):239–250

Vis AV, van der Gaast AG, van Rhijn BR, Catsburg TC, Schmidt CS, Mickisch GM (2002) A phase II trial of methotrexate-human serum albumin (MTX-HSA) in patients with metastatic renal cell carcinoma who progressed under immunotherapy. Cancer Chemother Pharmacol 49(4):342–345

Wachtel M, Bolliger MF, Ishihara H, Frei K, Bluethmann H, Gloor SM (2001) Down-regulation of occludin expression in astrocytes by tumour necrosis factor (TNF) is mediated via TNF type-1 receptor and nuclear factor-kappaB activation. J Neurochem 78(1):155–162

Wachters FM, Groen HJ, Maring JG, Gietema JA, Porro M, Dumez H, de Vries EG, van Oosterom AT (2004) A phase I study with MAG-camptothecin intravenously administered weekly for 3 weeks in a 4-week cycle in adult patients with solid tumours. Br J Cancer 90(12):2261–2267

Wang S, Hung M (2000) Transcriptional targeting of the HER-2/neu oncogene. Drugs Today (Barc) 36(12):835–843

Wang W, Dentler WL, Borchardt RT (2001) VEGF increases BMEC monolayer permeability by affecting occludin expression and tight junction assembly. Am J Physiol Heart Circ Physiol 280(1):H434–H440

Wang A, Gu F, Zhang L, Chan J, Radovic-Moreno A, Shaikh M, Farokhzad O (2008a) Biofunctionalized targeted Nanoparticles for therapeutic applications. Expert Opin Biol Ther 8(8):1063–1070

Wang X, Yang L, Chen ZG, Shin DM (2008b) Application of Nanotechnology in cancer therapy and imaging. CA Cancer J Clin 58(2):97–110

Wei MH, Popescu NC, Lerman MI, Merrill MJ, Zimonjic DB (1996) Localization of the human vascular endothelial growth factor gene, VEGF, at chromosome 6p12. Hum Genet 97(6):794–797

Weidner N, Semple JP, Welch WR, Folkman J (1991) Tumor angiogenesis and metastasis–correlation in invasive breast carcinoma. N Engl J Med 324(1):1–8

Weidner N, Carroll PR, Flax J, Blumenfeld W, Folkman J (1993) Tumor angiogenesis correlates with metastasis in invasive prostate carcinoma. Am J Pathol 143(2):401–409

Weir EK, Archer SL (1995) The mechanism of acute hypoxic pulmonary vasoconstriction: the tale of two channels. FASEB J 9(2):183–189

West GW, Weichselbaum R, Little JB (1980) Limited penetration of methotrexate into human osteosarcoma spheroids as a proposed model for solid tumor resistance to adjuvant chemotherapy. Cancer Res 40(10):3665–3668

White SC, Lorigan P, Margison GP, Margison JM, Martin F, Thatcher N, Anderson H, Ranson M (2006) Phase II study of SPI-77 (sterically stabilised liposomal cisplatin) in advanced non-small-cell lung cancer. Br J Cancer 95(7):822–828

Wiggins DL, Granai CO, Steinhoff MM, Calabresi P (1995) Tumor angiogenesis as a prognostic factor in cervical carcinoma. Gynecol Oncol 56(3):353–356

Abbreviations

aPKC	Atypical protein kinase-C
BBB	Blood brain barrier
GTPase	Guanine nucleotide triphosphatase
Ig-SF	Immunoglobulin-superfamily
JAM	Junctional adhesion molecule
Par	Partitioning-defective
TGFβ	Transforming growth factor-β
TJ	Tight junction

5.1 Junctional Adhesion Molecules (JAMs)

5.1.1 JAMs at Tight Junctions

JAM-A, the founding member of the JAM family of adhesion molecules, has originally been identified on the basis of an antibody that stimulates platelet aggregation and which later turned out to recognize JAM-A (Kornecki et al. 1990). The generation of new JAM-A antibodies against endothelial antigens and the identification of the cognate antigen by expression cloning revealed JAM-A as an adhesion molecule which localizes to cell-cell contacts of epithelial and endothelial cells. Confocal immunofluorescence as well as immunoelectron microscopy indicated that JAM-A localizes specifically to the TJs in epithelial cells (Martin-Padura et al. 1998). Later studies confirmed the localization of JAM-A at the TJs but also indicated that JAM-A does not exclusively localize to TJs (Liang et al. 2000; Liu et al. 2000). In epithelial cells grown on semipermeable filter membranes to allow for polarization and also in epithelial cells of the small intestine, JAM-A is enriched at the TJs but can also be found along the entire lateral cell-cell contact area (Liang et al. 2000; Liu et al. 2000) (own unpublished observations). The enrichment of JAM-A at the TJs as well as the exclusive phosphorylation of this TJ-associated population of JAM-A by atypical PKC (aPKC) suggest a TJ-specific role of JAM-A (discussed below in more detail). As opposed to JAM-A, JAM-B has not been found to be localized specifically at TJs, and its expression seems to be much more restricted than the expression of JAM-A and JAM-C. JAM-B expression in epithelial cells has so far been demonstrated in seminifereous epithelial cells where it co-localizes with ZO-1 at Sertoli cell TJs (Gliki et al. 2004; Liang et al. 2000). Upon transfection in MDCK epithelial cells, JAM-B does not co-distribute with ZO-1 suggesting that JAM-B does not localize to TJs in polarized epithelial cells (Aurrand-Lions et al. 2001). JAM-C on the other hand is exclusively localized at TJs when ectopically expressed in MDCK cells (Aurrand-Lions et al. 2001), and immunolocalization studies in native retinal pigment epithelial cells as well as in small intestine sections confirm a TJ-restricted localization of JAM-C in vivo (Economopoulou et al. 2009; Mandicourt et al. 2007) (own unpublished observations).

5.1.2 JAMs and the Regulation of Tight Junctions and Apico-Basal Polarity

The loss of cell polarity in epithelial cells is a contributing factor to the development of cancer (Wodarz and Nathke 2007). The TJs are important for the development of apico-basal polarity since they provide an intramembrane diffusion barrier which prevents the intermixing of membrane proteins and lipids which are vectorially delivered to the different membrane compartments. The molecular nature of this intramembrane diffusion barrier – also called the "fence" function of TJs – is not understood. It is retained in the absence of TJ strands which were considered before to provide a barrier for both the free diffusion of solutes along the paracellular pathway (named the "barrier" function) and the diffusion of intramembrane components. These two functions of TJs are probably regulated by separate molecular mechanisms. Proteins which localize to TJs could therefore be important for only one but also for both of the two TJ functions.

Among the three JAM proteins, the strongest evidence for a role in the establishment of epithelial TJs exists for JAM-A. JAM-A directly interacts with the cell polarity protein PAR-3 (Ebnet et al. 2001; Itoh et al. 2001). PAR-3 is part of the PAR – aPKC complex consisting of PAR-3, aPKC and PAR-6 (Suzuki and Ohno 2006). In vertebrate epithelial cells, this complex localizes to TJs, and knockdown of individual components of the complex or overexpression of kinase-dead aPKC mutant blocks or delays TJ formation (Gao et al. 2002; Yamanaka et al. 2001; Suzuki et al. 2002; Suzuki et al. 2001). The development of apico-basal polarity is regulated by aPKC which promotes the development of specific membrane domains by targeting specific proteins for removal from the aPKC-containing membrane domains (Hurov et al. 2004; Suzuki et al. 2004). One major role of JAM-A is most likely to recruit the PAR – aPKC complex to early sites of cell-cell adhesion. JAM-A is among the first proteins detectable at so-called primordial adherens junctions (pAJs) which are characterized by spots of cell-cell interactions between two mesenchymal-like cells (Ebnet et al. 2004). PAR-3, aPKC and PAR-6 appear later than JAM-A (Suzuki et al. 2002). Once recruited by JAM-A to pAJs, the binding of Rho family small GTPases to PAR-6 leads to the activation of aPKC which regulates the further maturation of pAJs into linear cell-cell contacts (Yamanaka et al. 2001). Consistent with the notion that aPKC regulates the maturation but not the formation of cell-cell contacts is the observation that cells expressing a dominant-negative mutant of aPKC are able to form pAJs but fail to form mature junctions with linear AJs and TJs (Suzuki et al. 2001).

More recent evidence indicates that JAM-A not only serves as positional cue to regulate the correct positioning of the PAR-3 – aPKC – PAR-6 complex, but is also part of the molecular machinery which regulates the subsequent junctional maturation. JAM-A is phosphorylated at Ser285 by aPKC, and this phosphorylation occurs at sites of pAJs simultanously with the recruitment of the PAR – aPKC complex by JAM-A (Iden et al. 2012). Ectopic expression of a phosphorylation-deficient JAM-A mutant (JAM-A/S285A) significantly delays junctional maturation suggesting that the normal maturation of pAJs into linear cell junctions requires JAM-A/S285 phosphorylation. The molecular mechanism underlying this junction formation-promoting role of JAM-A/S285 phosphorylation is unclear; it might be mediated through the recruitment of proteins, which bind selectively to S285-phosphorylated JAM-A, to sites of junctional maturation.

JAM-A is probably also required for the maintenance of functional TJs. In fully polarized epithelial cells, JAM-A/S285 phosphorylation is observed exclusively at the TJs, despite the localization of JAM-A along the entire lateral cell-cell contact (Iden et al. 2012). Ectopic expression of JAM-A/S285A increases the permeability for TRITC-labelled dextran and decreases the electrical resistance of the monolayer. In addition, upon Ca^{2+}-switch-induced disruption of epithelial cell junctions, JAM-A is dephosphorylated at S285 suggesting that JAM-A phosphorylation at Ser285 is involved in the maintenance of the integrity of cell junctions.

The role of JAM-C in TJ formation and maintenance is less well characterized. Ectopic expression of JAM-C in KLN205 promotes the formation of a functional barrier as indicated by the increase in transepithelial electrical resistance (TER) (Mandicourt et al. 2007). Like JAM-A, JAM-C can directly interact with PAR-3 (Ebnet et al. 2003) which opens the possibility that JAM-C is also involved in regulating the localization of the PAR – aPKC complex at TJs. Since JAM-C has not been demonstrated yet to exist in a complex with PAR proteins in cells, it is still unclear how JAM-C contributes to TJ formation and maintenance.

5.1.3 The PAR-aPKC Complex at Tight Junctions as Mediator of Oncogenic and EMT Signalling

The association of JAM-A with the PAR – aPKC complex is further intriguing since this complex has been implicated in oncogenic ErbB2 signalling. MCF-10A mammary epithelial cells, grown in a reconstituted, three-dimensional basement membrane, form spheroids which consist of a monolayer of epithelial cells surrounding a single lumen (Debnath and Brugge 2005). The spheroids resemble the glandular structures in the milk ducts, the so-called acini. ErbB2 signalling leads to multiacinar structures resulting from hyperproliferation and from filling of the luminal space due to a block of apoptosis of the cells in the acini. The block in apoptosis is mediated through the PAR-3 – aPKC – PAR-6 complex (Aranda et al. 2006). In response to ErbB2 signalling, the ErbB2 receptor outcompetes PAR-3 from the PAR-3 – aPKC – PAR-6 complex, and the newly formed ErbB2 – aPKC – PAR-6 complex disrupts apico-basal polarity and blocks apoptosis of the inner acinar cells (Aranda et al. 2006).

During epithelial-mesenchymal transition (EMT) cells lose some epithelial characteristics and adopt a mesenchymal-like phenotype, which is characterized by E-cadherin repression, reduced cell-cell adhesion and loss of cell polarity (Thiery 2002). Cancer cells share many characteristics with mesenchymal cells, and inappropriate activation of EMT pathways in the adult organism can therefore contribute to cancerogenesis. Transforming growth factor (TGF)-β is an important EMT regulator (Huber et al. 2005), and recent evidence indicates that TGFβ targets the PAR – aPKC complex at TJs. The TGFβ receptor TβRI associates with occludin and PAR-6, and TGFβ signals induce the heterodimerization of TβRI with TβRII leading to the assembly of an active TGFβ receptor complex which phosphorylates PAR-6 (Ozdamar et al. 2005). Through the recruitment of the ubiquitin ligase Smurf1 by phosphorylated PAR-6, and the Smurf1-mediated local degradation of RhoA which is critical for junctional integrity, TGFβ signals can lead to TJ disintegration. At the same time TGFβ downregulates PAR-3 (Wang et al. 2008), which might further lead to TJ destabilization. Since both TJ-associated JAMs, i.e. JAM-A and JAM-C, can interact with PAR-3 (Ebnet et al. 2003) it is possible that changes in the levels of JAM-A/-C expression or structural changes as a result of mutations influence the integrity of the PAR – aPKC complex and as a consequence to a loss in the functional integrity of TJs.

5.1.4 A Role for JAMs in Promoting Tumor Growth and Dissemination

Recent evidence suggests that JAMs are involved in tumor development. Depending on the tumor type, JAM levels positively as well as negatively correlate with tumor progression. In breast cancer patients, high mRNA and protein levels of JAM-A correlate with poor prognosis (McSherry et al. 2009; Murakami et al. 2011). This suggests that some functions of JAM-A provide a selective advantage for tumor cells. In support of this observation, various breast cancer-derived tumor cell lines downregulate micro RNA (miR)-145 which among other molecules targets JAM-A (Gotte et al. 2010). One function of JAM-A which is probably exploited by tumor cells, is its ability to influence cell migration by regulating the expression of β1-integrins (Mandell et al. 2005). By forming homodimers, JAM-A brings together AF-6, a scaffold protein for Rap1, and PDZ-GEF-2, a guanine nucleotide exchange factor for Rap1, which allows the activation of Rap1. Through this mechanism JAM-A regulates β1-integrin levels and cell adhesion to β1-integrin-dependent extracellular matrix proteins (Mandell et al. 2005; Severson et al. 2009). This mechanism operates also in breast carcinoma cells (McSherry et al. 2011). Knockdown of JAM-A in breast carcinoma cell lines impairs cell migration in vitro as well as invasion in matrigel (Gotte et al. 2010; McSherry et al. 2009). Therefore, high levels of JAM-A most likely provide an advantage to metastasizing tumor cells when they migrate through the extracellular matrix. JAM-A might provide an additional advantage to breast cancer cells. As observed in a mouse breast tumor model, the absence of JAM-A results in a decrease in tumor cell proliferation and an increase in tumor cell apoptosis (Murakami et al. 2011) suggesting that JAM-A can act as survival factor in these cells. It should be noted that some conflicting data exist regarding the role of JAM-A in breast cancer cells. In one report, an inverse correlation was described between (a) the levels of JAM-A and disease progression, and (b) the levels of

JAM-A and the capacity of the cells to migrate on collagen and invade collagen gels (Naik et al. 2008). However, in light of the clinical data sets provided in two studies (McSherry et al. 2009; Murakami et al. 2011), it appears that breast cancer cells benefit from high levels of JAM-A expression in vivo.

Similar to JAM-A, JAM-C levels were also found to correlate positively with tumorigenicity of certain tumors. For example, the metastatic potentials of different subclones derived from a parental melanoma cell line correlate with the levels of JAM-C (Fuse et al. 2007; Langer et al. 2011). Interestingly, in the same cells the levels of JAM-A and JAM-B do not correlate with their metastatic potential suggesting that the different JAM family members differentially contribute to tumorigenicity in tumors derived from different tissues. The tumor cells might benefit from the function of JAM-C to regulate cell adhesion to the ECM and to promote invasion through matrigel (Fuse et al. 2007).

It should be noted that tumor cells could exploit additional functions of JAMs. Both JAM-A and JAM-C are expressed in endothelial cells and are involved in leukocyte – endothelial cell interaction by mediating trans-heterophilic interactions with leukocyte integrins (Weber et al. 2007). However, JAMs are also expressed by leukocytes (Ebnet et al. 2004). JAM-A for example is expressed by neutrophils, and it is JAM-A expressed by neutrophils but not JAM-A expressed by endothelial cells which is responsible for the recruitment of neutrophils to the inflamed peritoneum (Corada et al. 2005). In this situation, JAM-A regulates neutrophil recruitment at the level of transendothelial migration as suggested by the observation that the absence of JAM-A in neutrophils increases the number of cells adhering to the vessel wall but decreases the number of cells that have invaded the tissue (Corada et al. 2005). Recent evidence indicates that melanoma cell-expressed JAM-C contributes to the transendothelial migration of melanoma cells, probably by mediating trans-homophilic JAM-C – JAM-C interactions (Ghislin et al. 2011; Langer et al. 2011). In addition, JAM-C expressed by some tumor cells can also regulate the adhesion of tumor cells to endothelial cells (Santoso et al. 2005). These observations indicate that JAM functions used by leukocyte to extravasate at sites of inflammation are exploited by metastasizing tumor cells to invade target tissues. Finally, JAMs could also contribute to tumor development in a non cell-autonomous manner. All JAMs are expressed by endothelial cells, and evidence exists for both JAM-A and JAM-C that they regulate angiogenesis (Cooke et al. 2006; Lamagna et al. 2005; Naik et al. 2003). Antibodies against JAM-C were shown to reduce tumor growth and vascularisation (Lamagna et al. 2005). Therefore, it is possible that aberrant expression of JAMs by endothelial cells could support tumor growth by influencing neo-angiogenesis.

5.2 The Blood-Brain Barrier and Tumor Metastasis

The blood-brain barrier (BBB) is a barrier formed by endothelial cells of the brain microvasculature and the surrounding astrocytes and pericytes (Abbott et al. 2010; Abbott et al. 2006). The barrier function is mediated predominantly by TJs of the endothelial cells which differ from the TJs found in other vascular beds by their very low permeability and their high electrical resistance. The barrier function of the BBB thus resembles more closely to the barrier function of epithelial cells rather than to the barrier function of endothelial outside of the central nervous system. In line with this, the TJs of the BBB also morphologically resemble more closely the TJ of epithelial cells. The TJ strands of BBB endothelial cells are predominantly P-face-associated whereas the TJ strands of endothelial TJs are normally E-face-associated (Wolburg et al. 2006). The low permeability of the BBB for ions and polar solutes protects the brain from toxic substances present in the circulation and might also help to restrict the invasion of the brain by inflammatory cells.

The brain is a preferred target organ for several metastasizing tumors with the majority of brain metastases originating from primary cancers in the lung, in the breast or from melanomas (Eichler et al. 2011). The metastasizing cells derived from these tumors must have developed mechanism to specifically home to the brain and to overcome the BBB. Several genes have been identified which mediate metastasis to organs that contain non-fenestrated endothelial as they exist in the brain or in the lung. These include the prostaglandin-synthesizing enzyme cyclooxygenase-2 (*COX2*), the heparin-binding epidermal growth factor (*HBEGF*) and the α2,6-sialyltransferase *ST6GALNAC5* (Bos et al. 2009). Interestingly, whereas COX2 and HBEGF mediate tropism of breast cancer cells to both lung and brain, ST6GALNAC5 mediates selective infiltration of the brain. ST6GALNAC5 acts at the level of tumor cell – BBB endothelial cell interaction as indicated by the positive correlation between the expression of ST6GALNAC5 in tumor cells and their capacity to transmigrate through an in vitro BBB model (Bos et al. 2009). Carbohydrate modifications of cell adhesion molecules which mediate leukocyte – endothelial interactions during lymphocyte homing are known to mediate tissue-specificity (Butcher and Picker 1996; Springer 1995). Therefore, it is likely that ST6GALNAC5 modifies surface cell adhesion molecules by sialylation which thereby acquire the ability to selectively interact with surface molecules expressed by BBB endothelial cells, which in turn promotes the transmigration through the BBB endothelium.

After overcoming the BBB endothelium, additional factors which are not brain-derived may contribute to brain-specific homing of metastasizing tumor cells. Tumor cells coopt pre-existing blood vessels for growth within the brain (Carbonell et al. 2009). Once transmigrated through the endothelium, tumor cells adhere in an integrin-β1-dependent manner predominantly to the vascular basement membrane implying the neurovasculature as a “niche” that promotes progression of malignant cells (Carbonell et al. 2009). It remains to be established if this is a specific property of the basement membrane of the BBB endothelium.

Interestingly, the BBB becomes structurally and functionally compromised when metastatic tumors grow beyond a diameter of 1–2 mm, which might be explained by a disruption of the interaction between astrocytes and endothelial cells (Eichler et al. 2011; Zhang et al. 1992). This suggests that the mechanisms required for metastatic tumor cells to overcome the BBB might operate only during early dissemination into the brain.

5.3 JAMs at the Blood-Brain Barrier

As outlined in the introduction, all three JAMs are expressed by endothelial cells. Immunogold electron microscopy performed in various regions of the mouse, rat and human brain has identified JAM-A at cell-cell contacts of BBB endothelial cells (Dobrogowska and Vorbrodt 2004; Vorbrodt and Dobrogowska 2003, 2004). In addition, JAM-A, -B, and -C are expressed in primary cultures of porcine BBB endothelial cells where they localize to interendothelial cell-cell contacts (Nagasawa et al. 2006). If changes in the expression levels or functional alterations of JAMs are involved in the tropism of brain-targeting metastatic tumor cells remains to be established. Interestingly, during cold injury-induced breakdown of the BBB in vivo, JAM-A expression levels decrease (Yeung et al. 2008) indicating a correlation between BBB integrity and JAM-A expression. Whether downregulation of JAM-A is a primary event and causative to BBB breakdown is still open. Evidence for a BBB-specific role of JAM-A has been provided by the observation that primary human BBB endothelial cells behave differently from non brain-derived primary endothelial cells (Haarmann et al. 2010). During inflammatory conditions JAM-A is redistributed away from cell-cell contacts and appears at the apical surface to mediate adhesion of lymphocytes in a $\alpha L\beta 2$-integrin-dependent manner (Ostermann et al. 2002; Ozaki et al. 1999). In addition, pro-inflammatory cytokines induce JAM-A shedding from the cell surface in vitro and in vivo, resulting in increased JAM-A serum levels (Hou et al. 2010; Koenen et al. 2009). Increased serum levels of JAM-A were found in patients with coronary artery disease, arterial hypertension, or systemic sclerosis (Cavusoglu et al. 2007; Hou et al. 2010; Ong et al. 2009). Brain-derived human microvascular endothelial cells, however, do not seem to respond to pro-inflammatory cytokines with increased JAM-A shedding, and JAM-A serum levels are not increased in patients with multiple sclerosis and ischemic stroke (Haarmann et al. 2010). These observations indicate that JAM-A shedding does not necessarily correlate with a loss in BBB function in all cases of BBB breakdown. However, it should be noted that the function of JAM-A is not exclusively regulated by the levels of surface expression. Its specific subcellular localization or its ability to form homodimers can also influence its functional activity. It will be important to understand these functions of JAM-A. In addition, it will be necessary to investigate how JAM-A regulates the integrity of the BBB to understand if JAM-A and perhaps JAM-B and JAM-C contribute to the brain-specific homing of metastasizing tumor cells.

References

Abbott NJ, Ronnback L, Hansson E (2006) Astrocyte-endothelial interactions at the blood-brain barrier. Nat Rev 7:41–53

Abbott NJ, Patabendige AA, Dolman DE, Yusof SR, Begley DJ (2010) Structure and function of the blood-brain barrier. Neurobiol Dis 37:13–25

Aranda V, Haire T, Nolan ME, Calarco JP, Rosenberg AZ, Fawcett JP, Pawson T, Muthuswamy SK (2006) Par6-aPKC uncouples ErbB2 induced disruption of polarized epithelial organization from proliferation control. Nat Cell Biol 8:1235–1245

Aurrand-Lions M, Johnson-Leger C, Wong C, Du Pasquier L, Imhof BA (2001) Heterogeneity of endothelial junctions is reflected by differential expression and specific subcellular localization of the three JAM family members. Blood 98:3699–3707

Bos PD, Zhang XH, Nadal C, Shu W, Gomis RR, Nguyen DX, Minn AJ, van de Vijver MJ, Gerald WL, Foekens JA, Massague J (2009) Genes that mediate breast cancer metastasis to the brain. Nature 459:1005–1009

Butcher EC, Picker LJ (1996) Lymphocyte homing and homeostasis. Science 272:60–66

Carbonell WS, Ansorge O, Sibson N, Muschel R (2009) The vascular basement membrane as "soil" in brain metastasis. PLoS One 4:e5857

Cavusoglu E, Kornecki E, Sobocka MB, Babinska A, Ehrlich YH, Chopra V, Yanamadala S, Ruwende C, Salifu MO, Clark LT, Eng C, Pinsky DJ, Marmur JD (2007) Association of plasma levels of F11 receptor/junctional adhesion molecule-A (F11R/JAM-A) with human atherosclerosis. J Am Coll Cardiol 50:1768–1776

Cooke VG, Naik MU, Naik UP (2006) Fibroblast growth factor-2 failed to induce angiogenesis in junctional adhesion molecule-A-deficient mice. Arterioscler Thromb Vasc Biol 26:2005–2011

Corada M, Chimenti S, Cera MR, Vinci M, Salio M, Fiordaliso F, De Angelis N, Villa A, Bossi M, Staszewsky LI, Vecchi A, Parazzoli D, Motoike T, Latini R, Dejana E (2005) Junctional adhesion molecule-A-deficient polymorphonuclear cells show reduced diapedesis in peritonitis and heart ischemia-reperfusion injury. Proc Natl Acad Sci USA 102:10634–10639

Debnath J, Brugge JS (2005) Modelling glandular epithelial cancers in three-dimensional cultures. Nat Rev Cancer 5:675–688

Dobrogowska DH, Vorbrodt AW (2004) Immunogold localization of tight junctional proteins in normal and osmotically-affected rat blood-brain barrier. J Mol Histol 35:529–539

Ebnet K, Suzuki A, Horikoshi Y, Hirose T, Meyer Zu Brickwedde MK, Ohno S, Vestweber D (2001) The cell polarity protein ASIP/PAR-3 directly associates with junctional adhesion molecule (JAM). EMBO J 20:3738–3748

Ebnet K, Aurrand-Lions M, Kuhn A, Kiefer F, Butz S, Zander K, Meyer Zu Brickwedde MK, Suzuki A, Imhof BA, Vestweber D (2003) The junctional adhesion molecule (JAM) family members JAM-2 and JAM-3 associate with the cell polarity protein PAR-3: a possible role for JAMs in endothelial cell polarity. J Cell Sci 116:3879–3891

Ebnet K, Suzuki A, Ohno S, Vestweber D (2004) Junctional adhesion molecules (JAMs): more molecules with dual functions? J Cell Sci 117:19–29

Economopoulou M, Hammer J, Wang F, Fariss R, Maminishkis A, Miller SS (2009) Expression, localization, and function of junctional adhesion molecule-C (JAM-C) in human retinal pigment epithelium. Invest Ophthalmol Vis Sci 50:1454–1463

Eichler AF, Chung E, Kodack DP, Loeffler JS, Fukumura D, Jain RK (2011) The biology of brain metastases-translation to new therapies. Nat Rev Clin Oncol 8:344–356

Fuse C, Ishida Y, Hikita T, Asai T, Oku N (2007) Junctional adhesion molecule-C promotes metastatic potential of HT1080 human fibrosarcoma. J Biol Chem 282:8276–8283

Gao L, Joberty G, Macara IG (2002) Assembly of epithelial tight junctions is negatively regulated by Par6. Curr Biol 12:221–225

Ghislin S, Obino D, Middendorp S, Boggetto N, Alcaide-Loridan C, Deshayes F (2011) Junctional adhesion molecules are required for melanoma cell lines transendothelial migration in vitro. Pigment Cell Melanoma Res 24:504–511

Gliki G, Ebnet K, Aurrand-Lions M, Imhof BA, Adams RH (2004) Spermatid differentiation requires the assembly of a cell polarity complex downstream of junctional adhesion molecule-C. Nature 431:320–324

Gotte M, Mohr C, Koo CY, Stock C, Vaske AK, Viola M, Ibrahim SA, Peddibhotla S, Teng YH, Low JY, Ebnet K, Kiesel L, Yip GW (2010) miR-145-dependent targeting of junctional adhesion molecule A and modulation of fascin expression are associated with reduced breast cancer cell motility and invasiveness. Oncogene 29:6569–6580

Haarmann A, Deiss A, Prochaska J, Foerch C, Weksler B, Romero I, Couraud PO, Stoll G, Rieckmann P, Buttmann M (2010) Evaluation of soluble junctional adhesion molecule-A as a biomarker of human brain endothelial barrier breakdown. PLoS One 5:e13568

Hou Y, Rabquer BJ, Gerber ML, Del Galdo F, Jimenez SA, Haines GK 3rd, Barr WG, Massa MC, Seibold JR, Koch AE (2010) Junctional adhesion molecule-A is abnormally expressed in diffuse cutaneous systemic sclerosis skin and mediates myeloid cell adhesion. Ann Rheum Dis 69:249–254

Huber MA, Kraut N, Beug H (2005) Molecular requirements for epithelial-mesenchymal transition during tumor progression. Curr Opin Cell Biol 17:548–558

Hurov JB, Watkins JL, Piwnica-Worms H (2004) Atypical PKC Phosphorylates PAR-1 Kinases to regulate localization and activity. Curr Biol 14:736–741

Iden S, Misselwitz S, Peddibhotla SS, Tuncay H, Rehder D, Gerke V, Robenek H, Suzuki A, Ebnet K (2012) aPKC phosphorylates JAM-A at Ser285 to promote cell contact maturation and tight junction formation. J Cell Biol 196:623–639

Itoh M, Sasaki H, Furuse M, Ozaki H, Kita T, Tsukita S (2001) Junctional adhesion molecule (JAM) binds to PAR-3: a possible mechanism for the recruitment of PAR-3 to tight junctions. J Cell Biol 154:491–498

Koenen RR, Pruessmeyer J, Soehnlein O, Fraemohs L, Zernecke A, Schwarz N, Reiss K, Sarabi A, Lindbom L, Hackeng TM, Weber C, Ludwig A (2009) Regulated release and functional modulation of junctional adhesion molecule A by disintegrin metalloproteinases. Blood 113:4799–4809

Kornecki E, Walkowiak B, Naik UP, Ehrlich YH (1990) Activation of human platelets by a stimulatory monoclonal antibody. J Biol Chem 265:10042–10048

Lamagna C, Hodivala-Dilke KM, Imhof BA, Aurrand-Lions M (2005) Antibody against junctional adhesion molecule-C inhibits angiogenesis and tumor growth. Cancer Res 65:5703–5710

Langer HF, Orlova VV, Xie C, Kaul S, Schneider D, Lonsdorf AS, Fahrleitner M, Choi EY, Dutoit V, Pellegrini M, Grossklaus S, Nawroth PP, Baretton G, Santoso S, Hwang ST, Arnold B, Chavakis T (2011) A novel function of junctional adhesion molecule-C in mediating melanoma cell metastasis. Cancer Res 71:4096–4105

Liang TW, DeMarco RA, Mrsny RJ, Gurney A, Gray A, Hooley J, Aaron HL, Huang A, Klassen T, Tumas DB, Fong S (2000) Characterization of huJAM: evidence for involvement in cell-cell contact and tight junction regulation. Am J Physiol Cell Physiol 279:C1733–C1743

Liu Y, Nusrat A, Schnell FJ, Reaves TA, Walsh S, Pochet M, Parkos CA (2000) Human junction adhesion molecule regulates tight junction resealing in epithelia. J Cell Sci 113:2363–2374

Mandell KJ, Babbin BA, Nusrat A, Parkos CA (2005) Junctional adhesion molecule 1 regulates epithelial cell morphology through effects on beta1 integrins and Rap1 activity. J Biol Chem 280:11665–11674

Mandicourt G, Iden S, Ebnet K, Aurrand-Lions M, Imhof BA (2007) JAM-C regulates tight junctions and integrin-mediated cell adhesion and migration. J Biol Chem 282:1830–1837

Martin-Padura I, Lostaglio S, Schneemann M, Williams L, Romano M, Fruscella P, Panzeri C, Stoppacciaro A, Ruco L, Villa A, Simmons D, Dejana E (1998) Junctional adhesion molecule, a novel member of the immunoglobulin superfamily that distributes at intercellular junctions and modulates monocyte transmigration. J Cell Biol 142:117–127

McSherry EA, McGee SF, Jirstrom K, Doyle EM, Brennan DJ, Landberg G, Dervan PA, Hopkins AM, Gallagher WM (2009) JAM-A expression positively correlates with poor prognosis in breast cancer patients. Int J Cancer 125:1343–1351

McSherry EA, Brennan K, Hudson L, Hill AD, Hopkins AM (2011) Breast cancer cell migration is regulated through junctional adhesion molecule-A-mediated activation of Rap1 GTPase. Breast Cancer Res 13:R31

Murakami M, Giampietro C, Giannotta M, Corada M, Torselli I, Orsenigo F, Cocito A, d'Ario G, Mazzarol G, Confalonieri S, Di Fiore PP, Dejana E (2011) Abrogation of junctional adhesion molecule-a expression induces cell apoptosis and reduces breast cancer progression. PLoS One 6:e21242

Nagasawa K, Chiba H, Fujita H, Kojima T, Saito T, Endo T, Sawada N (2006) Possible involvement of gap junctions in the barrier function of tight junctions of brain and lung endothelial cells. J Cell Physiol 208:123–132

Naik MU, Mousa SA, Parkos CA, Naik UP (2003) Signaling through JAM-1 and {alpha}v{beta}3 is required for the angiogenic action of bFGF: dissociation of the JAM-1 and {alpha}v{beta}3 complex. Blood 102:2108–2114

Naik MU, Naik TU, Suckow AT, Duncan MK, Naik UP (2008) Attenuation of junctional adhesion molecule-A is a contributing factor for breast cancer cell invasion. Cancer Res 68:2194–2203
Ong KL, Leung RY, Babinska A, Salifu MO, Ehrlich YH, Kornecki E, Wong LY, Tso AW, Cherny SS, Sham PC, Lam TH, Lam KS, Cheung BM (2009) Elevated plasma level of soluble F11 receptor/junctional adhesion molecule-A (F11R/JAM-A) in hypertension. Am J Hypertens 22:500–505
Ostermann G, Weber KS, Zernecke A, Schroder A, Weber C (2002) JAM-1 is a ligand of the beta(2) integrin LFA-1 involved in transendothelial migration of leukocytes. Nat Immunol 3:151–158
Ozaki H, Ishii K, Horiuchi H, Arai H, Kawamoto T, Okawa K, Iwamatsu A, Kita T (1999) Cutting edge: combined treatment of TNF-alpha and IFN-gamma causes redistribution of junctional adhesion molecule in human endothelial cells. J Immunol 163:553–557
Ozdamar B, Bose R, Barrios-Rodiles M, Wang HR, Zhang Y, Wrana JL (2005) Regulation of the polarity protein Par6 by TGFbeta receptors controls epithelial cell plasticity. Science 307:1603–1609
Santoso S, Orlova VV, Song K, Sachs UJ, Andrei-Selmer CL, Chavakis T (2005) The homophilic binding of junctional adhesion molecule-C mediates tumor cell-endothelial cell interactions. J Biol Chem 23:23
Severson EA, Lee WY, Capaldo CT, Nusrat A, Parkos CA (2009) Junctional adhesion molecule A interacts with Afadin and PDZ-GEF2 to activate Rap1A, regulate beta1 integrin levels, and enhance cell migration. Mol Biol Cell 20:1916–1925
Springer TA (1995) Traffic signals on endothelium for lymphocyte recirculation and leukocyte emigration. Annu Rev Physiol 57:827–872
Suzuki A, Ohno S (2006) The PAR-aPKC system: lessons in polarity. J Cell Sci 119:979–987
Suzuki A, Yamanaka T, Hirose T, Manabe N, Mizuno K, Shimizu M, Akimoto K, Izumi Y, Ohnishi T, Ohno S (2001) Atypical protein kinase C is involved in the evolutionary conserved PAR protein complex and plays a critical role in establishing epithelia-specific junctional structures. J Cell Biol 152:1183–1196
Suzuki A, Ishiyama C, Hashiba K, Shimizu M, Ebnet K, Ohno S (2002) aPKC kinase activity is required for the asymmetric differentiation of the premature junctional complex during epithelial cell polarization. J Cell Sci 115:3565–3573
Suzuki A, Hirata M, Kamimura K, Maniwa R, Yamanaka T, Mizuno K, Kishikawa M, Hirose H, Amano Y, Izumi N, Miwa Y, Ohno S (2004) aPKC acts upstream of PAR-1b in both the establishment and maintenance of mammalian epithelial polarity. Curr Biol 14:1425–1435
Thiery JP (2002) Epithelial-mesenchymal transitions in tumour progression. Nat Rev Cancer 2:442–454
Vorbrodt AW, Dobrogowska DH (2003) Molecular anatomy of intercellular junctions in brain endothelial and epithelial barriers: electron microscopist's view. Brain Res Brain Res Rev 42:221–242
Vorbrodt AW, Dobrogowska DH (2004) Molecular anatomy of interendothelial junctions in human blood-brain barrier microvessels. Folia Histochem Cytobiol 42:67–75
Wang X, Nie J, Zhou Q, Liu W, Zhu F, Chen W, Mao H, Luo N, Dong X, Yu X (2008) Downregulation of Par-3 expression and disruption of Par complex integrity by TGF-beta during the process of epithelial to mesenchymal transition in rat proximal epithelial cells. Biochim Biophys Acta 1782:51–59
Weber C, Fraemohs L, Dejana E (2007) The role of junctional adhesion molecules in vascular inflammation. Nat Rev Immunol 7:467–477
Wodarz A, Nathke I (2007) Cell polarity in development and cancer. Nat Cell Biol 9:1016–1024
Wolburg H, Lippoldt A, Ebnet K (2006) Tight junctions and the blood-brain barrier. In: Mariscal L (ed) Selected topics on tight junctions. RG Landes Bioscience, Georgetown, pp 175–195
Yamanaka T, Horikoshi Y, Suzuki A, Sugiyama Y, Kitamura K, Maniwa R, Nagai Y, Yamashita A, Hirose T, Ishikawa H, Ohno S (2001) Par-6 regulates aPKC activity in a novel way and mediates cell-cell contact-induced formation of epithelial junctional complex. Genes Cells 6:721–731
Yeung D, Manias JL, Stewart DJ, Nag S (2008) Decreased junctional adhesion molecule-A expression during blood-brain barrier breakdown. Acta Neuropathol 115:635–642
Zhang RD, Price JE, Fujimaki T, Bucana CD, Fidler IJ (1992) Differential permeability of the blood-brain barrier in experimental brain metastases produced by human neoplasms implanted into nude mice. Am J Pathol 141:1115–1124

FAK Focal adhesion kinase
GAG Glycosaminoglycan
HGF Hepatocyte growth factor
TCC Transitional cell carcinoma
TJ Tight junction
TUR Transurothelial resistance
ZO Zonula occludens

6.1 Introduction

6.1.1 Bladder Cancer

Bladder cancer is the second most common urological cancer after prostate cancer. Although the term bladder cancer includes transitional cell carcinoma (TCC), squamous cell carcinoma and adenocarcinoma, TCC represents over 90% of cases. TCC often behaves as a field change disease in which the entire urothelium, from the renal pelvis to the urethra may undergo malignant transformation and is responsible for the high incidence of multiple occurrences and the recurrences that are typical of urothelial tumours treated by local resection. However transitional carcinoma cells can also migrate and implant at other sites lined by urothelium. It is thus difficult to determine whether a recurrent tumour represents an inadequately treated initial tumour, a multifocal tumour or tumour implantation. In 2002, there were 10,199 new cases of bladder cancer in the UK representing 4% of all new cancers diagnosed annually (http://info.cancerresearchuk.org/cancerstats/types/bladder/?a=5441) (ISD 2005; Northern Ireland Cancer 2005; Office for National Statistics 2005; Welsh Cancer Intelligence and Surveillance 2005).

The most important determinant of survival in patients with bladder cancer is the depth of tumour invasion. Tumours that do not invade muscularis propria are classified as superficial with an 80–90% 5-year survival. However, patients with muscle-invasive bladder cancer have a 5-year survival of less than 50%. Malignant tumours are graded as low or high grade. Superficial tumours are classified as low risk or high risk according to the risk of progression to muscle invasive disease, which progresses to metastatic disease and eventually death. Low grade Ta lesions are at low risk for progression. High grade lesions including carcinoma in-situ have about 30–50% risk of progression. Although the disease can be controlled locally by radical treatment, many patients still die from metastatic disease. Thus early detection and treatment, when the disease is still superficial is the key to improving the outcome in bladder cancer.

6.1.2 Bladder Cancer and Metastasis

The spread of cancer involves a number of steps. Initially, increases in cell proliferation and increased cell motility is followed by invasion of surrounding stroma, angiogenesis, intravasation of cells into the circulation, and finally extravasation at distant sites to form metastatic deposits (Martin and Jiang 2001; Martin et al. 2011). Abnormal expression of growth factors or their receptors can stimulate tumour cell growth. There is increased expression of the EGF receptor that correlates with the invasive phenotype (Messing 1990; Neal et al. 1990). A large number of malignant transitional epithelial cells express collagenases, especially type IV, which help digest the basal lamina and lamina propria and allow cells to invade the stroma (Messing 2002). Small tumours (<2 mm diameter) can survive on nutrition obtained by local diffusion. However, larger tumours (>2 mm) need to establish a connection to the circulatory system to continue to grow (Martin and Jiang 2001). Some angiogenic factors such as fibroblast growth factor are excreted in the urine of bladder cancer patients (Chodak et al. 1988). Tumour cells can enter the circulation through these newly developed blood vessels, and when in the circulatory system must survive the mechanical stresses of blood flow and immunosurveillance defence mechanisms of the body before adhering to and extravasating through the endothelium of the blood vessels of the target organ, establish themselves and grow and progress as metastasis (Liotta and Kohn 2001).

6.1.3 Cell Adhesion and Bladder Cancer

It is generally accepted that decreased adhesion between adjacent cells is an early step in the invasive and metastatic process. Cell adhesion is maintained by a number of specialised membrane structures which include the TJs (TJ), adherens junctions (AJ) and demososmes. Reduced expression of adhesion molecules such as E-cadherin and β4 integrin as well as the cytoskeletal molecules α, β, γ catenins has been observed in invasive tumours (Messing 2002). The serum proteases urokinase and tissue plasminogen activator (uPA) convert plasminogen into plasmin, which in turn degrades laminin, found within the basement membrane. The u-PA also activates type IV collagenase. The expression of uPA and its receptor is related to invasiveness of bladder cancer, thus reflecting its proteolytic activity (Messing 2002). The urinary excretion of Hepatocyte Growth Factor (HGF), a known motogen is increased in TCC of the bladder, particularly in invasive or high grade tumours (Rosen et al. 1997). There is evidence to suggest that HGF induces disruption of the TJ structure and function, which is accompanied by relocation of the TJ protein occludin. These changes in the molecular structure of TJs elicited by HGF results in a decrease in the trans-epithelial resistance, promotes invasion and causes scattering of bladder cancer cells (Brown et al. 2003). In summary, the inter-cellular junctions, particularly TJ may represent the first barrier to metastasis (Martin and Jiang 2001).

6.2 TJs and Bladder Cancer

Epithelial and endothelial surfaces are primarily responsible for the separation of fluid compartments in the body by acting as permeability barriers. Transport of substances across these structures may occur through the transcellular or paracellular route. The paracellular flux of water-soluble molecules between adjacent cells is controlled by the apical junctional complex, which is composed of the TJ and adherens junction (AJ) (Fig. 6.1a). TJs are the most apical of these structures and separate the apical plasma membrane from the baso-lateral membrane. TJs occur between adjacent umbrella cells of the urothelium and play an important role in maintaining urothelial function.

In the bladder, the urothelium is thought to play a central role in maintaining the urinary composition by acting as a physiological barrier between urine and blood, and to play a major role in the storage and voiding of urine. The urothelium is a multi-layer structure. In humans, the urothelium comprises 6–7 rows of cells arranged into three distinct layers, viz. the basal cells which are germinal in nature, a middle layer of intermediate cells, which are larger than the basal cells and the outermost layer which is formed by the umbrella cells, which are the largest epithelial cells in the body (Fig. 6.1b). The surface of the umbrella cells is covered by a glycosaminoglycan (GAG) layer.

TJs appear as a series of fusion points between the outer leaflets of the plasma membrane of adjacent cells on ultra-thin section electron micrographs. The intercellular space is completely obliterated at these "kissing points" (Gonzalez-Mariscal et al. 2003) whereas in AJs and desmosomes there is a gap of 15–30 nm between adjacent membranes (Tsukita et al. 2001). Freeze fracture electron micrographs however reveal TJs to appear as a network of continuous intramembranous strands (known as TJ strands) that form continuous seals around cells. Each TJ strand is made up of protein that associates laterally with its pair in the plasma membrane of the adjacent cell. In the mammalian bladder TJs consist of 4–6 interconnecting strands that obliterate the intercellular space (Peter 1978). It is proposed that there are 'pores' located within these paired strands. These pores (which may be charged) act as gates that control the paracellular movement of solutes (Claude 1978; Colegio et al. 2002). The resistance barrier of the urothelium is thus formed by: (1) The uroplakins, which are the major site of resistance to transcellular ionic Permeability (Clausen et al. 1979) and (2) TJs, which represent sites of resistance to paracellular ion fluxes.

A single TJ strand of protein can comprise various combinations of claudin molecules. Differences in claudin composition in TJ strands produces distinct patterns in freeze-fracture studies (Furuse et al. 1999). For example, claudin 1 in combination with claudin 3 forms continuous TJ strands on the P (protoplasmic) surface of the cell membrane, whereas combination of claudins 1 and 2 or 3 and 2 produces particulate strands on the E (exoplasmic) surface of the cell. In addition, TJ strands in the CNS myelin and Sertoli cells are composed of parallel filaments of claudin 11 which are unlike the anastomosing strands observed in other TJs (Southwood and Gow 2001). TJs of neighbouring cells connect in the paracellular space through interactions between the extracellular domains of their constituent claudins (Furuse et al. 1999). This interaction may be selective, since studies with transfected L- fibroblasts reported that whereas TJ strands expressing claudin 3 associated with TJ strands expressing claudin 1 and 2, there was no interaction between claudin 1 and claudin 2 strands (Furuse et al. 1999). These observations suggest that selective interaction between different claudins may help explain the differential distribution of claudins in different tissues, and hence the resistance and permeability requirements of that tissue.

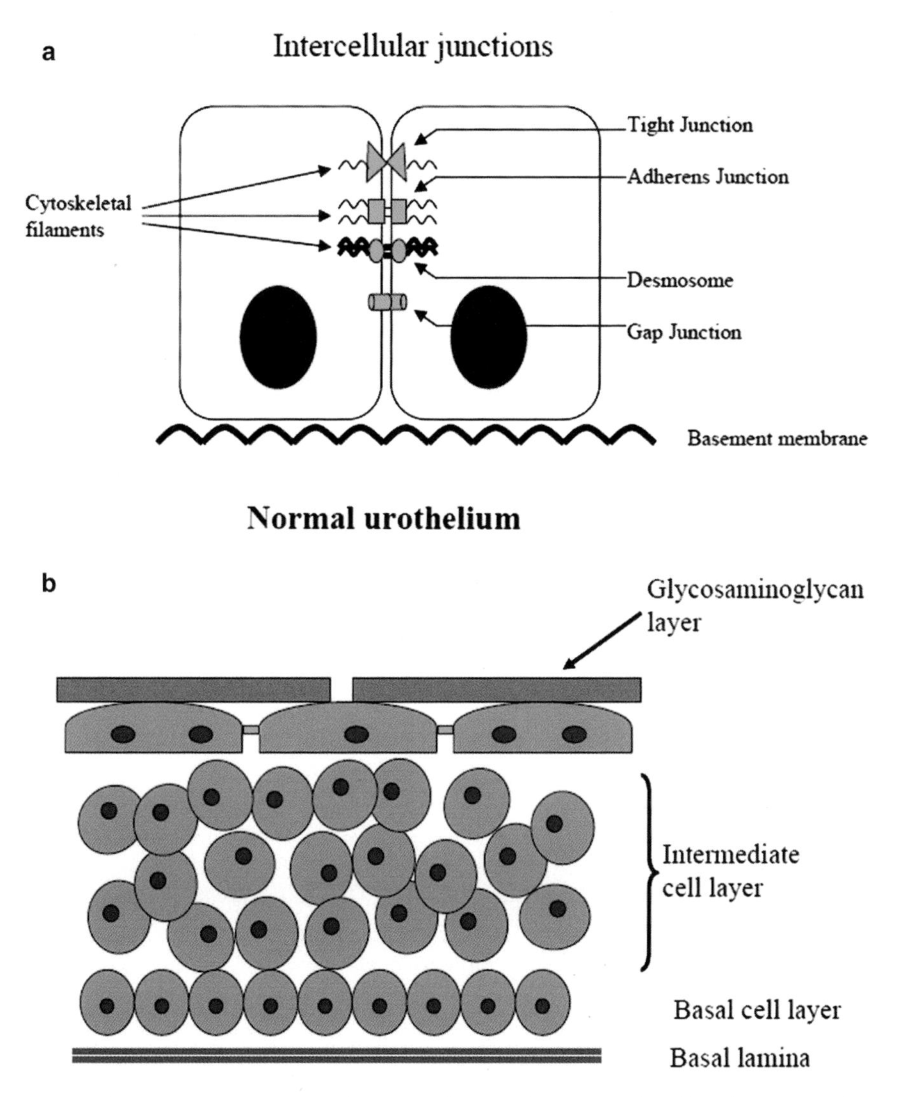

Fig. 6.1 (**a**) Location of intercellular junctions; (**b**) normal urothelium (Modified from Chancellor and Yoshimura (2002))

6.2.1 Tissue Distribution and Permeability Characteristics of Claudins

A number of reports suggest that claudins are distributed in a differential manner in various tissues. For example, in the kidney claudins 1, 3 and 8 are expressed in TJs in the distal tubule, claudins 1, 3, 4 and 8 in TJs in the collecting tubules, claudins 2, 10 and 11 in TJs in the proximal segment and claudins 5 and 15 in the endothelium (Kiuchi-Saishin et al. 2002; Reyes et al. 2002). Thus, claudin-1 is present in high resistance epithelia, such as the collecting tubules, and absent in leaky epithelia such as the proximal tubule (Reyes et al. 2002). The converse is true of claudin 2, which is present in the low resistance proximal tubule epithelium, and not detected in the collecting tubule. Claudins-4 (Van Itallie et al. 2001), -8 (Yu et al. 2003) and -14 (Ben-Yosef et al. 2003) reduce conductance across epithelia by selectively discriminating against cations (Table 1.6). Claudin-6 is expressed in embryonic epithelia (Turksen and Troy 2001). Claudin-1 knock-out mice (Furuse et al. 2002) and transgenic mice over-expressing claudin-6 (Turksen and Troy 2002) show defective skin barrier function. Patients with hypomagnesaemia hypercalciuria syndrome demonstrate a mutant form of claudin 16. In these patients, there is a selective defect in the reabsorption of calcium and magnesium, with intact sodium and chloride reabsorption in the thick ascending limb of the loop of Henle (Blanchard et al. 2001). Claudin-16 thus seems to show selective paracellular conductance for calcium and magnesium. In summary, it has been suggested that a tightness function can be demonstrated for claudins-1, 3, 4, 5, 8, 11, 14 and 19, whereas claudins-2, 7, 10, 15 and 16 are thought to function as pore forming claudins (Krause et al. 2008).

6.2.2 *Alteration of Claudin Expression in Cancer*

Aberrant expression of various claudins, including claudins 1, 2, 3, 4, 5, 7, 10, 16 and 23 has been described in a variety of immortalised cancer cell lines and in human cancer tissue. For example, claudin 1 expression is downregulated in breast cancer (Kramer et al. 2000). Similarly claudin 7 expression is down-regulated in breast (Kominsky et al. 2003) and head and neck cancers (Al Moustafa et al. 2002). This pattern of expression supports the generally accepted theory that there is disruption of TJs during the process of tumourigenesis, invasion and metastasis, with associated down-regulation of the expression of claudins, which may have a putative role in tumour suppression (Swisshelm et al. 2005). Indeed, studies have demonstrated that TJs of human and rat colonic tumours were leakier than TJs of normal colon and that increased TJ permeability of the colon epithelium and a decreased epithelial barrier function preceded the development of colonic cancers in an experimental carcinogenesis model (Soler et al. 1999). In addition to altered levels of expression, phosphorylation of TJ proteins such as claudins is thought to disrupt TJ structure and function. Thus, phosphorylation of claudin 3 in ovarian cancer cells disrupts TJs (D'Souza et al. 2005). Conversely, some claudin proteins are over-expressed in cancer. Thus, claudins 3 and 4 are over-expressed in ovarian cancer (Rangel et al. 2003), breast (Kominsky et al. 2004) and prostate adenocarcinoma (Long et al. 2001). Similarly, claudin 10 is overexpressed in thyroid papillary cancer (Aldred et al. 2004) and claudin 16 is overexpressed in ovarian cancer (Rangel et al. 2003), suggesting that claudins may have a role in tumourigenesis. There are reports that suggest a role for claudins in cell survival and invasiveness.

Expression of claudin 7 inversely correlates with the histological grade in breast cancer (Kominsky et al. 2003). Although claudin 4 is overexpressed in 54 pancreatic cancer tissues (Michl et al. 2003), the expression of claudin 4 is also inversely related to the invasiveness of pancreatic cells (Michl et al. 2003). Conversely, the expression of claudins 3 and 4 has been reported to positively correlate with the invasiveness of ovarian cancer cells (Agarwal et al. 2005). These unusual expression patterns of claudins in cancer could serve as potential indicators of invasiveness and prognosis and could be used as therapeutic targets (Morin 2005). There is a substantial body of evidence to suggest that cells, including prostatic (Long et al. 2001), pancreatic (Michl et al. 2003), breast (Kominsky et al. 2004) and ovarian cancer cells (Santin et al. 2005) that express claudins- 3 and 4 are targets for the *Clostridium perfringens* enterotoxin (CPE); which binds to these claudins and induces cytolysis. CPE-based therapy has been suggested as a novel treatment of these cancers (Hewitt et al. 2006). However, claudin-3 and 4 are also expressed in several normal tissues, and therefore systemic administration of CPE may have significant toxicity (Hewitt et al. 2006). Similarly, claudin-5 which is expressed at high levels in vascular endothelial cells, is also expressed in the brain and therefore systemic anti-angiogenic therapy targeting claudin-5 can have potentially serious side-effects (Hewitt et al. 2006). On the other hand, the urinary bladder lends itself to easy access via the urethral route for the intravesical instillation of novel therapies for bladder cancer such as CPE, with a decreased risk of systemic side effects.

6.3 Expression of TJ Molecules in Human Bladder and Bladder Cancer Tissues/Cells

6.3.1 Expression of TJs Molecular in Urethelial Cells and Tissues

A broad spectrum of TJ molecules are detectable in urothelial and urinary bladder tissues. Of the conventional TJ molecules, occludin, claudin-1 and ZO-1 are reported (Nakanishi et al. 2008). Subcoat molecules linked to TJ, namely ZO-1, ZO-2 and ZO-3, JAM1 and δ-catenin are also widely expressed in bladder cells and tissues (Fig. 6.2) (Sánchez Freire et al. 2011). It has also been shown that other claudins, namely claudin-4, claudin-5, claudin-7, claudin-11, claudin-12, claudin-16 are also present in bladder tissues (Boireau et al. 2007; Awsare et al. 2011) (Fig. 6.3). More information is available on cell lines from urethelial cells. In bladder cancer derived cell lines, claudins-1, 2, 4, 5, 6, 7, 8, 12, 14, 15 and 22 are detectable (Southgate et al. 2007; Rickard et al. 2008). However, claudins-1, 3, 9, and 20 are not commonly seen in these cells. Coxsackievirus and adenovirus receptor (CAR), a viral receptor protein and a known TJ proteins are also present in the urothelial cells and is in close proximity with occludin and ZO-1 in tissue (Gye et al. 2011).

are almost exclusively seen at the lateral junctions in normal cells but become reduce or lost in the lateral junctions in cancer tissues (Boireau et al. 2007).

Occludin has so far not shown a clear clinical relevance in bladder cancer (Nakanishi et al. 2008). This is interesting and may partly reflect the observation that occludin, despite abundantly expressed in urinary bladder, may play a lesser role or no role in the control of the barrier function of the bladder in a study using occludin knockout mice (Schulzke et al. 2005). It has been shown that the reduction of claudin-4 expression in urinary bladder cancer is likely to be via the methylation mechanism of CL4 promoters (Boireau et al. 2007). This is obviously a fertile area to explore in future.

6.4 Role of Claudin-11 in Human Bladder Cancer Cells

We recently investigated the possible role of Claudin-11 in bladder cancer tissues and cells (Awsare et al. 2011). The expression of claudin-11 in benign and malignant bladder tissue and the effect of forced expression of claudin-11 on TJ function and invasiveness of bladder cancer cells were studied. Benign bladder tissue demonstrated equal expression of claudin-11 mRNA as carcinoma, but displayed more intense staining than malignant tissue on immunohistochemistry (Fig. 6.4). Claudin-11 expression was tested in bladder cancer cell lines (T24/83, RT112/84 and EJ138) using reverse transcription-polymerase chain reaction (RT-PCR) and in benign and malignant bladder tissue by quantitative RT-PCR and immunohistochemistry. T24/83 cells were transfected with the full-length human claudin-11 sequence. Stable-transfected cells overexpressing claudin-11 (T24Cl-11Ex), wild-type cells (T24WT) and the empty plasmid control clone (T24GFP) were compared using transurothelial resistance (TUR), in vitro adhesion, invasion and growth assays. Claudin-11 was strongly expressed in the non-invasive RT112/84 cell line compared to the invasive T24/83 and EJ138 bladder cancer cell lines. Forced-expression of claudin-11 in T24/83 cells was confirmed by PCR, immunoprecipitation and by immunofluorescence, which demonstrated increased perinuclear claudin-11 staining (Fig. 6.5). Forced expression of claudin-11 did not affect TUR($p = 0.243$), but significantly reduced invasion ($p = 0.001$) while increasing cell matrix adhesion ($p = 0.001$) and growth rates ($p = 0.001$) (Awsare et al. 2011). The greater expression of claudin-11 in benign versus malignant tissue and non-invasive versus invasive cell lines, and its effect in reducing bladder cancer cell invasiveness suggests that Claudin-11 may have a role in preventing cancer progression and may serve as a therapeutic target in reducing metastasis. This study demonstrates the role of claudin-11 in bladder cancer cells. Claudin-11 expression could potentially be used as a biomarker to differentiate between non-invasive and aggressive bladder cancer. In vivo efficacy and safety studies must be performed to explore if claudin-11 gene therapy could be used to modify the course of the disease since this therapy can be delivered urethrally directly into the bladder.

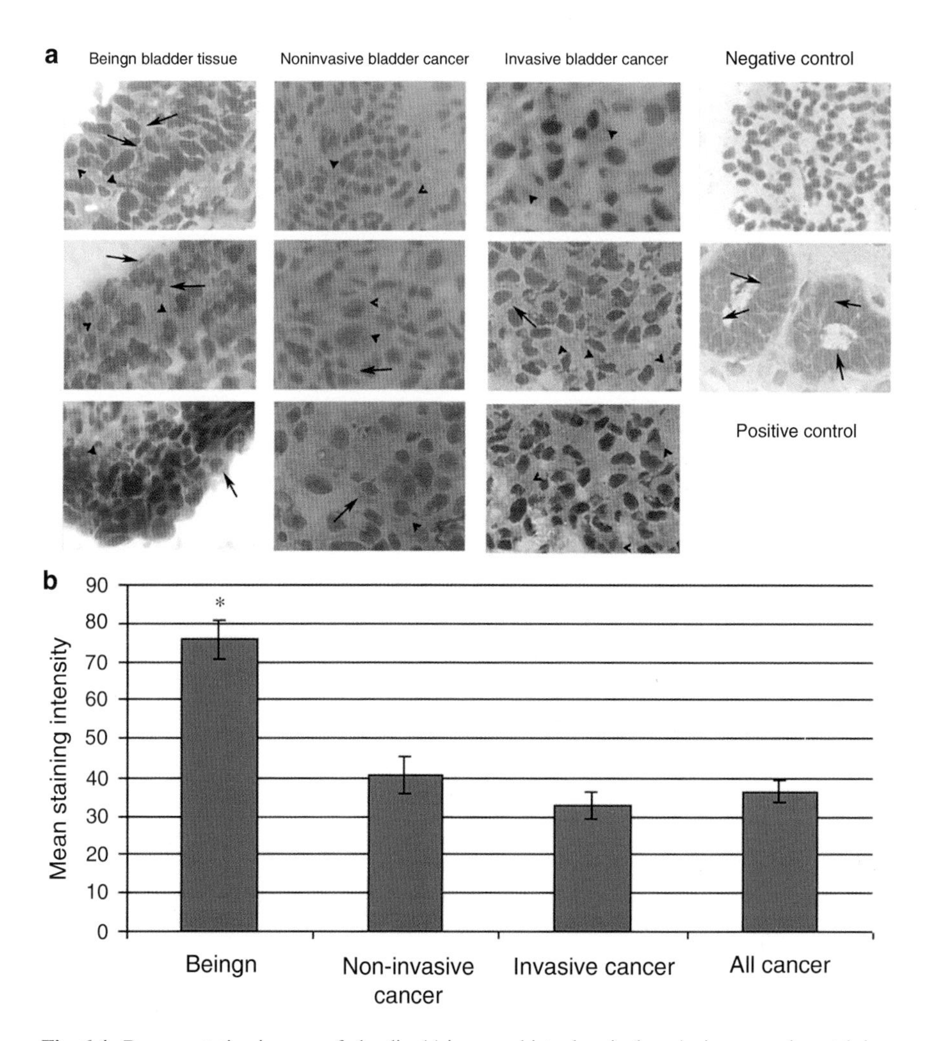

Fig. 6.4 Representative images of claudin-11 immunohistochemical analysis comparing staining patterns and intensity in benign, non-invasive and invasive bladder cancer tissue. (**a**) There was greater uptake in benign tissue as compared to TCC. Claudin-11 located to the intercellular region and cytoplasm of cells (×400 magnification). (**b**) Claudin-11 staining intensity was greater in benign bladder tissue, compared to non-invasive, invasive and all cancer tissue. *$p<0.05$, Error bars represent SEM

6.5 Regulation of TJs by Experimental Agents

It has been reported that the barrier functions of bladder cells can be regulated by protamine sulfate in vivo. Exposure to protamine resulted in a marked increase in the permeability to water and urea, accompanied by a reduction in transepithelial resistance (Lavelle et al. 2002). It is interesting to observe that the changes of the

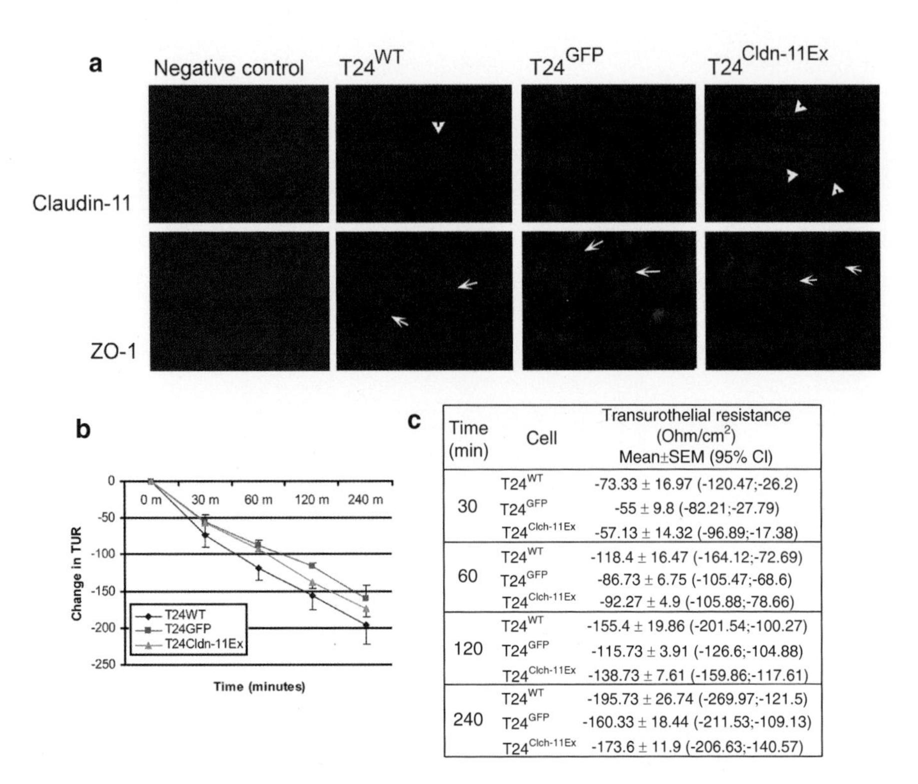

Time (min)	Cell	Transurothelial resistance (Ohm/cm^2) Mean±SEM (95% CI)
30	T24^{WT}	-73.33 ± 16.97 (-120.47;-26.2)
	T24^{GFP}	-55 ± 9.8 (-82.21;-27.79)
	T24$^{Clch-11Ex}$	-57.13 ± 14.32 (-96.89;-17.38)
60	T24^{WT}	-118.4 ± 16.47 (-164.12;-72.69)
	T24^{GFP}	-86.73 ± 6.75 (-105.47;-68.6)
	T24$^{Clch-11Ex}$	-92.27 ± 4.9 (-105.88;-78.66)
120	T24^{WT}	-155.4 ± 19.86 (-201.54;-100.27)
	T24^{GFP}	-115.73 ± 3.91 (-126.6;-104.88)
	T24$^{Clch-11Ex}$	-138.73 ± 7.61 (-159.86;-117.61)
240	T24^{WT}	-195.73 ± 26.74 (-269.97;-121.5)
	T24^{GFP}	-160.33 ± 18.44 (-211.53;-109.13)
	T24$^{Clch-11Ex}$	-173.6 ± 11.9 (-206.63;-140.57)

Fig. 6.5 (**a**) Representative images of immunofluorescence staining for claudin-11 and ZO-1 in T24/83 cells (×40 magnification). There was a stronger signal for claudin-11 in T24Cldn-11Ex compared to T24WTand T24GFPcells, thus confirming its forced expression. ZO-1 located to TJs (*white arrows*), whereas claudin-11 located to cytoplasm (*white arrowheads*). (**b**) Effect of forced-expression of claudin-11 on transurothelial resistance (change from baseline) of T24/83 cells over 240 min. There was no significant difference in the change in TUR in T24Cldn-11Ex cells compared to T24WTor T24GFPcells ($p=0.243$, ANOVA test). Error bars represent SEM. (**c**) The statistical information of the data set

permeability is sustained but recoverable. Urea and Proteins/peptides present in the urine may also influence the TJs functions (Lewis and Kleine 2000; Gallardo et al. 2002). Ca^{2+} plays an pivotal role in the regulation of TJ functions. Calcium regulators thus have been shown to be useful tools in regulating the TJs in urothelial cells (Lacaz-Vieira and Kachar 1996)

6.5.1 Regulation of TJs by Potential Therapeutic Agents

In bladder urethelial cells, troglitizone a PPAR-gamma agonist and PD153035 a EGF receptor kinase inhibitor are able to upregulate the expression of claudins-4 and 5 in bladder cell (Southgate et al. 2007). It has also been shown that Antiproliferative factor is able to decrease the expression of ZO-1 and occludin in the bladder whereas, Neu5Acα2-3Galβ1-3GalNAcα-O-TVPAAVVVA derivatives are able to upregulate expression of claudins (1, 4, 8 and 12), occludin and ZO-1 expression and change the function of bladder cells (Zhang et al. 2005; Keay et al. 2011). COX-2 inhibitor, for example Celecoxib has been shown to protect the function of TJs of urinary bladder in in vivo models (Celik-Ozenci et al. 2005). An antibiotics, Colistin, or polymyxin E, is also known to have have high affinity to uritherial cells and regulate the paracellular permeability (Lewis and Lewis 2004). Chitsan, a linear plysaccharide derived from Chitin, is a product that has been used as a hemostatic agent and for drug delivery. It is adhesive to mucosa and has been developed for cross skin and mucosa drug delivery. It has a marked impact on the permeability of the bladder urethelial cells by affecting the function of TJs in the tissue (Kos et al. 2006). Manipulation of the paracellular permeability may have therapeutic implication in topically applied medicine. For example, HPG-C(8/10)-MePEG-NH(2) nanoparticles has been reported to increase the permeability of the bladder tissue and increase the uptake of docetaxal (Mugabe et al. 2012)

6.5.2 TJ Modulation by Cytokines

A number of cytokines regulate TJ function. Proinflammatory cytokines such as TNF-α and IFN-γ downregulate the expression of the TJ protein occludin (Mankertz et al. 2000). Similarly, IFN-γ induces loss of ZO-1 and disrupts TJ structure and function (Youakim and Ahdieh 1999). Other cytokines such as EGF (Van Itallie et al. 1995), VEGF, and HGF (Martin et al. 2004) are also implicated in the disruption of TJ function.

6.5.3 Hepatocyte Growth Factor (HGF)

HGF was originally discovered to be a powerful stimulator of the growth and proliferation of hepatocytes (Nakamura et al. 1984). HGF is a multifunctional cytokine which elicits a number of effects on cell function (Jiang et al. 2005). HGF induces cytoskeletal changes through the Rho/Rac/ROCK and Grb-2 (growth factor receptor bound protein) pathways and leads to breakdown of cell-cell adhesion and increased motility. HGF also activates the focal adhesion kinase (FAK) pathway and coordinates the adhesion and migration of cancer cells over extracellular matrix by regulating the expression of integrins in cancer cells (Jiang et al. 2005). HGF/SF regulates TJ function in both normal epithelial cells (Hollande et al. 2001) and cancer cells (Kominsky et al. 2003; Martin et al. 2004). Treatment with HGF has been shown to cause a loss of claudin-7 (Kominsky et al. 2003), and phosphorylation of ZO-1 in breast cancer cells (Martin et al. 2004), that leads to disruption of TJ structure. It has recently been shown that HGF/SF disrupts TJs in bladder cancer cells, with relocation of occludin away from the TJ (Brown 2005). Since TJs play an important role in cell-cell adhesion and the prevention of trans-endothelial intravasation of cancer cells, HGF may have an important role in cancer metastasis (Jiang et al. 2005).

6.6 Discussion

The TJ of the umbrella cell layer of the urothelium create the paracellular barrier protecting the underlying detrusor against exposure from urinary toxins, with the claudins thought to play a central role in the maintenance of this barrier (Colegio et al. 2002). Disruptions of TJs with loss of TJ proteins including claudins are thought to reduce cell-cell adhesion and lower the diffusion barrier resulting in an abnormal influx of growth factors and nutrients, which offers a selective advantage to developing tumour cells (Mullin 1997). Altered expression of various claudins has been reported in several cancer types (Swisshelm et al. 2005) with an inverse correlation reported with histological grade (Kominsky et al. 2003) and invasiveness (Michl et al. 2003).

Conversely, expression of claudin-3 and -4 positively correlates with invasiveness of ovarian cancer cells (Agarwal et al. 2005). Claudin-11 has been reported to be expressed in a number of tissues including the nephron, oligodendrocytes, Sertoli cells and basal cells of the inner ear (Gow et al. 2004; 1999; Kiuchi-Saishin et al. 2002) but not previously noted in the urothelium. In addition, claudin-11 expression was decreased in invasive compared to non-invasive cell lines (Awsare et al. 2011). Claudin-4 was identified as a target of the pro-invasive cytokine transforming growth factor beta and Ras/Raf/extracellular signal-regulated kinase pathway (Michl et al. 2003). In contrast, the over-expression of claudins-3 and 4 in human ovarian surface epithelial (HOSE) cells increased cell invasion, motility and cell survival, probably by activating matrix metalloproteinase- 2 (Agarwal et al. 2005). Conversely, small interfering (siRNA) mediated knockdown of claudin-3 and 4 expression in ovarian cancer cell lines reduced invasion (Agarwal et al. 2005).

Claudins can therefore promote or inhibit tumourigenesis and metastasis in various cancers by altering the malignant phenotype and this appears to be highly tissue specific. A number of signalling mechanisms, not directly involved in TJ formation seem to mediate these actions. Further studies are needed to identify the signalling pathways mediating the effects of claudin-11 and 20 in reducing the invasive phenotype of TCC cells. The expression of claudins in tissues has also been explored as a potential therapeutic target. Claudins- 3 and 4 are receptors for CPE, which upon binding to its receptors causes cytolysis. Experimental studies in prostate (Long et al. 2001), breast (Kominsky et al. 2004), ovarian (Santin et al. 2005) and pancreatic (Michl et al. 2003) cancer cells have shown that cells expressing claudin-3 and 4 are sensitive to CPE-mediated cytolysis, while those that lack their expression are unaffected by CPE.

Thus, claudins could have a significant role to play in the regulation of bladder cancer progression.

6.7 Conclusions

Claudins could have a potential for use in gene therapy to modify the course of the disease in individual patients in the future. Detailed studies of local and systemic side-effects and toxicities will have to be conducted before gene therapy becomes a reality. However, the easy accessibility of the bladder per urethra could be exploited for the intravesical application of such therapy; potentially minimising systemic side-effects as has been the case with intravesical chemotherapy and BCG treatments for bladder TCC.

We can conclude that claudins appear to have a potential role as prognostic markers of tumour aggressiveness and may become therapeutic targets in the control of invasion and metastasis in bladder cancer.

References

Acharya P, Beckel J, Ruiz WG, Wang E, Rojas R, Birder L, Apodaca G (2004) Distribution of the TJ proteins ZO-1, occludin, and claudin-4, -8, and -12 in bladder epithelium. Am J Physiol Renal Physiol 287(2):F305–F318

Agarwal R, D'Souza Tand Morin PJ (2005) Claudin-3 and claudin-4 expression in ovarian epithelial cells enhances invasion and is associated with increased matrix metalloproteinase-2 activity. Cancer Res 65:7378

Al Moustafa AE, Alaoui-Jamali MA, Batist G, Hernandez-Perez M, Serruya C, Alpert L, Black MJ, Sladek R, Foulkes WD (2002) Identification of genes associated with head and neck carcinogenesis by cDNA microarray comparison between matched primary normal epithelial and squamous carcinoma cells. Oncogene 21:2634–2640

Aldred MA, Huang Y, Liyanarachchi S, Pellegata NS, Gimm O, Jhiang S, Davuluri RV, de la Chapelle A, Eng C (2004) Papillary and follicular thyroid carcinomas show distinctly different microarray expression profiles and can be distinguished by a minimum of five genes. J Clin Oncol 22:3531–3539

Awsare NS, Martin TA, Haynes MD, Matthews PN, Jiang WG (2011) Claudin-11 decreases the invasiveness of bladder cancer cells. Oncol Rep 25(6):1503–1509

Ben-Yosef T, Belyantseva IA, Saunders TL, Hughes ED, Kawamoto K, Van Itallie CM, Beyer LA, Halsey K, Gardner DJ, Wilcox ER, Rasmussen J, Anderson JM, Dolan DF, Forge A, Raphael Y, Camper SA, Friedman TB (2003) Claudin 14 knockout mice, a model for autosomal recessive deafness DFNB29, are deaf due to cochlear hair cell degeneration. Hum Mol Genet 12:2049–2061

Blanchard A, Jeunemaitre X, Coudol P, Dechaux M, Froissart M, May A, Demontis R, Fournier A, Paillard M, Houillier P (2001) Paracellin-1 is critical for magnesium and calcium reabsorption in the human thick ascending limb of Henle. Kidney Int 59:2206–2215

Boireau S, Buchert M, Samuel MS, Pannequin J, Ryan JL, Choquet A, Chapuis H, Rebillard X, Avancès C, Ernst M, Joubert D, Mottet N, Hollande F (2007) DNA-methylation-dependent alterations of claudin-4 expression in human bladder carcinoma. Carcinogenesis 28(2):246–258

Brown GM (2005) Tight junction structure in human bladder cancer: effects of hepatocyte growth factor. Thesis, University of Wales College of Medicine, Cardiff, 2005

Brown GM, Martin TA, Matthews PN, Jiang WG (2003) Hepatocyte growth factor (HGF) and its impact on the tight junctions of human bladder cancer cells. BJU Int 92:836–837

Chancellor MB, Yoshimura N (2002) Physiology and pharmacology of the bladder and urethra. In: Walsh PC, Retik AB, Vaughan ED, Wein AJ (eds) Campbell's urology, 8th edn. Saunders, Philadelphia

Chodak GW, Hospelhorn V, Judge SM, Mayforth R, Koeppen H, Sasse J (1988) Increased levels of fibroblast growth factor-like activity in urine from patients with bladder or kidney cancer. Cancer Res 48:2083–2088

Claude P (1978) Morphological factors influencing transepithelial permeability: a model for the resistance of the zonula occludens. J Membr Biol 39:219–232

Clausen C, Lewis SA, Diamond JM (1979) Impedance analysis of a tight epithelium using a distributed resistance model. Biophys J 26:291–317

Rosen EM, Joseph A, Jin L, Yao Y, Chau MH, Fuchs A, Gomella L, Hastings H, Goldberg ID, Weiss GH (1997) Urinary and tissue levels of scatter factor in transitional cell carcinoma of bladder. J Urol 157:72–78

Sánchez Freire V, Burkhard FC, Schmitz A, Kessler TM, Monastyrskaya K (2011) Structural differences between the bladder dome and trigone revealed by mRNA expression analysis of cold-cut biopsies. BJU Int 108(2 Pt 2):E126–E135

Santin AD, Cane S, Bellone S, Palmieri M, Siegel ER, Thomas M, Roman JJ, Burnett A, Cannon MJ, Pecorelli S (2005) Treatment of chemotherapy-resistant human ovarian cancer xenografts in C.B-17/SCID mice by intraperitoneal administration of Clostridium perfringens enterotoxin. Cancer Res 65:4334–4342

Schulzke JD, Gitter AH, Mankertz J, Spiegel S, Seidler U, Amasheh S, Saitou M, Tsukita S, Fromm M (2005) Epithelial transport and barrier function in occludin-deficient mice. Biochim Biophys Acta 1669(1):34–42

Soler AP, Miller RD, Laughlin KV, Carp NZ, Klurfeld DM, Mullin JM (1999) Increased tight junctional permeability is associated with the development of colon cancer. Carcinogenesis 20:1425–1431

Southgate J, Varley CL, Garthwaite MA, Hinley J, Marsh F, Stahlschmidt J, Trejdosiewicz LK, Eardley I (2007) Differentiation potential of urothelium from patients with benign bladder dysfunction. BJU Int 99(6):1506–1516

Southwood CM, Gow A (2001) Functions of OSP/claudin-11-containing parallel tight junctions: implications from the knockout mice. In: Cereijido M, Anderson JM (eds) Tight junctions. CRC Press, Boca Raton, pp 723–745

Swisshelm K, Macek Rand Kubbies M (2005) Role of claudins in tumorigenesis. Adv Drug Deliv Rev 7:919–928

Székely E, Törzsök P, Riesz P, Korompay A, Fintha A, Székely T, Lotz G, Nyirády P, Romics I, Tímár J, Schaff Z, Kiss A (2011) Expression of claudins and their prognostic significance in noninvasive urothelial neoplasms of the human urinary bladder. J Histochem Cytochem 59(10):932–941

Tsukita S, Furuse M, Itoh M (2001) Multifunctional strands in tight junctions. Nat Rev Mol Cell Biol 2:285–293

Turksen K, Troy TC (2001) Claudin-6: a novel tight junction molecule is developmentally regulated in mouse embryonic epithelium. Dev Dyn 222:292–300

Turksen K, Troy TC (2002) Permeability barrier dysfunction in transgenic mice overexpressing claudin 6. Development (Cambridge, England) 129(7):1775–1784. http://www.ncbi.nlm.nih.gov/pubmed/11923212

Van Itallie CM, Balda MS, Anderson JM (1995) Epidermal growth factor induces tyrosine phosphorylation and reorganization of the tight junction protein ZO-1 in A431 cells. J Cell Sci 108:1735–1742

Van Itallie C, Rahner C, Anderson JM (2001) Regulated expression of claudin-4 decreases paracellular conductance through a selective decrease in sodium permeability. J Clin Invest 107:1319–1327

Welsh Cancer Intelligence and Surveillance (2005) http://www.wales.nhs.uk/sites3/page.cfm?orgid=242&pid=18140

Youakim A, Ahdieh M (1999) Interferon-gamma decreases barrier function in T84 cells by reducing ZO-1 levels and disrupting apical actin. Am J Physiol 276:G1279–G1288

Yu AS, Enck AH, Lencer WI, Schneeberger EE (2003) Claudin-8 expression in Madin-Darby canine kidney cells augments the paracellular barrier to cation permeation. J Biol Chem 278:17350–17359

Zhang CO, Wang JY, Koch KR, Keay S (2005) Regulation of TJ proteins and bladder epithelial paracellular permeability by an antiproliferative factor from patients with interstitial cystitis. J Urol 174(6):2382–2387

Chapter 7
Tight Junctions in Colorectal Cancer

Frédéric Hollande and Marina Papin

Abstract The intestine plays a major role in the absorption of nutrients, water, and electrolytes during digestion and protects the inner organs from the external environment. One of the major characteristics of this organ is that, while maintaining its function, it undergoes constant renewal at a very high rate, with the entire intestine renewed every 5–7 days in humans. This process occurs without disrupting the epithelial barrier, thus guaranteeing that the separation between external (lumen) and internal compartments is maintained, and that cell polarity is preserved. Tight junctions (TJ) are among the multi-protein complexes that play an essential role in the maintenance of this epithelial barrier. A number of observations have been made that the structure of tight junctions can be disrupted from early stages of neoplastic development in the intestine, and alterations of tight junction protein expression and/or localization have been reported in epithelial cancers, which significantly impacts on early tumorigenesis as well as tumor progression. In the present review, we summarize the current knowledge concerning alterations of tight junction protein expression in colorectal tumors, before discussing some of the likely consequences of these disruptions for tumorigenesis and the potential clinical use of tight junction proteins as prognostic markers or as targets for therapy in this type of tumors.

F. Hollande (✉)
CNRS, UMR-5203, Institut de Génomique Fonctionnelle, Montpellier F-34000, France

INSERM, U661, Montpellier F-34000, France

Universités de Montpellier 1 & 2, UMR-5203, Montpellier F-34000, France
e-mail: frederic.hollande@unimelb.edu.au

M. Papin
Department of Pathology, The University of Melbourne, Australia

CNRS, UMR-5203, Institut de Génomique Fonctionnelle, Montpellier F-34000, France

INSERM, U661, Montpellier F-34000, France

Universités de Montpellier 1 & 2, UMR-5203, Montpellier F-34000, France

T.A. Martin and W.G. Jiang (eds.), *Tight Junctions in Cancer Metastasis*,
Cancer Metastasis - Biology and Treatment 19, DOI 10.1007/978-94-007-6028-8_7,

cancer, was shown to promote metastasis development and to increase claudin-1 (and claudin-4) levels with, again, a prominent cytoplasmic and nuclear localization for claudin-1 (Halder et al. 2008). Mutations in the *APC* or *CTNNB1* gene as well as Smad4 down-regulation are frequent in colorectal cancer, providing a convincing mechanistic rationale for the increased claudin-1 expression detected by several groups in large numbers of colorectal carcinoma samples.

In stark contrast, other studies point to the fact that loss of claudin-1 expression is associated with increased invasion and metastasis as well as poorer clinical outcome. Thus, Resnick et al. (2005) found that claudin-1 expression was robust in 75% of CRC patient tumor sections in a retrospective IHC study on 129 stage II tumors, but that loss of claudin-1 expression represented a very strong independent predictor of increased recurrence and poor survival (Resnick et al. 2005). Another recent study found that claudin-1 expression decreased significantly in the presence of lymph node metastasis, seemingly pointing to a negative correlation between Claudin-1 expression and neoplastic progression (Ersoz et al. 2011). Similarly, another group described the sharp decrease of claudin-1 expression during tumor progression, as detected in stage IV primary CRC and liver metastasis samples (Georges et al. 2011). These authors went on to use an allogenic graft model to demonstrate that siRNA-mediated claudin-1 and claudin-4 knockdown increased migration of CC531 rat colon adenocarcinoma cells in vitro.

The discrepancy between studies demonstrating overexpression and down-regulation of claudin-1 during CRC progression remains unexplained. Further studies are required to demonstrate whether these differences reflect more than mere technical issues and have an underlying biological significance. They could for example reflect the existence of subclasses of colorectal cancer in terms of claudin-1 expression levels. Alternatively, although the study of Resnick and colleagues (2005) analysed claudin-1 expression in these areas, the presence of locally high claudin-1 levels towards the invasive front of those tumors where the overall claudin-1 expression is low would partially reconcile both sets of results, since cells located in these areas are the most likely to play an active part in the metastasis process. In view of recent results (Singh et al. 2011) showing that claudin-1 overexpression is likely to promote Epithelial to Mesenchyme Transition (EMT) via the activation of ZEB-1, a precise characterization of the claudin-1 expression status in cells displaying EMT-like phenotypes at the leading/invasive edge may resolve this issue. Finally, a degree of uncertainty also remains concerning the sub-cellular pool of claudin-1 that may play a role during tumor development. Indeed, far from the tight junctional localization expected for this protein in the healthy intestine (Holmes et al. 2006; Rahner et al. 2001), multiple studies in CRC patient samples report the detection of claudin-1 along the whole cell membrane, in the cell nucleus, as well as within the cytoplasm of colorectal tumor cells, as diffuse staining or in intracellular vesicles (Dhawan et al. 2005; Grone et al. 2007; Kinugasa et al. 2007; Mees et al. 2009; Miwa et al. 2001; Resnick et al. 2005). No conclusive evidence has been brought forward so far to correlate specific sub-cellular localizations of claudin-1 with selective tumor-promoting (or inhibiting) functions. Yet, one could argue that the underlying mechanisms could be partly different in pre-neoplastic/

early neoplastic lesions and in more established/advanced carcinomas. In the former, alterations of membrane claudin-1 expression at the tight junction is likely to participate in the disruption of junction stability and tissue permeability, thus allowing increased pro-mitotic signalling to occur (Mullin 2004). In contrast, tight junctional structure is already altered in a large proportion of stage II/IV carcinomas, albeit sometimes in selective regions only, and the impact of claudin-1 alterations in this case may be more related to a (direct or indirect) signalling role of claudin-1, in a sub-cellular compartment yet to be specified.

7.2.2 Other Tight Junction Proteins

Over the last 15 years, an increasing number of studies have assessed the expression status of other tight junction proteins in colorectal cancer. Several reports on medium to large patient populations concur to say that claudin-2 mRNA and/or protein is overexpressed in a significant proportion of CRC samples (Aung et al. 2006; Huo et al. 2009; Kinugasa et al. 2007), following stimulatory signals induced by inflammation (Mima et al. 2008; Weber et al. 2008) and by increased EGF signalling during tumor progression (Dhawan et al. 2011). An overexpresion of claudin-2 transcripts in CRC, albeit moderate, was also detected by Hewitt et al., who analyzed gene expression data for all 21 human claudins across 266 tissues using serial analysis of gene expression (SAGE) Genie database and validation by RT-qPCR on a small number of human tumor samples (Hewitt et al. 2006). In addition, our group showed that transcription of the CLDN2 gene in CRC was under the control of the symplekin/ZONAB complex, the former protein being overexpressed in colorectal tumors (Buchert et al. 2010). In contrast, while Mees et al. found no significant change in claudin-2 immunofluorescent detection in 16 CRC samples (Mees et al. 2009), one study reported a down-regulation of claudin-2 in comparison with normal colon tissue on a panel of 33 human CRC samples using immunohistochemistry, without finding any correlation with clinical parameters such as tumor location and size, stage, lymph node involvement, and degree of differentiation (Hahn-Stromberg et al. 2009).

The dynamics of expression during the course of tumorigenesis may be more complex in the case of other claudins, such claudin-3, -4 and -7. Thus, expression of claudin-4 appeared to be marginally elevated in 58% of samples in a large-scale immunohistochemistry study of stage II tumors (Resnick et al. 2005). In addition, a moderate increase of claudin-3 and 4 expression was identified in a Western blot-based study of 12 CRC samples and matched normal epithelia, with tumor tissues also demonstrating a higher level of paracellular leakiness (permeability to ruthenium red) (de Oliveira et al. 2005). A larger study on 205 samples from untreated patients also reported increased expression of claudin-3 and -4 mRNA (Oshima et al. 2008). In addition, the relationship between inflammation and CRC development is clearly not as straightforward concerning claudin-4 expression as it was for claudin-1, with one study reporting a strong increase in membrane expression of claudin-4 in grade II inflammatory bowel disease but not in other inflammatory grades or derived tumors (Weber et al. 2008) and another detecting increased expression in ulcerative colitis-derived CRC samples (Mees et al. 2009). Using the in silico analysis and RT-PCR approach described above for claudin-2, P. Morin's group also reported that claudin-3 and -4 were overexpressed (Hewitt et al. 2006). In contrast, other studies point to non-significant increases of claudins-3 and -4 expression (Kinugasa et al. 2007) or to a decreased expression of claudin-4 during progression to stage IV primary and liver metastasis samples (Ersoz et al. 2011; Georges et al. 2011; Ueda et al. 2007).

The situation is similarly complex in the case of claudin-7. While our laboratory demonstrated increased claudin-7 expression in 12 samples from stage I–III tumors, larger studies have suggested that expression of claudin-7 mRNA decreases in correlation with invasion and lymph node metastasis (Oshima et al. 2008; Tang et al. 2011), maybe due to hypermethylation of its promoter (Nakayama et al. 2008). Yet, convincing data was obtained by Kuhn and colleagues (2007) on 104 primary CRC and 66 liver metastasis samples, demonstrating that the expression of claudin-7, along with that of interacting proteins EpCAM, CO-029 and CD44v6, was frequently up-regulated to the same extent in both types of samples, arguing against a relationship between metastatic progression and claudin-7 expression. Interestingly, Hewitt and colleagues (2006) also found an overexpression of claudin-7 in colon cancer samples and suggested that the expression of claudins-3, -4, and -7, which was frequently elevated in several cancers and appeared to show a tight association, may reflect a coordinated regulation of these genes. Finally, a frequent up-regulation of claudin-12 and down-regulation of claudin-8 mRNA was detected in a panel of 30 microdissected CRC samples of all stages (Grone et al. 2007).

As the original transmembrane tight junction protein to be discovered in 1993 (Furuse et al. 1993), occludin has also been subject to intense scrutiny in the cancer setting. Most results concur in saying that occludin expression is stable or locally decreased in human CRC samples (Kinugasa et al. 2007; Resnick et al. 2005; Tobioka et al. 2002), with a clear trend towards down-regulation in poorly differentiated primary carcinomas (Kimura et al. 1997), followed by reexpression in liver metastasis samples (Orban et al. 2008). Similarly, although very little is known about the expression of TJ cytoplasmic plaque proteins in colorectal cancer, the expression of ZO-1 was often reported to be down-regulated during tumor progression (Kaihara et al. 2003; Kimura et al. 1997; Resnick et al. 2005), but the protein appears to be strongly re-expressed in liver metastasis samples (Kaihara et al. 2003; Orban et al. 2008). Kinugasa and colleagues (2007) also reported a moderate overexpression of ZO-2 using RT-qPCR and immunohistochemistry techniques on a small population of patient CRC samples.

7.2.3 Altered Localization of TJ Proteins in CRC

As briefly mentioned above, it is important to remember that, if the quantification of expression levels obviously provides essential information concerning tight junction proteins, precise analysis of their localisation should always be taken into consideration.

Indeed, alterations of mRNA/protein expression are not an essential requirement to explain disruptions in the functionality of tight junctions or their involvement in tumor progression. The localization of proteins from the TJ cytoplasmic plaque is particularly sensitive to subtle phenotypic changes in tumor cells. Several of these proteins are known to shuttle between the TJ and the cytoplasm or nucleus under physiological conditions (Anderson and Van Itallie 2009; Polette et al. 2007), and their preferential localization may be tipped away from the TJ in tumor cells (Harhaj and Antonetti 2004). Such variations can occur in a subtle way in early, preneoplastic stages, as demonstrated by our group in the case of ZO-1 (Pannequin et al. 2007), while the localization of shuttling proteins such as symplekin can become exclusively nuclear in human CRC (Buchert et al. 2010). ZO-1 and ZO-2 can also display a quite prominent nuclear localization (Gottardi et al. 1996; Islas et al. 2002), although this has not been demonstrated as yet in CRC. Finally, in a study on 12 colon adenocarcinoma samples using immunoperoxidase-based staining, peculiar cingulin immunoreactivity was detected at the interface between tumor tissue and the surrounding stroma in poorly differentiated CRC, while the profile of cingulin staining appeared similar in well to moderately differentiated samples, (Citi et al. 1991). In addition, as briefly mentioned above, mislocalization of TJ transmembrane proteins is also frequently encountered in CRC histological samples, as demonstrated in the case of claudin-1, which can be found in the cell nucleus (Dhawan et al. 2005). The localization of claudin-4 is also altered in CRC (Resnick et al. 2005), and this was shown to disrupt paracellular permeability (Tanaka et al. 2005). Interestingly, tyrosine phosphorylation of claudin-4 by the Eph receptor EphA2 was shown to enhance paracellular permeability by reducing the cell-cell contact localization of claudin-4 and decreasing its association with ZO-1 (Tanaka et al. 2005). The overexpression of EphA2 in colorectal cancer correlates with invasion and metastasis development (Saito et al. 2004), suggesting that this protein could be involved in the transcription-independent disruption of tight junctional complexes during cancer progression. Another mechanism that leads to the loss of tight junction involves EphrinB1, which can compete with the association of Cdc42 with Par-6, thereby inactivating the Par complex (Lee et al. 2008). The occurrence of such mechanisms calls for a more thorough examination of TJ protein expression/localization relationships in tumors, as well as for a precise examination of their localization in multiple areas of the tumor, and particularly highly angiogenic and in invasive regions.

7.3 Consequences of TJ Alteration for Tumor Growth and Metastasis

Before analysing putative causal relationships between tight junction alterations and cancer development or progression, one should bear in mind that tumors are often extremely heterogeneous from a structural, genetic and phenotypic point of view. This is obviously true between different types of cancers, but also between cancers of the same type across different patients, at different stages of a given tumor's evolution, as well as between different regions of a given tumor at a given point in time. Apparent paradoxes between tight junction protein expression in different studies should be viewed in this context, and precise description of local and temporal variations of these (and most other) proteins will be necessary before we can gain a broader understanding of their real impact on tumor evolution, and of their actual interest for prognosis or therapy.

7.3.1 Disruption of Paracellular Permeability in Early-Stage Tumors

This being said, alterations of TJ protein expression and/or localization can clearly become important enough to disrupt the structure of tight junctions as well as their fence and barrier functions. Such alterations are likely to have a high impact during early stages of colon cancer development (Soler et al. 1999), and the resulting leakiness can promote inflammatory processes (Mullin 2004) and increase the permeabilization to various ions, thereby altering tissue homeostasis (Tsukita et al. 2008). Another significant consequence of this leakiness is that the passage of growth factors from the lumen through the paracellular space becomes possible, thereby allowing them to come in contact with their receptors previously segregated along the basolateral membrane of intestinal cells (Bishop and Wen 1994), resulting in an increased pro-mitotic signalling (Mullin 2004). As an example, Epidermal Growth Factor (EGF) is secreted into the gastrointestinal lumen while its receptors are mainly expressed on the basolateral membrane of epithelial cells (Playford et al. 1996; Playford and Wright 1996). In a physiological setting, this pool of EGF is only though to come in contact with its receptors upon mechanical or infectious damage to the epithelium, and the ensuing proliferating effect of EGF is thought to facilitate wound healing (Murphy 1998; Playford and Wright 1996). However, in premalignant tissue, constant proliferative effect of EGF is made possible by chronic TJ disruption, thus contributing to the onset of adenomas and/or carcinomas (Mullin 1997, 2004).

7.3.2 Modulation of Tumor-Promoting Cell Signalling by TJ Protein Alterations

A number of results also suggest that, beyond their consequences on tight junction structure, alterations in the expression and/or subcellular localization of individual tight junction proteins can results in tumor promotion, suggesting that these alterations can modulate intracellular signalling in cancer cells. A number of results obtained in vitro also corroborate the fact that TJ proteins modulate cell phenotype in sub-confluent conditions where no epithelial barrier is established. Thus, claudin-1 was shown to stimulate the transcriptional activity of the β-catenin/Tcf-4 complex, a key effector of the canonical Wnt pathway, in CRC cells (Dhawan et al. 2005; Miwa et al. 2001). This pathway has a well-described role in the aetiology of colorectal cancer, from the onset of intestinal tumorigenesis (Morin et al. 1997) to the maintenance (Scholer-Dahirel et al. 2011) and invasion (Brabletz et al. 1998; Hlubek et al. 2004) of more advanced tumors. Our laboratory also showed that claudin-7 overexpression in HT-29 CRC cells increases Tcf-4 target gene expression, proliferation, and tumorigenicity after injection in nude mice (Darido et al. 2008). Other groups also found that claudin-7 overexpression promotes resistance to apoptosis and tumor growth, particularly in tumors where EpCAM, CO-029, and CD44v6 are co-expressed with this protein (Kuhn et al. 2007). Overexpression of claudin-2 in CRC cells promotes cell proliferation and anchorage-independent growth, and claudin-2 may mediate some of the effects of EGF on tumor cell proliferation, raising the intriguing possibility that claudin-2 may be involved in EGF receptor transactivation (Dhawan et al. 2011). Intriguingly, an opposite effect was reported concerning occludin. Indded, decreased TJ expression of occludin following treatment with synthetic peptides homologous to the second extracellular domain of this protein was demonstrated to increase transcription by β-catenin/TCF and to up-regulate the expression of its downstream target gene c-Myc (Vietor et al. 2001).

Finally, modulation of the TJ cytoplasmic plaque, known to be a dynamic process under the tight regulation of multiple signalling pathways (Anderson and Van Itallie 2009), may also play a regulatory role on the behaviour or cancer cells. Thus, several proteins of the *Zonula Occludens* (ZO) family have been shown to modulate the cell cycle and proliferative capacities of epithelial cells (Anderson and Van Itallie 2009; Steed et al. 2010). These proteins are thought to have a 'tumor suppressor' effect and their mislocalization in CRC could therefore enhance the proliferative potential of tumor cells (Harhaj and Antonetti 2004). Thus, when expressed at the TJ, ZO-1 functions as an inhibitor of ZONAB nuclear accumulation by cytoplasmic sequestration (Balda and Matter 2000), and this ZO-1/ZONAB interaction regulates nuclear cdk-4 accumulation and cell proliferation (Balda et al. 2003). ZO-2 was also shown to localization at the TJ or in the nucleus, and nuclear ZO-2 can interact with c-Myc to bind the Cyclin D1 promoter, thereby repressing its activity (Huerta et al. 2007) and inhibiting cell cycle progression (Gonzalez-Mariscal et al. 2009). In addition, ZO-2 overexpression decreases the phosphorylation of GSK3β, increasing its ability to phosphorylate β-catenin and thus decreasing Wnt target gene expression (Tapia et al. 2009). In contrast, an interaction between ZO-3 and cyclin D1 was shown to stabilize the latter and thus to facilitate G1/S phase transition (Capaldo et al. 2011).

7.3.3 A Role for TJ Protein Alterations in CRC Invasion and Metastasis Development?

As tumors become more aggressive, Epithelial to Mesenchyme Transition (EMT) is a crucial step where epithelial cell acquire a pro-migratory mesenchymal phenotype and gain the ability to invade adjacent tissues and initiate metastasis development.

To gain this aggressive phenotype, tumor cells located towards the invasive edge must first lose their cell-cell adhesion and their polarity. The disruption of TJ is thus an essential step towards the acquisition of a more invasive state, and several EMT inducers like TGF-β, HGF, EGF, ZEB1, Snail or Slug are also negative regulators of TJ, promoting their disruption (Grotegut et al. 2006; Ikenouchi et al. 2003; Masuda et al. 2010; Ozdamar et al. 2005; Vincent et al. 2009; Wang et al. 2007).

In turn, a number of teams have obtained results demonstrating that additional pro-invasive signalling can be induced in tumor cells by overexpression or by down-regulation of TJ proteins, depending on the protein considered. Thus, increased claudin-1 expression was shown to disrupt the polarized nature of T84 colorectal cancer cells (Huo et al. 2009), and induced overexpression of Claudin-1 in human CRC cells increased liver metastasis following intrasplenic injection (Dhawan et al. 2005). This latter effect appears mediated by a strong ZEB-1-driven down-regulation of E-cadherin expression in these cells (Singh et al. 2011), a mechanism that seems to be independent from SNAI1 (Snail) and SNAI2 (Slug) (Dhawan et al. 2005). A recent work from the same group suggests that physical association of claudin-1 with the Src tyrosine kinase and subsequent modulation of anoikis could be a key regulatory mechanism in the promoting role of claudin-1 on CRC invasion (Singh et al. 2012). Claudin-2, often overexpressed in CRC, was shown to promote motility and invasion of gastric, lung (Mima et al. 2008) and colorectal carcinoma cells (Takehara et al. 2009) in vitro, an effect that appears mediated by Sp1-driven transcription of MMP-9 in non-small cell lung cancer (Ikari et al. 2011).

The precise role of claudin-7 on tumor progression is less well-characterized, with one group demonstrating that its overexpression, in correlation with that of EpCAM, CO-029, and CD44v6, facilitates tumor progression and metastasis formation (Kuhn et al. 2007) by regulating EpCAM-mediated functions (Nubel et al. 2009), while two other publications correlate loss of expression of claudin-7 with increased invasion (Nakayama et al. 2008; Tang et al. 2011).

Down-regulation of other TJ transmembrane proteins was shown to promote loss of polarity and invasive behavior. For example, decreased expression of occludin induced a loss in the polarized distribution of carcinoembryonic antigen in human colorectal adenocarcinoma samples. (Tobioka et al. 2002), and siRNA-mediated claudin-4 knockdown up-regulated the motility in SW480 CRC cells (Ueda et al. 2007).

Finally, as suggested above, more subtle regulation of TJ cytoplasmic plaque proteins may also be involved in local alterations of the invasive behaviour of tumor cells. Although to our knowledge no precise demonstration was performed in colorectal cancer, the report of Hirakawa and colleagues suggests that interaction of ZO-1 with cortactin, an actin filament (F-actin)-binding protein that promotes actin polymerization and stimulates cancer cell migration and invasion, promotes cancer progression (Hirakawa et al. 2009). This result is in accordance with studies reporting the formation of a ZO-1/alpha5-integrin complex involved in lamellae formation at the leading edge of migrating lung cancer cells (Tuomi et al. 2009). Finally, ZO-1 mutants that encode the PDZ domains but do not localize at the plasma membrane, strongly induce EMT and beta-catenin signalling in canine kidney (MDCK) cells (Reichert et al. 2000), again suggesting that observed changes in the localization of this protein will promote CRC progression.

7.3.4 *Permeabilization of Endothelial TJ During the Metastasis Process*

An important step in the formation of cancer metastasis is the penetration of the vascular endothelium by dissociated cancer cells. Tumor cells may have several weapons at their disposal to modulate the permeability of tight junctions and thus facilitate their intravasion and extravasion through endothelial walls. Although little data is available from colorectal cancer studies, lessons can be learned from work performed on other cancers. A first study showed that treatment of endothelial cell monolayer with conditioned medium prepared from highly invasive and metastatic melanoma cells, increased the endothelium permeability and damaged the TJ integrity (as assayed using TER, transendothelial resistance) (Utoguchi et al. 1995). Thus, matrix metalloproteinases may play a role in the degradation of transmembrane tight junction proteins in endothelial cells. Indeed, in Central Nervous System leukemia, matrix metalloproteinases MMP-2 and MMP-9 are secreted by leukemic cells and mediate the Blood-brain Barrier opening by disrupting transmembrane tight junction proteins (occludin and claudin-5), but also junctional plaque proteins such as ZO-1 (Feng et al. 2011a).

Moreover, secretion of VEGF by disseminating tumor cells may also play a role in enhancing the permeability of endothelia by disrupting tight junction proteins. Vascular endothelial growth factor A (VEGF) is a key angiogenic mediator that stimulates endothelial cell proliferation and regulates vascular permeability. Feng and colleagues hypothesized that leukemic cells could secrete VEGF to reduce the expression of occludin, claudin-5 and ZO-1 in order to disrupt the Blood-brain Barrier (Feng et al. 2011b). A first experimental proof of this hypothesis came from the work of Argaw and colleagues (2009), who describe the down-regulation of claudin-5 and occludin expression after VEGF treatment on brain microvascular endothelial cells or in mouse cerebral cortex. In addition, only recombinant claudin-5, not occludin, protected brain microvascular endothelial cells from a VEGF-induced increase in paracellular permeability (Argaw et al. 2009). The important role of claudin-5 in tumor progression is also suggested by observations made in situ on tumor vasculature. As an example, in hepatocellular carcinoma, poor differentiation and lower postoperative overall survival rate is associated with lower expression of claudin-5 in tumor vessels (Sakaguchi et al. 2008).

7.4 TJ Proteins in the Clinical Setting: Prognostic Markers and/or Targets for Therapy in CRC?

For a number of years, scientists have suggested that tight junction proteins could represent promising prognostic markers as well as putative targets for cancer therapy. Yet, as summarized above, the alterations of TJ proteins in colorectal cancer are complex and not entirely understood, and the possibility to use this information in the clinical setting therefore remains unproven.

7.4.1 TJ Proteins as Biomarkers

Thus, a small number of investigators have attempted to determine whether measurement of expression and/or localization of selective claudin isoforms in primary CRC could generate valuable information to help predict disease evolution. Although a number of promising avenues have been uncovered and should be explored further, opposite results obtained by different groups still make for a somewhat fuzzy picture of how these markers could be used for the benefit of patients. As an example, while most groups reported the overexpression of claudin-1 in CRC of all stages (see above), and despite the fact that molecular data points towards a pro-invasive role of overexpressed claudin-1 (Singh et al. 2011), the two main studies to have specifically tackled this issue found a significant a negative correlation between claudin-1 expression and good prognosis (Resnick et al. 2005) or lymph node metastasis (Ersoz et al. 2011). Similarly, overexpression of claudin-7, in association with that of EpCAM, CO-029, and CD44v6, inversely correlated with disease-free survival in a study analysing a large number of primary colorectal cancer ($n=104$) and liver metastasis ($n=66$) samples (Kuhn et al. 2007). Yet, using quantitative real-time RT-PCR on a population of tumor and adjacent normal mucosa samples from 205 untreated patients, Oshima and colleagues found that reduced expression of the claudin-7 gene may be a useful predictor of liver metastasis in patients with CRC (Oshima et al. 2008). Finally, only one large study has been performed so far concerning the potential of claudin-4 as a biomarker in CRC, and in this case the molecular data fits well with in vivo correlation data between expression and clinicopathological parameters. Thus, in a population of 129 primary and 44 metastasis tumor samples, Ueda and colleagues (2007) found that claudin-4 immunoreactivity was significantly decreased in moderately and poorly differentiated adenocarcinomas, particularly at the invasive front and at regions of vessel infiltration. Decreased expression of claudin-4 was significantly correlated with depth of invasion, lymphatic vessel invasion and presence of metastases, suggesting that claudin-4 expression and metastatic development are inversely correlated in CRC. Using multivariate Cox analysis, this group also detected a strong but not quite significant trend towards increased local recurrence in stage II CRC tumors displaying decreased claudin-4 expression when compared with tumors of the same stage with high histological grade or detectable vessel invasion (Ueda et al. 2007). In accordance with this data, their in vitro results indicated that claudin-4 down-regulation significantly upregulated cell motility in colorectal cancer cells. Their conclusion was that decreased claudin-4 expression might be a good biomarker for diagnosing the risk of distant metastasis. Finally, it is very interesting to note that, in vitro, tumor cells from various cancers including CRC appear to shed full-length claudin molecules in the culture medium, and these molecules are incorporated into exosomes. As an example, due to the selective presence of claudin-4-containing exosomes in the bloodstream of women with ovarian cancer, this isoform is proposed as a good biomarker for this type of cancer (Li et al. 2009). These results suggest that detection of other claudins in vivo could signal the presence of different tumor types, including CRC.

In any case, a number of additional thorough studies are clearly required before any of these prospective markers can be forwarded towards clinical use in the diagnosis or the prognosis of CRC evolution.

7.4.2 TJ-Targeted Therapeutic Approaches

In terms of TJ-targeted therapy, *Clostridium perfringens* enterotoxin (CPE) stands out as the most promising candidate in a number of solid tumors, although the data obtained in CRC is so far minimal. CPE, produced by the food poisoning-inducing Clostridium perfringens type A strain, is known to induce cell death via the formation of large pores into the plasma membrane, which dramatically increases membrane permeability and leads to the loss of osmotic equilibrium. CPE strongly binds via its C-terminus to claudin-4 and, with lower affinity, to claudin-3 (for review, see for example (Kondoh et al. 2006)). Administration of various forms of full-length or C-terminal fragments of CPE has been demonstrated to exert potent in vitro and in vivo cytotoxic effects on several types of solid tumors, such as pancreatic endometrial, ovarian, lung, and prostate cancers (Morin 2005). Although the ubiquitous nature of claudin-3 and -4 expression suggests that systemic administration of CPE may trigger a number of toxic side effects, precise dosage of CPE as well as, whenever possible, direct intra-tumoral administration, could certainly provide therapeutic benefit for patients with tumors that strongly overexpress claudin-3 and/or claudin-4. New strategies such as the use non-viral intratumoral gene transfer of CPE cDNA may also provide a more selective targeting allowing to circumvent potential side effects (Walther et al. 2012). One should keep in mind however that only those tumors that express significant amount of claudin-3 and/or -4 at the plasma membrane will be targeted by CPE, implying that immunohistochemical detection of claudin localization in tumor samples is an essential prerequisite before using such approaches.

Various other approaches are currently under investigation in order to target claudins or other TJ proteins, although these strategies are not as still in its infancy. Selective antibodies were recently developed against claudin-4 (Suzuki et al. 2009), claudin-3/4 (Kato-Nakano et al. 2010), and claudin-18 (Sahin et al. 2008), with the former being able to mediate antibody-dependent cell cytotoxicity (ADCC) and in vivo anti-tumor activity on pancreatic and ovarian cells. No results are available as yet concerning the prospective use of these antibodies in the CRC setting. Another, more recent approach aims to enhance the permeability of tight junctions, mostly through the use of peptide modulators (Takahashi et al. 2011), in order to facilitate the passage of other drugs. A proof of concept experiment for such approach was provided on breast, gastric, ovarian and lung cancer cell xenografts with the use of a small, recombinant adenovirus serotype 3-derived protein, termed junction opener 1 (JO-1), which greatly enhanced penetration and/or efficiency of the anti-HER2 and anti-EGFR monoclonal antibodies Trastuzumab and Cetuximab (Beyer et al. 2011). The usefulness of such approach for CRC treatment remains to

Kaihara T, Kawamata H, Imura J, Fujii S, Kitajima K, Omotehara F, Maeda N, Nakamura T, Fujimori T (2003) Redifferentiation and ZO-1 reexpression in liver-metastasized colorectal cancer: possible association with epidermal growth factor receptor-induced tyrosine phosphorylation of ZO-1. Cancer Sci 94:166–172

Kato-Nakano M, Suzuki M, Kawamoto S, Furuya A, Ohta S, Nakamura K, Ando H (2010) Characterization and evaluation of the antitumour activity of a dual-targeting monoclonal antibody against claudin-3 and claudin-4. Anticancer Res 30:4555–4562

Kimura Y, Shiozaki H, Hirao M, Maeno Y, Doki Y, Inoue M, Monden T, Ando-Akatsuka Y, Furuse M, Tsukita S, Monden M (1997) Expression of occludin, tight-junction-associated protein, in human digestive tract. Am J Pathol 151:45–54

Kinugasa T, Huo Q, Higashi D, Shibaguchi H, Kuroki M, Tanaka T, Futami K, Yamashita Y, Hachimine K, Maekawa S, Nabeshima K, Iwasaki H (2007) Selective up-regulation of claudin-1 and claudin-2 in colorectal cancer. Anticancer Res 27:3729–3734

Kondoh M, Takahashi A, Fujii M, Yagi K, Watanabe Y (2006) A novel strategy for a drug delivery system using a claudin modulator. Biol Pharm Bull 29:1783–1789

Korinek V, Barker N, Morin PJ, van Wichen D, de Weger R, Kinzler KW, Vogelstein B, Clevers H (1997) Constitutive transcriptional activation by a beta-catenin-Tcf complex in APC–/– colon carcinoma. Science 275:1784–1787

Kuhn S, Koch M, Nubel T, Ladwein M, Antolovic D, Klingbeil P, Hildebrand D, Moldenhauer G, Langbein L, Franke WW, Weitz J, Zoller M (2007) A complex of EpCAM, claudin-7, CD44 variant isoforms, and tetraspanins promotes colorectal cancer progression. Mol Cancer Res 5:553–567

Lee HS, Nishanian TG, Mood K, Bong YS, Daar IO (2008) EphrinB1 controls cell-cell junctions through the Par polarity complex. Nat Cell Biol 10:979–986

Li J, Sherman-Baust CA, Tsai-Turton M, Bristow RE, Roden RB, Morin PJ (2009) Claudin-containing exosomes in the peripheral circulation of women with ovarian cancer. BMC Cancer 9:244

Maloy KJ, Powrie F (2011) Intestinal homeostasis and its breakdown in inflammatory bowel disease. Nature 474:298–306

Martin TA, Mason MD, Jiang WG (2011) Tight junctions in cancer metastasis. Front Biosci 16:898–936

Masuda R, Semba S, Mizuuchi E, Yanagihara K, Yokozaki H (2010) Negative regulation of the tight junction protein tricellulin by snail-induced epithelial-mesenchymal transition in gastric carcinoma cells. Pathobiology 77:106–113

Mees ST, Mennigen R, Spieker T, Rijcken E, Senninger N, Haier J, Bruewer M (2009) Expression of tight and adherens junction proteins in ulcerative colitis associated colorectal carcinoma: upregulation of claudin-1, claudin-3, claudin-4, and beta-catenin. Int J Colorectal Dis 24:361–368

Mima S, Takehara M, Takada H, Nishimura T, Hoshino T, Mizushima T (2008) NSAIDs suppress the expression of claudin-2 to promote invasion activity of cancer cells. Carcinogenesis 29:1994–2000

Miwa N, Furuse M, Tsukita S, Niikawa N, Nakamura Y, Furukawa Y (2001) Involvement of claudin-1 in the beta-catenin/Tcf signaling pathway and its frequent upregulation in human colorectal cancers. Oncol Res 12:469–476

Morin PJ (2005) Claudin proteins in human cancer: promising new targets for diagnosis and therapy. Cancer Res 65:9603–9606

Morin PJ, Sparks AB, Korinek V, Barker N, Clevers H, Vogelstein B, Kinzler KW (1997) Activation of beta-catenin-Tcf signaling in colon cancer by mutations in beta-catenin or APC. Science 275:1787–1790

Mullin JM (1997) Potential interplay between luminal growth factors and increased tight junction permeability in epithelial carcinogenesis. J Exp Zool 279:484–489

Mullin JM (2004) Epithelial barriers, compartmentation, and cancer. Sci STKE 2004:pe2

Mullin JM, Laughlin KV, Ginanni N, Marano CW, Clarke HM, Peralta Soler A (2000) Increased tight junction permeability can result from protein kinase C activation/translocation and act as a tumor promotional event in epithelial cancers. Ann N Y Acad Sci 915:231–236

Murphy MS (1998) Growth factors and the gastrointestinal tract. Nutrition 14:771–774
Nakayama F, Semba S, Usami Y, Chiba H, Sawada N, Yokozaki H (2008) Hypermethylation-modulated downregulation of claudin-7 expression promotes the progression of colorectal carcinoma. Pathobiology 75:177–185
Nubel T, Preobraschenski J, Tuncay H, Weiss T, Kuhn S, Ladwein M, Langbein L, Zoller M (2009) Claudin-7 regulates EpCAM-mediated functions in tumor progression. Mol Cancer Res 7:285–299
Orban E, Szabo E, Lotz G, Kupcsulik P, Paska C, Schaff Z, Kiss A (2008) Different expression of occludin and ZO-1 in primary and metastatic liver tumors. Pathol Oncol Res 14:299–306
Oshima T, Kunisaki C, Yoshihara K, Yamada R, Yamamoto N, Sato T, Makino H, Yamagishi S, Nagano Y, Fujii S, Shiozawa M, Akaike M, Wada N, Rino Y, Masuda M, Tanaka K, Imada T (2008) Reduced expression of the claudin-7 gene correlates with venous invasion and liver metastasis in colorectal cancer. Oncol Rep 19:953–959
Ozdamar B, Bose R, Barrios-Rodiles M, Wang HR, Zhang Y, Wrana JL (2005) Regulation of the polarity protein Par6 by TGFbeta receptors controls epithelial cell plasticity. Science 307:1603–1609
Pannequin J, Delaunay N, Darido C, Maurice T, Crespy P, Frohman MA, Balda MS, Matter K, Joubert D, Bourgaux JF, Bali JP, Hollande F (2007) Phosphatidylethanol accumulation promotes intestinal hyperplasia by inducing ZONAB-mediated cell density increase in response to chronic ethanol exposure. Mol Cancer Res 5:1147–1157
Playford RJ, Wright NA (1996) Why is epidermal growth factor present in the gut lumen? Gut 38:303–305
Playford RJ, Hanby AM, Gschmeissner S, Peiffer LP, Wright NA, McGarrity T (1996) The epidermal growth factor receptor (EGF-R) is present on the basolateral, but not the apical, surface of enterocytes in the human gastrointestinal tract. Gut 39:262–266
Polette M, Mestdagt M, Bindels S, Nawrocki-Raby B, Hunziker W, Foidart JM, Birembaut P, Gilles C (2007) Beta-catenin and ZO-1: shuttle molecules involved in tumor invasion-associated epithelial-mesenchymal transition processes. Cells Tissues Organs 185:61–65
Radtke F, Clevers H, Riccio O (2006) From gut homeostasis to cancer. Curr Mol Med 6:275–289
Rahner C, Mitic LL, Anderson JM (2001) Heterogeneity in expression and subcellular localization of claudins 2, 3, 4, and 5 in the rat liver, pancreas, and gut. Gastroenterology 120:411–422
Reichert M, Muller T, Hunziker W (2000) The PDZ domains of zonula occludens-1 induce an epithelial to mesenchymal transition of Madin-Darby canine kidney I cells. Evidence for a role of beta-catenin/Tcf/Lef signaling. J Biol Chem 275:9492–9500
Resnick MB, Konkin T, Routhier J, Sabo E, Pricolo VE (2005) Claudin-1 is a strong prognostic indicator in stage II colonic cancer: a tissue microarray study. Mod Pathol 18:511–518
Sahin U, Koslowski M, Dhaene K, Usener D, Brandenburg G, Seitz G, Huber C, Tureci O (2008) Claudin-18 splice variant 2 is a pan-cancer target suitable for therapeutic antibody development. Clin Cancer Res 14:7624–7634
Saito T, Masuda N, Miyazaki T, Kanoh K, Suzuki H, Shimura T, Asao T, Kuwano H (2004) Expression of EphA2 and E-cadherin in colorectal cancer: correlation with cancer metastasis. Oncol Rep 11:605–611
Sakaguchi T, Suzuki S, Higashi H, Inaba K, Nakamura S, Baba S, Kato T, Konno H (2008) Expression of tight junction protein claudin-5 in tumor vessels and sinusoidal endothelium in patients with hepatocellular carcinoma. J Surg Res 147:123–131
Scholer-Dahirel A, Schlabach MR, Loo A, Bagdasarian L, Meyer R, Guo R, Woolfenden S, Yu KK, Markovits J, Killary K, Sonkin D, Yao YM, Warmuth M, Sellers WR, Schlegel R, Stegmeier F, Mosher RE, McLaughlin ME (2011) Maintenance of adenomatous polyposis coli (APC)-mutant colorectal cancer is dependent on Wnt/β-catenin signaling. Proc Natl Acad Sci U S A 108(41):17135–17140
Segditsas S, Tomlinson I (2006) Colorectal cancer and genetic alterations in the Wnt pathway. Oncogene 25:7531–7537
Shin DY, Kim GY, Kim JI, Yoon MK, Kwon TK, Lee SJ, Choi YW, Kang HS, Yoo YH, Choi YH (2010) Anti-invasive activity of diallyl disulfide through tightening of tight junctions and inhibition of matrix metalloproteinase activities in LNCaP prostate cancer cells. Toxicol In Vitro 24:1569–1576

multifaceted and depend on specific cellular and molecular interactions will be presented. Additionally, some of the more recent studies which suggest that some members of the claudin family, particularly claudin 1, may be useful as diagnostic markers for some breast cancer subtypes, will be highlighted.

Keywords Mammary gland • Breast cancer • Breast cancer subtypes • Tight junction proteins • Claudins • Occludin • ZO-1 • JAMs • Methylation • Epigenetic regulation • Posttranslational modifications • Phosphorylation • Transcription factors • Ephrins

Abbreviations

+ve	Positive
−ve	Negative
AJ	Adherin junction
BC	Breast cancer
BLBC	Basal-like breast cancer
BM	Basement membrane
CK5/6	Cytokeratin 5,6
DCIS	Ductal carcinoma in situ
EGF	Epidermal growth factor
EGFR	Epidermal growth factor receptor
EMT	Epithelial-mesenchymal transition
Eph	Ephrin
ER	Estrogen receptor
HBC	Human breast cancer
HER2	Human epidermal growth factor receptor 2
HMEC	Human mammary epithelial cells
IHC	Immunohistochemical
JAMs	Junctional adhesion molecules
LCIS	Lobular carcinoma in situ
N/A	Not available
PKA	Protein kinase A
PKC	Protein kinase C
PR	Progesterone receptor
TJ	Tight junction
TPA	12-O-tetradecanoylphorbol-13-acetate
ZO-1	Zona occludin-1

8.1 Normal Mammary Epithelia

8.1.1 Structure: Organization and Cellular Subtypes

Generally, although glandular development is initiated during embryogenesis, much of the morphogenesis of the mammary gland occurs postnatally. Morphogenesis commences with ductal branching and elongation during puberty, and is followed by cycles of cellular proliferation, differentiation and apoptosis that accompany each round of pregnancy, lactation and involution (for review, (Daniel and Silberstein 1987; Johnson 2010)).

The adult human breast is an area of skin and underlying connective tissue containing a group of 15–20 modified sweat glands (referred to as lobes) that collectively make up the mammary gland. Within each lobe, lobules drain into a series of ducts that in turn drain into a single lactiferous duct (Fig. 8.1) that opens onto the surface of the nipple. The epithelium within the ductwork progressively thickens as its tributaries converge towards the nipple.

In essence, three subtypes of epithelial cells line the mammary gland; ductal luminal, alveolar luminal and myoepithelial cells (Fig. 8.1). Ductal and alveolar cells constitute the inner layer of the duct and the lobuloalveolar units respectively, and each is surrounded by a basal layer of myoepithelial cells. The luminal epithelial cells have little cytoplasm and oval central nuclei with one or more nucleoli and scattered or peripheral chromatin. They are cuboidal to columnar and each cell has a complete lateral belt of tight junction (TJs) near its apex. The smallest ducts are lined with simple cuboidal epithelium while the largest ducts are lined with stratified columnar epithelium (Junquiera and Carneiro 2003).

The entire tubuloalveolar system is surrounded by a basement membrane (BM), and in between the luminal epithelial cells and the BM is interposed in an incomplete layer of stellate myoepithelial cells which establish epithelial polarity by

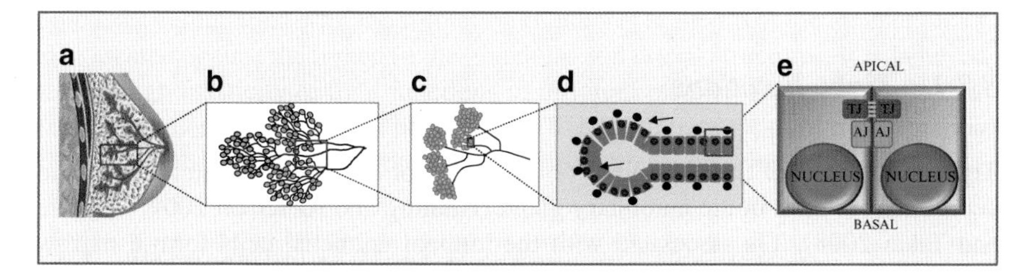

Fig. 8.1 Tight junctions in mammary gland epithelia. The adult human breast is comprised of 15–20 lobes (**a**). Within each lobe, lobules drain into a series of ducts (**b**) that in turn drain into a single lactiferous duct (**c**). Ductal and alveolar luminal cells constitute the inner layer of duct and the lobuloalveolar units, respectively, and are surrounded by a basal layer of myoepithelial cells (**d**). The luminal epithelial cells are cuboidal to columnar in morphology and each cell has a complete lateral belt of TJs near its apex (**e**). *TJ* tight junction, *AJ* adherin junction, *M*, myoepithelial cells, *E* epithelial cells

by a decrease in milk secretion rate, and conversely, a decrease in TJ permeability is accompanied by an increase in milk secretion rate (Stelwagen et al. 1998).

The permeability of the TJs can be regulated by a number of local and systemic stimuli. Key hormones, such as prolactin, hydrocortisone and progesterone, which play an integral part in milk production, maintenance and secretion, have also been shown to alter TJ permeability (Nguyen and Neville 1998; Itoh and Bissell 2003). Prolactin, which regulates the synthesis and secretion of milk as well as progesterone, the hormone that maintains pregnancy (Stelwagen et al. 1998), as well as the glucocorticoids (Stelwagen et al. 1999) which are essential for the maintenance of lactation, have all been shown to alter TJ permeability (Nguyen and Neville 1998). In addition, local stimuli such as intramammary pressures or growth factors, for example TGF beta, has also been shown to influence TJ permeability through its regulation of glucocorticoids (Nguyen and Neville 1998).

8.2.5 The Integral Tight Junction Proteins

The TJ is regarded as a complex array of peripheral and integral proteins. The peripheral proteins, also referred to as adapter proteins, connect the transmembrane proteins of the TJ with the actin cytoskeleton and numerous other proteins involved in cell signaling and vesicle trafficking (Gonzalez-Mariscal et al. 2007). The integral proteins that constitute the TJ are occludin, the junctional adhesion molecules (JAMs) and the claudins.

The two main molecular components in forming the TJ strand, are occludin and the claudins. Although both are tetraspan transmembrane proteins with four transmembrane domains they share no sequence similarity. Occludin has a long carboxy-terminal cytoplasmic domain and a short amino-terminal cytoplasmic domain and its expression is correlated with the number of TJ strands in various tissues. Its over expression has been shown to increase transepithelial resistance (TER) and at the same time increase paracellular flux (Balda et al. 1996). Occludin can regulate TJ barrier function and TJ protein interactions (Raleigh et al. 2011), but has been shown to be non-essential for strand assembly (Saitou et al. 1998).

The claudin family of proteins, (24 members to date (Myal et al. 2010)) share a common transmembrane topology. The C-terminus of most claudins end in a putative PDZ-binding domain, important for interaction with cytoskeleton proteins (Heiskala et al. 2001). The claudins are important players in strand formation and the over-expression of claudin 1 results in aberrant strand formation on the lateral surface of epithelial cells (McCarthy et al. 2000). The exact combination of claudin proteins within a given tissue determines the selectivity, strength and tightness of the TJ (Morin 2005). The expression pattern of claudins, which is tissue specific, determines the particular characteristic of its paracellular pathway (Gonzales-Mariscal et al. 2010). Aside from being indispensible for maintaining apical polarity and paracellular functions, the claudins are also involved in recruiting signaling proteins (Itoh et al. 1999; Zahraoui et al. 2000) thereby regulating various cellular processes including differentiation and cell growth and tumor development.

Unlike occludin and claudins, JAMs have an extracellular region that contains two variable type Ig-like domains, and a short cytoplasmic domain (Brennan et al. 2010; Martin-Padura et al. 1998). They are a family of glycosylated proteins that have been shown to be ubiquitously expressed in epithelia and interact with the PDZ domains of other TJ associated proteins. JAM proteins regulate numerous cellular adhesive processes including intercellular junction assembly (Liang et al. 2000) and cell morphology (Martin-Padura et al. 1998; Mandell and Parkos 2005; Ostermann et al. 2002). JAMS also show an intimate relationship with TJ strands. They do not form TJ strand-like structures as observed with claudins (Itoh et al. 2001) but function as the initial spatial clue for TJ formation as they begin to concentrate at cell-cell contact sites earlier than occludin and claudin (Martin-Padura et al. 1998; Forrest et al. 2003; Naik et al. 2001; Tsukita et al. 2001).

The structure of the integral TJ proteins share similarity, and appears to be important in how they interact with other signaling pathways. All three integral proteins, occludin, claudins and JAMs are linked to the actin cytoskeleton where they make contact with signal transducers such as kinases, phosphatases and G proteins (Fanning and Anderson 2009; Miyoshi and Takai 2008; Gonzalez-Mariscal et al. 2008; Paris and Bazzoni 2008). A comprehensive review of the interaction of the TJ proteins, in particular the claudins, with signal transduction pathways has recently been published (Nguyen and Neville 1998; Itoh and Bissell 2003).

8.2.6 Role in Cellular Proliferation

Carcinomas are malignancies that originate in epithelial tissues and account for 90% of all neoplasms including breast (Gonzalez-Mariscal et al. 2008). Proliferation and invasion are two major aspects of tumor development. TJs have been proposed to be involved in proliferation control more than two decades ago (Stevenson and Keon 1998), and have now been shown to be active participants in the regulation of gene transcription and cell proliferation. The role of TJs in cellular proliferation is well reviewed (Gonzalez-Mariscal et al. 2007).

8.3 Breast Cancer

8.3.1 Breast Cancer Subtypes and Hypothetical Models of Origin

Most breast cancers have their origin in the milk secreting luminal epithelial cells of the mammary gland. Control of cell proliferation, survival and polarity in the mammary epithelium is critical for normal mammary gland function and deregulation of these processes is thought to drive breast cancer initiation and progression (Itoh and Bissell 2003; Mareel and Leroy 2003). Furthermore, the acquisition of the invasive

Table 8.1 The five distinct subtypes of breast cancer and their corresponding phenotypes

Subtypes	Phenotype
Luminal A	ER+ve and/or PR+ve, Her2−ve
Luminal B	ER+ve and/or PR+ve, Her2+ve
Her2 overexpressing	ER−ve, PR−ve, Her2+ve
Basal-like	ER−ve, PR−ve, Her2−ve, CK5,6+ve and/or EGFR+ve
Normal-like	Not clearly defined, similar to normal epithelia, displays putative-initiating stem cell phenotype

BC breast cancer, *ER* estrogen receptor, *PR* progesterone receptor, *Her2* human epidermal growth factor receptor 2, *CK5,6* cytokeratin 5,6, *EGFR* epidermal growth factor receptor, *+ve* positive, *−ve* negative

phenotype is seen as the most significant change in breast cancer biology. As cells convert from the non-invasive to the invasive phenotype, they proliferate, become anchorage independent, more aggressive and exhibit enhanced motility.

In the last decade, a growing understanding of the heterogeneous nature of this disease has stemmed primarily from molecular and immunohistochemical (IHC) studies, leading to a redefinition of breast cancer subsets. To date, five distinct breast cancer subtypes have been identified based on estrogen receptor/progesterone receptor (ER/PR) status, human epidermal growth factor receptor 2 (HER2), cytokeratin 5,6 (CK5/6) and epidermal growth factor receptor (EGFR) expression. (Table 8.1, (Finak et al. 2006; Kwan et al. 2009; Nielsen et al. 2004; Perou et al. 2000; Sorlie et al. 2003; Yu et al. 2009)). These subtypes differ in their morphology, clinical course and display a wide variety of responses to different treatments (Kwan et al. 2009; Bild et al. 2009; Hugh et al. 2009). Whereas the luminal subtype is characterized by its epithelial phenotype, mildly invasive capacity and relatively good clinical outcome, the basal-like breast cancer (BLBC) subtype, which is not sensitive to hormones are more aggressive, characterized by enhanced invasiveness and formation of distant metastases and demonstrate the worst prognosis (Finak et al. 2006; Kwan et al. 2009). The enhanced metastatic capacity of the BLBC subtype is associated with their migratory, mesenchymal phenotype (Nielsen et al. 2004). These subtypes were found to be conserved across ethnic groups and are already evident at the ductal carcinoma in situ (DCIS) stages (Yu et al. 2004) suggesting distinct tumor progression pathways for each tumor type (Polyak 2007).

Notably, a sixth molecular subtype, a “claudin low” subtype, has been identified that exhibits characteristically low expression of claudin 3, 4, and 7 and high expression of stem cell markers and was recently shown to associate with poor survival (Herschkowitz et al. 2007). More recently, a large comprehensive study based on genetic analysis and exome mapping (Curtis et al. 2012) identified ten breast cancer subtypes, including four of the previously identified ones. It has been speculated that these genomically defined tumor subtypes may represent transformation of stem cells with arrest at specific stages of development, or, alternatively, direct transformation of various cell types (Prat and Perou 2009). One can also speculate that the high plasticity of the organization of the tight junctions that is observed during different physiological stages of morphogenesis may be characteristically unique for each of these subtypes.

Also in recent years a number of hypothetical models have been proposed to explain breast cancer progression and the origins of breast cancer subtypes. One hypothetical model outlines a continuum of lesions to describe the stepwise progression of breast cancer: initially epithelial hyperplasia develops, followed by atypical hyperplasia and ductal carcinoma in situ (DCIS) to invasive carcinoma and eventually metastatic disease (Polyak 2007). In this linear model, the cell of origin may be the same for the different tumor subtypes. Thus, the tumor subtype is determined by acquired genetic and epigenetic events (Polyak 2007). In more recent years other models have been proposed based on genomic studies which suggest that each tumor subtype is initiated in a different cell type (presumably stem cell or progenitor cell (Lim et al. 2009)). In support of such a hypothesis, recent work by Jeselsohn et al. (2010) has provided evidence pointing to two populations of progenitor cells, one giving rise to luminal-like breast cancers and another, to basal-like breast cancers.

8.3.2 Tight Junctions and Breast Cancer Progression

It is believed that the acquisition of the invasive phenotype represents the greatest threat for the development of metastatic disease. For cancer to progress to metastasis, individual cells must leave the primary tumor, invade surrounding normal tissues and enter the blood vessels to reach distant sites. For these cells, the breakdown of cell-cell interactions as well as decreased cellular polarity and differentiation are critical steps in invasion and metastasis (Kominsky et al. 2003). It has been observed that the disruption of TJs leads to loss of cohesion, invasiveness and the lack of differentiation (Martin and Jiang 2009). This resulting loss of TJ integrity may be particularly important in facilitating motility as well as allowing the diffusion of nutrients and other factors necessary for the survival and growth of the cancer cells thereby promoting tumorigenesis (Hoevel et al. 2002; Martin et al. 2002; Ren et al. 1990; Satoh et al. 1996).

8.3.3 Tight Junctions and Epithelial Mesenchymal Transition

Epithelial-mesenchymal transition (EMT) is the process by which tumor cells overcome fundamental cell controls that facilitate cell-cell contact and prevents migration and has been attributed to the reactivation of a dormant developmental process (Hay 1995; Valles et al. 1994). Studies show that some of the molecular programs of EMT might be involved in the development of the mammary gland, particularly at the terminal end buds or possibly, in the lateral branching (O'Brien et al. 2002; Vincent-Salomon and Thiery 2003).

Just as EMT has been implicated to be involved in mammary gland development, it has also been implicated to play a role in breast cancer progression. Many signaling pathways identified in the EMT process have been identified in breast cancer malignancy. EMT allows epithelial cells to acquire a mesenchymal migratory phenotype (for review, (Guarino et al. 2007)) and as mesenchymal cells, they can migrate as individual cells, penetrate into surrounding tissues and move to spread to distant sites (Hay 1995; Guarino et al. 2007; Boyer et al. 2000). One of the earliest events during EMT, involves disassembly of the lateral seals close to the apical surface. This disassembly is accompanied by a loss of cell-cell junctions and the reorganization of the cytoskeleton which results in the loss of apical-basal cell polarity and the acquisition of spindle-shaped morphology (Trimboli et al. 2008). The loss of proteins involved in epithelial cell-cell contact, as in the case of the TJ proteins is crucial for the development of a fibroblastic phenotype with invasive properties in cancerous tissues (Gonzalez-Mariscal et al. 2007). Thus, EMT involves the redistribution of key molecules in the junctional complex, the TJ proteins (Mareel and Leroy 2003; Mullin 1997). Recently, the first direct evidence that EMT plays a major role in breast tumor progression has been demonstrated (Trimboli et al. 2008). Reports on the phenotype of breast cancer micrometastases in lymph nodes and that of bone marrow, also suggest that EMT occurs within the primary tumors (Braun and Pantel 1999). Knudsen et al. (2012) has also shown that progression of DCIS to invasive breast cancer, is associated with EMT, and that the most significant alterations occur in the epithelial compartment (Braun and Pantel 1999).

8.4 Tight Junctions in Breast Cancer

8.4.1 Alterations in Tight Junction Protein/Gene Expression

An increasing number of studies have shown that the deregulated expression of the junctional proteins are directly or indirectly involved in cancer progression, including breast cancer (for review, (Itoh and Bissell 2003; Brennan et al. 2010; Martin and Jiang 2009)). Some of the recent developments in the understanding pertaining to the role of the integral tight junction proteins will be discussed in this section.

JAMS: To date, only a limited number of studies have addressed the role of the JAMS in breast cancer and a deregulation of JAMs has been implicated (McSherry et al. 2009; Naik et al. 2008). Naik et al. (2008) showed that over expression of JAM-A decreased migration and invasion in breast cancer cell lines, while knockdown of JAM-A enhanced invasiveness. Since JAM proteins regulate numerous cellular adhesive processes including intercellular junction assembly and cell morphology (Mandell and Parkos 2005) and loss of tissue architecture is a prerequisite for cancer invasion and metastasis, JAMs may play a key role in breast cancer progression. Conversely, McSherry et al. (2009) subsequently showed a significant association between high JAM gene and protein expression and poor survival in two large cohorts of human invasive breast cancer suggesting a dual role for the JAM proteins in breast cancer. It is difficult to explain why these results are so contradictory; however it is likely that each cohort may have different representations of the different breast cancer subtypes. Thus if JAMS havedifferent functions in different molecular subtypes of breast cancer (Table 8.1), this will impact greatly on the outcome of such studies

Occludin: A relationship between occludin silencing and oncogenic transformation has been reported for several cancers including breast cancer. As in the case of the JAMs, over expression of occludin in HBC cells has been shown to decrease cancer cell migration and invasion both in vitro and in vivo (Osanai et al. 2006). As well, the expression of occludin decreases with disease progression (Tobioka et al. 2004). Occludin has also been shown to be down regulated in murine breast cancers (Osanai et al. 2007a). In murine breast cancer cells, occludin over expression was found to induce premature senescence (Osanai et al. 2007a) as well as promote detachment-induced apoptosis (anoikis) (Osanai et al. 2006). As a result of such observations in both human and mice, occludin has been considered a putative tumor suppressor gene.

Claudins: Of the integral junctional proteins, the claudins are the most widely studied in breast cancer, in particular claudin 1, 3, 4, 6, 7 and to some extent, claudin 16. Some of these claudins have demonstrated an increase, or a decrease or both, in gene expression in breast cancer (Gonzalez-Mariscal et al. 2007; Kominsky et al. 2003; Blanchard et al. 2009; Hewitt et al. 2006; Kim et al. 2008; Lanigan et al. 2009; Martin et al. 2008; Morohashi et al. 2007; Sauer et al. 2005; Soini 2005; Tokes et al. 2005). An increase in the expression of claudin 3 and 4 has been frequently observed in breast cancer, both at the mRNA and the protein level (Blanchard et al. 2009; Lanigan et al. 2009; Soini 2005; Tokes et al. 2005). However, breast tumors exhibiting low levels of claudin 3, 4 and 7 expression and high expression of stem cell and epithelial markers (designated 'claudin low'), believed to represent a rare and aggressive breast cancer subtype (Herschkowitz et al. 2007), has recently been shown to be associated with poor overall survival (Perou 2010). Conversely, some claudins are frequently down-regulated or lost in invasive breast cancer (Hoevel et al. 2004; Michl et al. 2003; Swisshelm et al. 1999). Suppression of claudin 6 has been shown to increase MMP2 activity (Osanai et al. 2007b) and promote resistance to apoptosis and anoikis. Its reintroduction into HBC cells in vitro, increased cellular adhesion and abrogated enhanced invasion and migration (Osanai et al. 2007b). Expression of claudin 7 has also been shown to be reduced or lost in invasive ductal carcinoma compared to normal breast (Kominsky et al. 2003; Hoevel et al. 2004). Furthermore, loss of claudin 7 expression also correlated with increasing tumor grade and metastatic disease (Kominsky et al. 2003; Sauer et al. 2005). As well, the over expression of claudin 16 in HBC cells was shown to reduce aggressiveness and motility (Martin et al. 2008).

In breast cancer, the role of claudin 1 is not fully understood. However, studies from our laboratory and others, demonstrate that claudin 1 is down regulated or lost in invasive HBC (Blanchard et al. 2009; Tokes et al. 2005; Kramer et al. 2000; Kulka and Tokes 2005; Swisshelm et al. 2005). These observations, as well as a recent study reporting a correlation between the down regulation of claudin 1 and disease recurrence (Morohashi et al. 2007) suggest that it may function as a tumor suppressor in breast cancer. Claudin 1 is also generally expressed in the membranes of normal

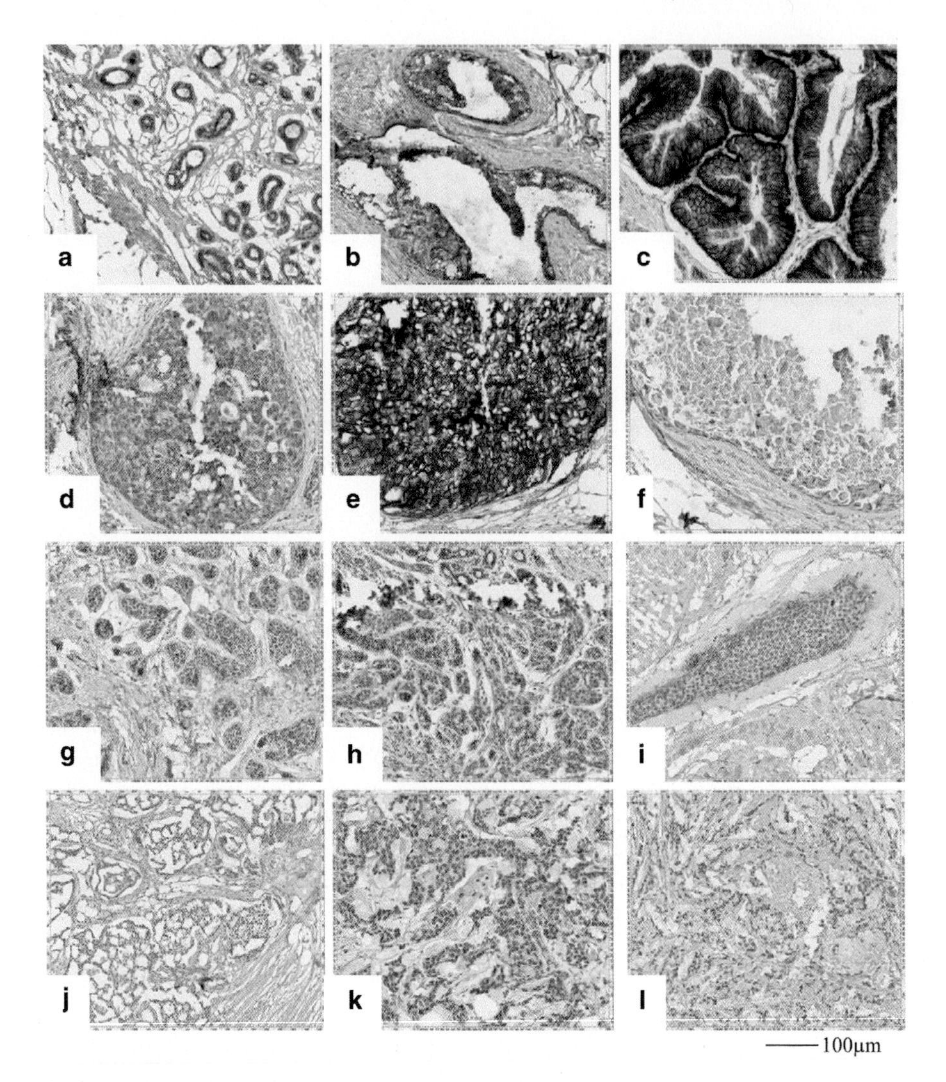

Fig. 8.2 Claudin 1 is highly expressed in some pre-invasive breast cancers. IHC analysis, showing claudin 1 staining in (**a**) normal ducts; (**b**) atypical ductal hyperplasia (ADH); (**c**) apocrine; (**d**, **e**, **f**) ductal carcinoma in situ (DCIS); (**g**, **h**) lobular neoplasia; (**i**) lobular carcinoma in situ (LCIS); (**j**, **k**) infiltrating ductal carcinoma; (**l**) infiltrating lobular carcinoma

ductal epithelial cells and in some pre-invasive breast cancers including DCIS and LCIS (lobular carcinoma in situ; Fig. 8.2). As well, there is now mounting evidence that suggests claudin 1 may play a direct role in breast cancer progression. A re-expression of claudin 1 alone was sufficient to induce apoptosis in a HBC cell line (Hoevel et al. 2004). Claudin 1 has also been shown to singularly exert TJ mediated gate function (paracellular sealing) in metastatic breast cancer cells in the absence of other tight junction proteins (Hoevel et al. 2002). Furthermore, it has

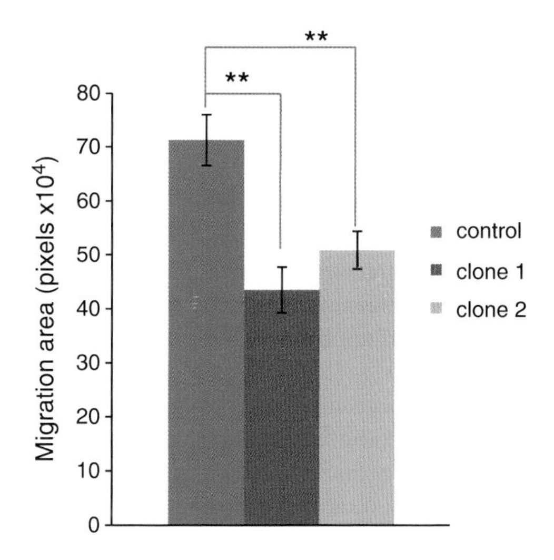

Fig. 8.3 Over expression of claudin 1 results in a decrease in cell migration in the human breast cancer cell line, MDA-MB231. Cells were grown to confluency and a scratch made through the cell monolayer. Measurements of the wound areas at time 0 and 16 h were compared using the Image-J program. MDA-MB231 cells over expressing claudin 1 (clone 1, 2) migrated slower than the cells stably transfected with the empty vector (control; mean ± S.E; ANOVA $p = 0.0008$, control $n = 10$; clone 1 $n = 5$; clone 2 $n = 8$). $^{**}p < 0.01$, Bonferroni's Multiple Comparison Test

been demonstrated that the down regulation of claudin 1 gene expression can lead to neoplastic transformation of mammary epithelial cells in vitro (Kulawiec et al. 2008).

However, recent work from our laboratory suggests that the role of claudin 1 in breast cancer may be more complex than originally thought. While studies from our laboratory have shown that in invasive HBC, the frequency of claudin 1 expression is low (Blanchard et al. 2009), paradoxically, we have also observed that high levels of claudin 1 protein was significantly associated with the estrogen receptor negative (ER−ve) basal-like breast cancer (BLBC) subtype (Blanchard et al. 2009), a very invasive and aggressive form of breast cancer, suggesting that, in addition to being a putative tumor suppressor, claudin 1 may have other functions as well depending on the subtype of breast cancer. We have also now shown that over expression of claudin 1 increase cell migration and motility in the HBC cell line MDA231 which is phenotypically basal-like (Fig. 8.3).

Interestingly, the possibility of a dual role for claudin 1 in the same cancer has previously been speculated. For example, the over expression of claudin 1 in type II seropapillary endometrial carcinoma distinguishes it from type I endometriod carcinoma (Sobel et al. 2006), highlighting its potential as a diagnostic marker to differentiate subtypes of cancers. There are a number of plausible explanations why in some breast cancers claudin 1 is down regulated or absent and in others it is over expressed. Claudin 1 may behave differently in different breast cancer subtypes. Thus, if claudin 1 expression is breast cancer subtype specific as it appears to be, then specificity of expression could be dictated by the cell type of origin of the

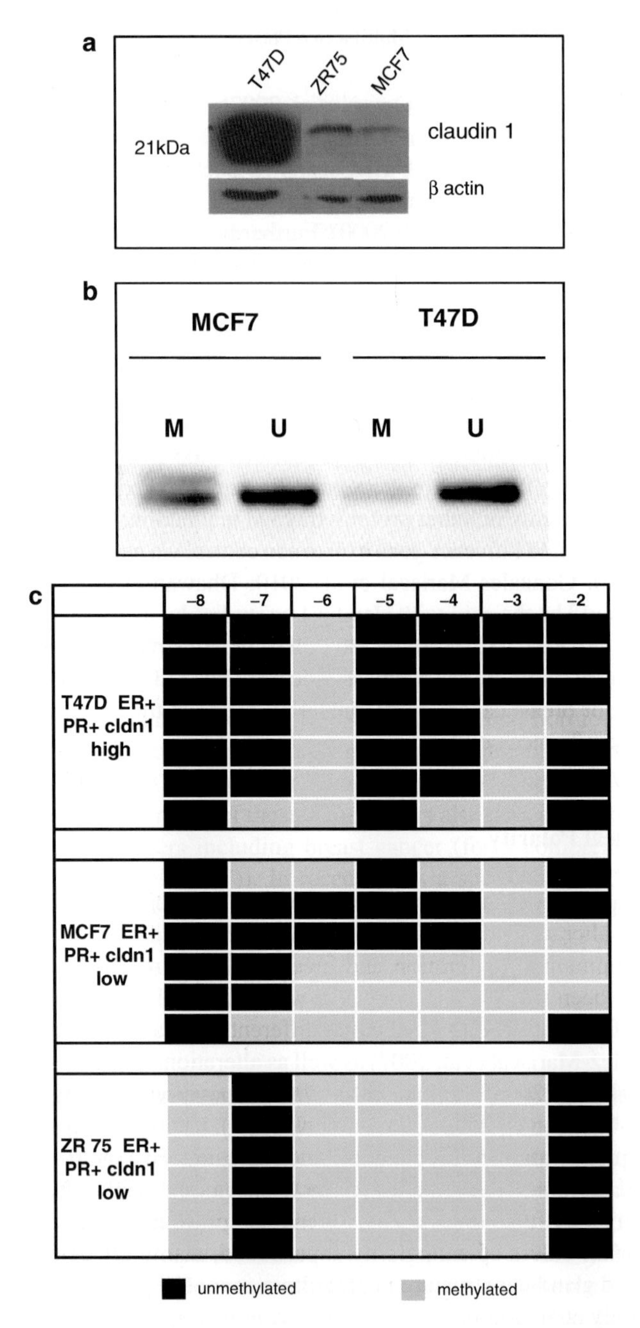

Fig. 8.4 Methylation of the claudin 1 promoter in human breast cancer cell lines. (**a**) Western blot showing relative amount of endogenous claudin 1 levels in the T47D, ZR75 and MCF7 HBC cell lines. (**b**) Bisulphite converted DNA methylation specific PCR was carried out on two HBC cell lines. Hypermethylation of claudin 1 was observed in the MCF7 cell line which expresses low levels of endogenous claudin 1. Hypomethylation was observed in the T47D cells which express high levels of endogenous claudin 1. As well, an additional band was associated with the MCF7 cells. *M* methylated, *U* unmethylated. (**c**) Genomic DNA from cell lines was extracted, and following

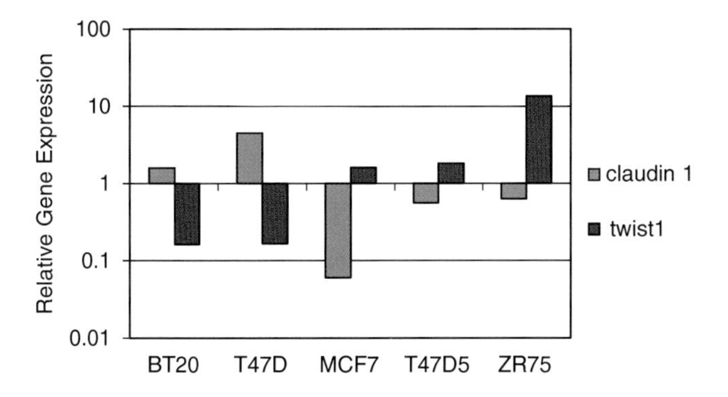

Fig. 8.5 The EMT marker, twist, and claudin 1 are inversely regulated in human breast cancer. Twist gene expression is up regulated in HBC cell lines that exhibit low claudin 1 expression, and down regulated in those lines that exhibit higher levels of claudin 1 expression of twist and claudin 1. Gene expression in each cell line was determined by RT-qPCR and was compared to gene expression in the MCF10A normal mammary cell line

other categories of progenitor cells because of their different chromatin organization (Lelievre 2010; Boheler 2009). Although there is no direct evidence reported to date to validate any of these hypotheses, the ever increasing identification of new molecular subtypes of breast cancer provides new opportunities for delineating the influences of the TJs on apical polarity as it relates to breast cancer progression.

8.4.2.3 Transcriptional Factors

The transcriptional factors snail (Moody et al. 2005) and twist (Yang et al. 2004) have been unveiled as key regulators in induction of EMT in breast cancer and other cancers, by suppressing the expression of the epithelial specific adhesion molecule, E cadherin. Both snail and twist binds directly to the E box of the E cadherin promoter and represses its expression (Lopez et al. 2009). Moreover, studies have now shown that these transcription factors may also control the gene expression of TJ proteins. Snail has been shown to bind to the E-box motifs in the human claudin 1 promoter and inhibit claudin 1 expression (Martinez-Estrada et al. 2006). Also, the same E box motif in the E cadherin promoter to which twist binds is also present in the claudin promoter (unpublished observation). Interestingly, we have observed an inverse relationship between claudin 1 and twist in HBC cell lines (Fig. 8.5

Fig. 8.4 (continued) bisulphite conversion, PCR amplification of a 169 bp sequence, 5 of the claudin 1 translational start site was carried out. The PCR fragments were ligated into a cloning vector and clones were sequenced. Each row represents a single clone and the numbers above each column indicate the location of the CpG relative to the ATG translational start site. Some methylated sites were common to all three cell lines (e.g. –6, –3). Chi-squared analysis of the number of methylated versus unmethylated sites, showed a significantly higher number of methylated sites in the MCF7 and ZR75 cell line as compared with T47D ($p=0.0002$)

2005; Lippoldt et al. 2000). PKC/PKA activity is important in regulating claudin 1 expression and localization in several cancers (Dhawan et al. 2005; French et al. 2009; Leotlela et al. 2007; Oku et al. 2006). In hepatoma cells a decrease in PKA resulted in sublocalization of claudin 1 to the cytoplasm (Farquhar et al. 2008). In melanoma, PKC activity has been shown to increase the transcription of claudin 1 and phosphorylation modification by PKA/PKC further resulted in the retention of claudin 1 in the cytoplasm (French et al. 2009). Furthermore, mutation of PKA/PKC phosphorylation sites on the claudin 1 protein that led to its constitutive phosphorylation resulted in claudin 1 retention in the cytoplasm. In particular, the mutation of one PKA site contained within the C-terminal tail was alone responsible for retaining claudin 1 in the cytoplasm. In melanomas, as well as in oral and colon cancers, the observed mislocalization of the claudin 1 protein was shown to be associated with enhanced metastatic capacity (Dhawan et al. 2005; Leotlela et al. 2007; Oku et al. 2006).

8.5.3 Ephrins

Ephrin (Eph) receptors are frequently over expressed in cancerous tissues (Tanaka et al. 2005) and the over expression of this receptor and its ligand has been reported in various tumors of epithelial origin, but their significance is not well understood. In neural cells as well as epithelial cells, the interaction of the Eph family of receptor protein tyrosine kinase and its ephrin ligand family can induce bidirectional signaling via cell-cell contact (Tanaka et al. 2005). Ephrin B1 becomes phosphorylated on tyrosine residues following contact with the Eph receptor ecodomain (Bruckner et al. 1997).

Recent studies show that there exists a relationship between the ephrin B1 ligand and claudin 1. Claudin 1 has been shown to facilitate the phosphorylation of ephrin B1. Ephrin B1 creates an in vivo complex with adjacent claudin 1 via the extracellular domains. (Tanaka et al. 2005). Claudin 1 was shown to induce phosphorylation of ephrin B1 resulting in formation of cell-cell contact. Depending on the level of phosphorylation the over expression of ephrin B1 resulted in a consequent change in the level of cell-cell adhesion. Furthermore, the cell-cell contact formation enhanced the tyrosine phosphorylation of the cytoplasmic region of ephrin B1 in a manner dependent on claudin 1.

A search of common phosphorylation sites among claudins reveals that the tyrosine present at position −1 (from the C-terminus) of all claudins (except 11, 12, 13 16, 22 and 23), is a conserved putative Eph phosphorylation site (Gonzales-Mariscal et al. 2010). The revelation that the claudins have conserved Eph phosphorylation sites suggest that the claudins in turn may also be phosphorylated by members of the Eph family.

The relationship between claudin 1 and the ephrin family is not straightforward. However the interactions between the ephrin family and the claudins are intriguing and more extensive investigations are warranted and understanding the interactions of the Eph family and the claudins will be an important novel area of study with regards to breast cancer progression.

8.6 Conclusion and Future Perspectives

As many as ten different breast cancer subtypes have now been identified to date, underscoring the heterogeneous nature of this disease. Unfortunately, most of these are poorly characterized. As a result, this void in our knowledge presents an insurmountable challenge to finding effective strategies for managing this disease.

Research into the development of approaches to modulate barrier function for efficient drug delivery continues to receive much attention (Itoh and Bissell 2003). Since the loss of cell-cell adhesion is a crucial step in EMT, strategies to overcome the altered expression of TJ proteins in cancerous tissues are attractive as they could eventually lead to the development of effective therapeutic management for treating and possibly preventing human cancers. It is well established that changes in the TJ permeability can be reproduced readily in response to a number of physiological stimuli in the healthy mammary gland of a number of species (Freeman et al. 2000) Enhancing our understanding in the regulation of TJ permeability pathological states as breast cancer, is critical. At the same time we also need to further our understanding of how TJ interact with signaling cascades. To address such interactions, established and available mouse models with genetic variation in many types of signaling molecules can be utilized. The combination of studies in these mice with careful biochemical studies of TJ components may provide powerful insight into TJ regulation.

The claudin family of TJ proteins are of particular interest in the quest to identify new targets for breast cancer therapy as (1) an increasing number of claudin family members are being shown to have direct involvement in the migration and proliferation rates in cancer (Agarwal et al. 2009; Dhawan et al. 2005; French et al. 2009; Leotlela et al. 2007) and (2) some claudin family members (Claudin 3 and 4) are known receptors for the enterotoxins (*Clostridium perfringens)* that can elicit rapid and specific cell lysis in numerous cancer cells (Long et al. 2001; Michl et al. 2001; Santin et al. 2005), including breast (Kominsky et al. 2004).

The pattern of expression of TJ proteins in normal breast and breast cancer may also serve as a tool for predicting disease progression. We have shown a significant association between claudin 1 and the BLBC subtype (Blanchard et al. 2009). Also, the recent identification of a new "claudin low" subtype (Herschkowitz et al. 2007) that characteristically exhibits low levels of claudin 3 and 4, and which has been shown to be associated with a particularly aggressive form of breast cancer, is one example. Additional support regarding the utility of TJ proteins as potential biomarkers for breast cancer comes from a recent report which showed that when claudin 1 was used in conjunction with four other markers, this cohort of markers was a useful predictive indicator for breast cancer patients (Charpin et al. 2012).

Altogether, these observations present new opportunities for developing effective therapeutic strategies for managing breast cancer.

Itoh M, Bissell MJ (2003) The organization of tight junctions in epithelia: implications for mammary gland biology and breast tumorigenesis. J Mammary Gland Biol Neoplasia 8:449–462
Itoh M, Furuse M, Morita K, Kubota K, Saitou M, Tsukita S (1999) Direct binding of three tight junction-associated MAGUKs, ZO-1, ZO-2, and ZO-3, with the COOH termini of claudins. J Cell Biol 147:1351–1363
Itoh M, Sasaki H, Furuse M, Ozaki H, Kita T, Tsukita S (2001) Junctional adhesion molecule (JAM) binds to PAR-3: a possible mechanism for the recruitment of PAR-3 to tight junctions. J Cell Biol 154:491–497
Jeselsohn R, Brown NE, Arendt L, Klebba I, Hu MG, Kuperwasser C, Hinds PW (2010) Cyclin D1 kinase activity is required for the self-renewal of mammary stem and progenitor cells that are targets of MMTV-ErbB2 tumorigenesis. Cancer Cell 17:65–76
Johnson MC (2010) Anatomy and physiology of the breast. In: Jatio I, Kaufmann M (eds) Management of breast diseases. Springer, Berlin, pp 1–36
Junquiera L, Carneiro J (2003) Basic histology text and atlas. Lange Medical Books/McGraw-Hill, New York
Kaihara T, Kawamata H, Imura J, Fujii S, Kitajima K, Omotehara F, Maeda N, Nakamura T, Fujimori T (2003) Redifferentiation and ZO-1 reexpression in liver-metastasized colorectal cancer: possible association with epidermal growth factor receptor-induced tyrosine phosphorylation of ZO-1. Cancer Sci 94:166–172
Kim TH, Huh JH, Lee S, Kang H, Kim GI, An HJ (2008) Down-regulation of claudin-2 in breast carcinomas is associated with advanced disease. Histopathology 53:48–55
Knudsen ES, Ertel A, Davicioni E, Kline J, Schwartz GF, Witkiewicz AK (2012) Progression of ductal carcinoma in situ to invasive breast cancer is associated with gene expression programs of EMT and myoepithelia. Breast Cancer Res Treat 133:1009–1024
Koizumi J, Kojima T, Ogasawara N, Kamekura R, Kurose M, Go M, Harimaya A, Murata M, Osanai M, Chiba H, Himi T, Sawada N (2008) Protein kinase C enhances tight junction barrier function of human nasal epithelial cells in primary culture by transcriptional regulation. Mol Pharmacol 74:432–442
Kominsky SL, Argani P, Korz D, Evron E, Raman V, Garrett E, Rein A, Sauter G, Kallioniemi OP, Sukumar S (2003) Loss of the tight junction protein claudin-7 correlates with histological grade in both ductal carcinoma in situ and invasive ductal carcinoma of the breast. Oncogene 22:2021–2033
Kominsky SL, Vali M, Korz D, Gabig TG, Weitzman SA, Argani P, Sukumar S (2004) Clostridium perfringens enterotoxin elicits rapid and specific cytolysis of breast carcinoma cells mediated through tight junction proteins claudin 3 and 4. Am J Pathol 164:1627–1633
Konska G, Guillot J, L.M d, Fonck Y (1998) Expression of Tn antigen and N-acetyllactosamine residues in malignant and benign human breast tumors detected by lectins and monoclonal antibody 83D4. Int J Oncol 12:361–367
Kramer F, White K, Kubbies M, Swisshelm K, Weber BH (2000) Genomic organization of claudin-1 and its assessment in hereditary and sporadic breast cancer. Hum Genet 107:249–256
Kulawiec M, Safina A, Desouki MM, Still I, Matsui SI, Bakin A, Singh KK (2008) Tumorigenic transformation of human breast epithelial cells induced by mitochondrial DNA depletion. Cancer Biol Ther 7:1732–1743
Kulka J, Tokes AM (2005) Claudin expression in breast tumors. Hum Pathol 36:859
Kwan ML, Kushi LH, Weltzien E, Maring B, Kutner SE, Fulton RS, Lee MM, Ambrosone CB, Caan BJ (2009) Epidemiology of breast cancer subtypes in two prospective cohort studies of breast cancer survivors. Breast Cancer Res 11:R31
Lanigan F, McKiernan E, Brennan DJ, Hegarty S, Millikan RC, McBryan J, Jirstrom K, Landberg G, Martin F, Duffy MJ, Gallagher WM (2009) Increased claudin-4 expression is associated with poor prognosis and high tumour grade in breast cancer. Int J Cancer 124:2088–2097
Lelievre SA (2010) Tissue polarity-dependent control of mammary epithelial homeostasis and cancer development: an epigenetic perspective. J Mammary Gland Biol Neoplasia 15:49–63
Lelievre SA, Bissell MJ (1998) Communication between the cell membrane and the nucleus: role of protein compartmentalization. J Cell Biochem Suppl 30–31:250–263

Leotlela PD, Wade MS, Duray PH, Rhode MJ, Brown HF, Rosenthal DT, Dissanayake SK, Earley R, Indig FE, Nickoloff BJ, Taub DD, Kallioniemi OP, Meltzer P, Morin PJ, Weeraratna AT (2007) Claudin-1 overexpression in melanoma is regulated by PKC and contributes to melanoma cell motility. Oncogene 26:3846–3856

Liang TW, DeMarco RA, Mrsny RJ, Gurney A, Gray A, Hooley J, Aaron HL, Huang A, Klassen T, Tumas DB, Fong S (2000) Characterization of huJAM: evidence for involvement in cell-cell contact and tight junction regulation. Am J Physiol Cell Physiol 279:C1733–C1743

Lim E, Vaillant F, Wu D, Forrest NC, Pal B, Hart AH, Asselin-Labat ML, Gyorki DE, Ward T, Partanen A, Feleppa F, Huschtscha LI, Thorne HJ, Fox SB, Yan M, French JD, Brown MA, Smyth GK, Visvader JE, Lindeman GJ (2009) Aberrant luminal progenitors as the candidate target population for basal tumor development in BRCA1 mutation carriers. Nat Med 15:907–913

Lippoldt A, Liebner S, Andbjer B, Kalbacher H, Wolburg H, Haller H, Fuxe K (2000) Organization of choroid plexus epithelial and endothelial cell tight junctions and regulation of claudin-1, -2 and -5 expression by protein kinase C. Neuroreport 11:1427–1431

Long H, Crean CD, Lee WH, Cummings OW, Gabig TG (2001) Expression of Clostridium perfringens enterotoxin receptors claudin-3 and claudin-4 in prostate cancer epithelium. Cancer Res 61:7878–7881

Lopez D, Niu G, Huber P, Carter WB (2009) Tumor-induced upregulation of Twist, Snail, and Slug represses the activity of the human VE-cadherin promoter. Arch Biochem Biophys 482:77–82

Mandell KJ, Parkos CA (2005) The JAM family of proteins. Adv Drug Deliv Rev 57:857–867

Mareel M, Leroy A (2003) Clinical, cellular, and molecular aspects of cancer invasion. Physiol Rev 83:337–376

Martin TA, Jiang WG (2009) Loss of tight junction barrier function and its role in cancer metastasis. Biochim Biophys Acta 1788:872–891

Martin TA, Mansel RE, Jiang WG (2002) Antagonistic effect of NK4 on HGF/SF induced changes in the transendothelial resistance (TER) and paracellular permeability of human vascular endothelial cells. J Cell Physiol 192:268–275

Martin TA, Harrison GM, Watkins G, Jiang WG (2008) Claudin-16 reduces the aggressive behavior of human breast cancer cells. J Cell Biochem 105:41–52

Martinez-Estrada OM, Culleres A, Soriano FX, Peinado H, Bolos V, Martinez FO, Reina M, Cano A, Fabre M, Vilaro S (2006) The transcription factors Slug and Snail act as repressors of Claudin-1 expression in epithelial cells. Biochem J 394:449–457

Martin-Padura I, Lostaglio S, Schneemann M, Williams L, Romano M, Fruscella P, Panzeri C, Stoppacciaro A, Ruco L, Villa A, Simmons D, Dejana E (1998) Junctional adhesion molecule, a novel member of the immunoglobulin superfamily that distributes at intercellular junctions and modulates monocyte transmigration. J Cell Biol 142:117–127

McCarthy KM, Francis SA, McCormack JM, Lai J, Rogers RA, Skare IB, Lynch RD, Schneeberger EE (2000) Inducible expression of claudin-1-myc but not occludin-VSV-G results in aberrant tight junction strand formation in MDCK cells. J Cell Sci 113(Pt 19):3387–3398

McSherry EA, McGee SF, Jirstrom K, Doyle EM, Brennan DJ, Landberg G, Dervan PA, Hopkins AM, Gallagher WM (2009) JAM-A expression positively correlates with poor prognosis in breast cancer patients. Int J Cancer 125:1343–1351

Michl P, Buchholz M, Rolke M, Kunsch S, Lohr M, McClane B, Tsukita S, Leder G, Adler G, Gress TM (2001) Claudin-4: a new target for pancreatic cancer treatment using Clostridium perfringens enterotoxin. Gastroenterology 121:678–684

Michl P, Barth C, Buchholz M, Lerch MM, Rolke M, Holzmann KH, Menke A, Fensterer H, Giehl K, Lohr M, Leder G, Iwamura T, Adler G, Gress TM (2003) Claudin-4 expression decreases invasiveness and metastatic potential of pancreatic cancer. Cancer Res 63:6265–6271

Miyoshi J, Takai Y (2008) Structural and functional associations of apical junctions with cytoskeleton. Biochim Biophys Acta 1778:670–691

Moody SE, Perez D, Pan TC, Sarkisian CJ, Portocarrero CP, Sterner CJ, Notorfrancesco KL, Cardiff RD, Chodosh LA (2005) The transcriptional repressor Snail promotes mammary tumor recurrence. Cancer Cell 8:197–209

Morgan G, Wooding FB (1982) A freeze-fracture study of tight junction structure in sheep mammary gland epithelium during pregnancy and lactation. J Dairy Res 49:1–11

Morin PJ (2005) Claudin proteins in human cancer: promising new targets for diagnosis and therapy. Cancer Res 65:9603–9606

Morohashi S, Kusumi T, Sato F, Odagiri H, Chiba H, Yoshihara S, Hakamada K, Sasaki M, Kijima H (2007) Decreased expression of claudin-1 correlates with recurrence status in breast cancer. Int J Mol Med 20:139–143

Mullin JM (1997) Potential interplay between luminal growth factors and increased tight junction permeability in epithelial carcinogenesis. J Exp Zool 279:484–489

Myal Y, Leygue E, Blanchard AA (2010) Claudin 1 in breast tumorigenesis: revelation of a possible novel "claudin high" subset of breast cancers. J Biomed Biotechnol 2010:956897

Naik UP, Naik MU, Eckfeld K, Martin-DeLeon P, Spychala J (2001) Characterization and chromosomal localization of JAM-1, a platelet receptor for a stimulatory monoclonal antibody. J Cell Sci 114:539–547

Naik MU, Naik TU, Suckow AT, Duncan MK, Naik UP (2008) Attenuation of junctional adhesion molecule-A is a contributing factor for breast cancer cell invasion. Cancer Res 68:2194–2203

Neville MC, Daniel CW (1987) The mammary gland: development, regulation, and function. Plenum Press, New York

Nguyen DA, Neville MC (1998) Tight junction regulation in the mammary gland. J Mammary Gland Biol Neoplasia 3:233–246

Nielsen TO, Hsu FD, Jensen K, Cheang M, Karaca G, Hu Z, Hernandez-Boussard T, Livasy C, Cowan D, Dressler L, Akslen LA, Ragaz J, Gown AM, Gilks CB, van de RM, Perou CM (2004) Immunohistochemical and clinical characterization of the basal-like subtype of invasive breast carcinoma. Clin Cancer Res 10:5367–5374

O'Brien LE, Zegers MM, Mostov KE (2002) Opinion: building epithelial architecture: insights from three-dimensional culture models. Nat Rev Mol Cell Biol 3:531–537

Oku N, Sasabe E, Ueta E, Yamamoto T, Osaki T (2006) Tight junction protein claudin-1 enhances the invasive activity of oral squamous cell carcinoma cells by promoting cleavage of laminin-5 gamma2 chain via matrix metalloproteinase (MMP)-2 and membrane-type MMP-1. Cancer Res 66:5251–5257

Osanai M, Murata M, Nishikiori N, Chiba H, Kojima T, Sawada N (2006) Epigenetic silencing of occludin promotes tumorigenic and metastatic properties of cancer cells via modulations of unique sets of apoptosis-associated genes. Cancer Res 66:9125–9133

Osanai M, Murata M, Nishikiori N, Chiba H, Kojima T, Sawada N (2007a) Occludin-mediated premature senescence is a fail-safe mechanism against tumorigenesis in breast carcinoma cells. Cancer Sci 98:1027–1034

Osanai M, Murata M, Chiba H, Kojima T, Sawada N (2007b) Epigenetic silencing of claudin-6 promotes anchorage-independent growth of breast carcinoma cells. Cancer Sci 98:1557–1562

Ostermann G, Weber KS, Zernecke A, Schroder A, Weber C (2002) JAM-1 is a ligand of the beta(2) integrin LFA-1 involved in transendothelial migration of leukocytes. Nat Immunol 3:151–158

Paris L, Bazzoni G (2008) The protein interaction network of the epithelial junctional complex: a system-level analysis. Mol Biol Cell 19:5409–5421

Perou CM (2010) Molecular stratification of triple-negative breast cancers. Oncologist 15(Suppl 5):39–48

Perou CM, Sorlie T, Eisen MB, van de RM, Jeffrey SS, Rees CA, Pollack JR, Ross DT, Johnsen H, Akslen LA, Fluge O, Pergamenschikov A, Williams C, Zhu SX, Lonning PE, Borresen-Dale AL, Brown PO, Botstein D (2000) Molecular portraits of human breast tumours. Nature 406:747–752

Polak-Charcon S, Shoham J, Ben-Shaul Y (1980) Tight junctions in epithelial cells of human fetal hindgut, normal colon, and colon adenocarcinoma. J Natl Cancer Inst 65:53–62

Polyak K (2007) Breast cancer: origins and evolution. J Clin Invest 117:3155–3163

Prat A, Perou CM (2009) Mammary development meets cancer genomics. Nat Med 15:842–844

Raleigh DR, Boe DM, Yu D, Weber CR, Marchiando AM, Bradford EM, Wang Y, Wu L, Schneeberger EE, Shen L, Turner JR (2011) Occludin S408 phosphorylation regulates tight junction protein interactions and barrier function. J Cell Biol 193:565–582

Ren J, Hamada J, Takeichi N, Fujikawa S, Kobayashi H (1990) Ultrastructural differences in junctional intercellular communication between highly and weakly metastatic clones derived from rat mammary carcinoma. Cancer Res 50:358–362

Resnick MB, Konkin T, Routhier J, Sabo E, Pricolo VE (2005) Claudin-1 is a strong prognostic indicator in stage II colonic cancer: a tissue microarray study. Mod Pathol 18:511–518

Ruffer C, Gerke V (2004) The C-terminal cytoplasmic tail of claudins 1 and 5 but not its PDZ-binding motif is required for apical localization at epithelial and endothelial tight junctions. Eur J Cell Biol 83:135–144

Saitou M, Fujimoto K, Doi Y, Itoh M, Fujimoto T, Furuse M, Takano H, Noda T, Tsukita S (1998) Occludin-deficient embryonic stem cells can differentiate into polarized epithelial cells bearing tight junctions. J Cell Biol 141:397–408

Sallee JL, Burridge K (2009) Density-enhanced phosphatase 1 regulates phosphorylation of tight junction proteins and enhances barrier function of epithelial cells. J Biol Chem 284: 14997–15006

Santin AD, Cane S, Bellone S, Palmieri M, Siegel ER, Thomas M, Roman JJ, Burnett A, Cannon MJ, Pecorelli S (2005) Treatment of chemotherapy-resistant human ovarian cancer xenografts in C.B-17/SCID mice by intraperitoneal administration of Clostridium perfringens enterotoxin. Cancer Res 65:4334–4342

Satoh H, Zhong Y, Isomura H, Saitoh M, Enomoto K, Sawada N, Mori M (1996) Localization of 7H6 tight junction-associated antigen along the cell border of vascular endothelial cells correlates with paracellular barrier function against ions, large molecules, and cancer cells. Exp Cell Res 222:269–274

Sauer T, Pedersen MK, Ebeltoft K, Naess O (2005) Reduced expression of Claudin-7 in fine needle aspirates from breast carcinomas correlate with grading and metastatic disease. Cytopathology 16:193–198

Sobel G, Nemeth J, Kiss A, Lotz G, Szabo I, Udvarhelyi N, Schaff Z, Paska C (2006) Claudin 1 differentiates endometrioid and serous papillary endometrial adenocarcinoma. Gynecol Oncol 103:591–598

Soini Y (2005) Expression of claudins 1, 2, 3, 4, 5 and 7 in various types of tumours. Histopathology 46:551–560

Soler AP, Miller RD, Laughlin KV, Carp NZ, Klurfeld DM, Mullin JM (1999) Increased tight junctional permeability is associated with the development of colon cancer. Carcinogenesis 20:1425–1431

Sorlie T, Tibshirani R, Parker J, Hastie T, Marron JS, Nobel A, Deng S, Johnsen H, Pesich R, Geisler S, Demeter J, Perou CM, Lonning PE, Brown PO, Borresen-Dale AL, Botstein D (2003) Repeated observation of breast tumor subtypes in independent gene expression data sets. Proc Natl Acad Sci USA 100:8418–8423

Stelwagen K, van Espen DC, Verkerk GA, McFadden HA, Farr VC (1998) Elevated plasma cortisol reduces permeability of mammary tight junctions in the lactating bovine mammary epithelium. J Endocrinol 159:173–178

Stelwagen K, McFadden HA, Demmer J (1999) Prolactin, alone or in combination with glucocorticoids, enhances tight junction formation and expression of the tight junction protein occludin in mammary cells. Mol Cell Endocrinol 156:55–61

Stevenson BR, Keon BH (1998) The tight junction: morphology to molecules. Annu Rev Cell Dev Biol 14:89–109

Swisshelm K, Machl A, Planitzer S, Robertson R, Kubbies M, Hosier S (1999) SEMP1, a senescence-associated cDNA isolated from human mammary epithelial cells, is a member of an epithelial membrane protein superfamily. Gene 226:285–295

Swisshelm K, Macek R, Kubbies M (2005) Role of claudins in tumorigenesis. Adv Drug Deliv Rev 57:919–928

Takai E, Tan X, Tamori Y, Hirota M, Egami H, Ogawa M (2005) Correlation of translocation of tight junction protein Zonula occludens-1 and activation of epidermal growth factor receptor in the regulation of invasion of pancreatic cancer cells. Int J Oncol 27:645–651

Tan X, Egami H, Ishikawa S, Nakagawa M, Ishiko T, Kamohara H, Hirota M, Ogawa M (2004) Relationship between activation of epidermal growth factor receptor and cell dissociation in pancreatic cancer. Int J Oncol 25:1303–1309

Tanaka M, Kamata R, Sakai R (2005) Phosphorylation of ephrin-B1 via the interaction with claudin following cell-cell contact formation. EMBO J 24:3700–3711

Tay PN, LanY, Leung CH, Hooi SC (2008) Genes associated with metastasis and epithelial-mesenchymal transition (EMT)-like phenotype in human colon cancer cells. Palladin in colon cancer metastasis. AACR. http://www.aacr.org/home/scientists/meetings--workshops/aacr-annual-meeting-2013/previous-annual-meetings/annual-meeting-2008/abstracts.aspx

Tobioka H, Isomura H, Kokai Y, Tokunaga Y, Yamaguchi J, Sawada N (2004) Occludin expression decreases with the progression of human endometrial carcinoma. Hum Pathol 35:159–164

Tokes AM, Kulka J, Paku S, Szik A, Paska C, Novak PK, Szilak L, Kiss A, Bogi K, Schaff Z (2005) Claudin-1, -3 and -4 proteins and mRNA expression in benign and malignant breast lesions: a research study. Breast Cancer Res 7:R296–R305

Trimboli AJ, Fukino K, de BA, Wei G, Shen L, Tanner SM, Creasap N, Rosol TJ, Robinson ML, Eng C, Ostrowski MC, Leone G (2008) Direct evidence for epithelial-mesenchymal transitions in breast cancer. Cancer Res 68:937–945

Tsukita S, Furuse M, Itoh M (2001) Multifunctional strands in tight junctions. Nat Rev Mol Cell Biol 2:285–293

Valles AM, Thiery JP, Boyer B (1994) In vitro studies of epithelium-to-mesenchymal transitions. In: Celis J (ed) Cell biology: a laboratory handbook. Academic, New York, pp 232–242

Vincent-Salomon A, Thiery JP (2003) Host microenvironment in breast cancer development: epithelial-mesenchymal transition in breast cancer development. Breast Cancer Res 5:101–106

Yang J, Mani SA, Donaher JL, Ramaswamy S, Itzykson RA, Come C, Savagner P, Gitelman I, Richardson A, Weinberg RA (2004) Twist, a master regulator of morphogenesis, plays an essential role in tumor metastasis. Cell 117:927–939

Yu K, Lee CH, Tan PH, Tan P (2004) Conservation of breast cancer molecular subtypes and transcriptional patterns of tumor progression across distinct ethnic populations. Clin Cancer Res 10:5508–5517

Yu KD, Shen ZZ, Shao ZM (2009) The immunohistochemically "ER-negative, PR-negative, HER2-negative, CK5/6-negative, and HER1-negative" subgroup is not a surrogate for the normal-like subtype in breast cancer. Breast Cancer Res Treat 118(3):661–663

Zahraoui A, Louvard D, Galli T (2000) Tight junction, a platform for trafficking and signaling protein complexes. J Cell Biol 151:F31–F36

Chapter 9
Regulation of Tight Junctions for Therapeutic Advantages

Lorenza González-Mariscal, Mónica Díaz-Coránguez, and Miguel Quirós

Abstract Tight junctions (TJs) mediate cell-cell adhesion between epithelial cells, maintain cell polarity and regulate the paracellular transit of ions and molecules. Epithelial cells in order to become cancerous disassemble their TJs. Although many TJ proteins are down-regulated during cell transformation, others are in contrast overexpressed. The case is particularly relevant for claudins, where not only each tissue, but each tumor within a given organ expresses a specific set of claudins. In this chapter we will explain how the expression of TJ proteins in cancerous cells, has been used as a molecular tool to identify tumors and predict patient survival rates. In addition we will describe novel strategies that target TJ proteins for cancer treatment and to enhance the delivery of therapeutic drugs through the paracellular pathway of tumors.

Keywords Tight junctions • Cancer • Claudins • Paracellular pathway • Epithelia

9.1 Introduction

Tight junctions (TJs), located at the uppermost portion of the lateral membrane of epithelial cells, regulate the permeability of ions and molecules through the paracellular pathway, located between neighboring cells, and maintain a polarized distribution of lipids and proteins between the apical and basolateral plasma membranes (Gonzalez-Mariscal et al. 2003). Besides these canonical functions of barrier and fence, TJs have in more recent times, been recognized as the frontline structure that cells have to overcome to become cancerous and to metastasize.

These authors contributed equally to this manuscript

L. González-Mariscal (✉) • M. Díaz-Coránguez • M. Quirós
Department of Physiology, Biophysics and Neuroscience, Center for Research and Advanced Studies (Cinvestav), Ave IPN 2508, México D.F 07360, Mexico
e-mail: lorenza@fisio.cinvestav.mx

T.A. Martin and W.G. Jiang (eds.), *Tight Junctions in Cancer Metastasis*, Cancer Metastasis - Biology and Treatment 19, DOI 10.1007/978-94-007-6028-8_9,

Table 9.1 Expression of tight junction proteins in cancer

TJ protein / Carcinoma	Claudins 1	2	3	4	5	6	7	8	9	10	11	12	16	18	23	Occ	Tric	MARVEL	JAM	ZO-1	ZO-2	ZO-3	Par-3	MUPP1	Cin	AF6
Biliary tract																										
Bladder																										
Breast																										
Cervical																										
Colorectal																										
Endometrial																										
Epidermoid																										
Esophageal																										
Gastric																										
Head and neck																										
Intestinal																										
Liver																										
Lung																										
Melanoma																										
Meningioma																										
Mesothelioma																										
Nasopharyngeal																										
Oral																										
Ovarian																										
Pancreatic																										
Pleura																										
Prostate																										
Renal																										
Testis																										
Thyroid																										
Tongue																										
Uterus																										

downregulation

upregulation

up and down regulation

Table 9.2 Expression profile of TJ proteins in a variety of tumors of different organs

a

Breast		
TJ protein \ Tumor	Her2 overexpressing (ER–, PR–,Her2+)	Basal-like (ER–, PR–, Her2–)
ZO-1	+++	
ZO-2	+++	
PAR3	+++	
Cl 1		+++
Cl 4		+++
Cl 16	+++	+++

b

Lung					
TJ protein \ Tumor	NSCC			SCC	Met
	SQ	AC	LCC		
Occ	–		–	–	
ZO-1	+	+			
Cingulin	–				
Cl 1	+	–	–		+
Cl 2	+	+			+
Cl 3	–	+			+
Cl 4	+	+	+		
Cl 5	–	+			–
Cl 7	+	+	–		+

AC adenocarcinoma, *LCC* large cell carcinoma, *Met* metastasis, *NSCC* non-small cell carcinoma, *SCC* small cell carcinoma, *SQ* aquamous cell carcinoma

c

Pancreas							
TJ protein \ Tumor	IPMN	IIPMN	Endocrine	Exocrine			
				AC	SPT	PB	ACC
Cl 1			–	+			
Cl 2			–	+	+		
Cl 3			+		–		
Cl 4	+	++	–	+	–		
Cl 5					+		
Cl 7			+	+		+	+

AC ductal adenocarcinoma, *ACC* undifferentiated carcinoma with osteoclastic-like giant cells and acinar cell carcinomas, *IPMN* intraductal papillary mucinous neoplasm, *IIPMN* intestinal type IPMN, *PB* pancreatoblastoma, *SPT* solid pseudopapillary tumor

d

Stomach					
TJ protein \ Tumor	IM	Displasia	IAC	Difusse	PDAC
Cl 2	–	++	+++		
Cl 4				++	++
Cl 7	++	++	++	+	

IAC intestinal adenocarcinoma, *IM* intestinal metaplasia, *PDAC* poor differentiated adenocarcinoma

(continued)

Table 9.2 (continued)

e

Endometrium								
Tumor / TJ protein	Type I	Type II						
		SAC	Hyperplasia			EC	CCEC	EEC
			Simple	Complex	Atypical			
Cl 1	+	+++						
Cl 2	+++	+						
Cl 3		+++	+	+	+++	+++	+++	++
Cl 4		+++	+	+	+++	+++	+++	++

CCEC clear cell endometrial cancer, *EC* endometrioid carcinomas, *EEC* endometrioid endometrial cancer, *SAC* serouspapillary adenocarcinomas

f

Ovary			
Tumor / TJ protein	EOC	Serous	Adenoma
Cl 3	+++	++	++
Cl 4	+++	++	++
Cl 7	+++	++	++

EOC epitelial ovarian carcinoma

g

Liver and billiary tract				
Tumor / TJ protein	HCC		Cholangiocarcinoma	Gall bladder
	Conventional	Fibrolamelar		
Occ	–		+ (↓)	+ (↓)
Tric	–			
ZO-1	–		+ (↓)	+ (↓)
Cl 1	+ (↓)	+		
Cl 2		+++		
Cl 3	–	–	+++	
Cl 4	–	–	+++	+++
Cl 5	–	++	–	
Cl 7	–	–	+++	
Cl 10				+ (↓)

HCC hepatocellular cell carcinoma, () in comparison with normal tissue

h

Kidney				
Tumor / TJ protein	CCC	Papillary	Chromophobe	Oncocytoma
Cl 1	–	+		
Cl 7	+	+++	+++	++

CCC clear cell carcinoma

i

Bladder and upper urinary tract.					
Tumor / TJ protein	UP	PUNLMP	LG-UCC	HG-UCC	IUP
Cl 1	+	+	+		+++
Cl 4			+++	+	

HG-UCC invasive high grade urothelial cell carcinoma, *IUP* inverted UP, *LG-UCC* low-grade urothelial cell carcinomas, *PUNLMP* papillary urothelial neoplasms of low malignana potential, *UP* urothelial papilloma

(continued)

Table 9.2 (continued)

j

Esophageal						
Tumor / TJ protein	SCC	AC	Barret's esophagus	LGD	HGD	Met
Cl 1	+++					
Cl 2		+++	+			
Cl 3		+++	++		+++	+++
Cl 4		+++	+++		+++	+++
Cl 7	–	++	+++	+++	++	

AC adenocarcinoma, *HGD* high-grade dysplasia, *LGD* low-grade dysplasia, *Met* metastasis, *SCC* squamous cell carcinoma

k

Oral cavity and minor salivary glands			
Tumor / TJ protein	SCC	Low-MEC	High-MEC
Occ	–		
Cl 1	+++	+++	
Cl 3		+	+++
Cl 4	++		
Cl 5	+		
Cl 7	+++		

MEC mucoepidermoid carcinoma, *SCC* squamous cell carcinoma

l

Thyroid						
Tumor / TJ protein	Follicular	Papillary	PMC	Medullary	PDC	UDC
Cl 1	++	++	++	+	++	+
Cl 4	++	++	++	++	++	+
Cl 7	+	++	++	+	+	+

PDC poorly differentiated carcinoma, *PMC* papillary microcarcinoma, *UDC* undifferentiated carcinoma

m

Pediatric CNS				
Tumor / TJ protein	Teratoid/ Rhabdoid	Germinomas	Choroid plexus	Wilms
Cl6	+++	++	++	++

n

Mesenchymal tissue of the gastrointestinal tract					
Tumor / TJ protein	GIST	Angiosarcoma	Hemangiomas	Leiomyoma	Leiomyosarcoma
Cl 1					+
Cl 2	+	+	+	+	+
Cl 3	+				+
Cl 4	+				+
Cl 5	+	+			+
Cl 7	+				+

GIST gastrointesinal stromal tumor

(continued)

Table 9.2 (continued)

o

Brain				
Tumor / TJ protein	Meningioma	Fibrous	Hemangioperycitoma	Vestibular schwannoma
Cl 1	+	–	–	–

p

Sarcoma					
Tumor / TJ protein	AS	Bifasic synovial	Kaposi	LGFM	Perineurioma
Cl 1				+	+
Cl 5	+	+	+		

AS angiosarcoma, *LGFM* low-grade fibromyxoid

EGFR+) and Normal like (cells are somewhat similar to normal epithelia but yet display putative initiating stem cell phenotype). The luminal A subtype which is sensitive to hormones has the most favorable outcome, whereas those not sensitive to hormones are more aggressive and have fewer therapeutic options.

TJ proteins that have been identified as new biomarkers predictive of poor prognosis in breast cancer patients are:

9.2.1.1 Peripheral TJ Proteins with PDZ Domains: ZO-1, ZO-2, MUPP1 and PAR3

In ductal carcinomas, immunohistochemical studies have shown that as tumors become more undifferentiated, the expression of ZO-1 and E-cadherin decreases (Hoover et al. 1998), and in a similar fashion, PCR analysis demonstrated that the expression of ZO-1 and MUPP1 is significantly lower in samples derived from patients with metastatic disease than in those remaining disease free (Martin et al. 2004). Another study detected by PCR and immunocytochemistry, a decrease in ZO-1 and ZO-2 in invasive breast carcinoma samples in comparison with normal breast tissue (Tokes et al. 2012). Although these result indicate that loss of ZO-1, ZO-2 an MUPP1 is associated with poor patient prognosis, the subtype of tumor should also be taken into account, as ZO-1, ZO-2 and PAR3 genes are unexpectedly upregulated in Her2 overexpressing tumors (Tokes et al. 2012) (Fig. 9.1a and Table 9.2A).

9.2.1.2 Integral TJ Proteins: Occludin, JAM-A and Claudins

Occludin

PCR analysis and immunohistochemical staining show loss of occludin expression in human breast cancer tissue and correlation of this loss with a worse patient prognosis. In accordance, ZO-2 silencing increases the invasiveness of breast cancer cell lines (Martin et al. 2010) (Fig. 9.1b).

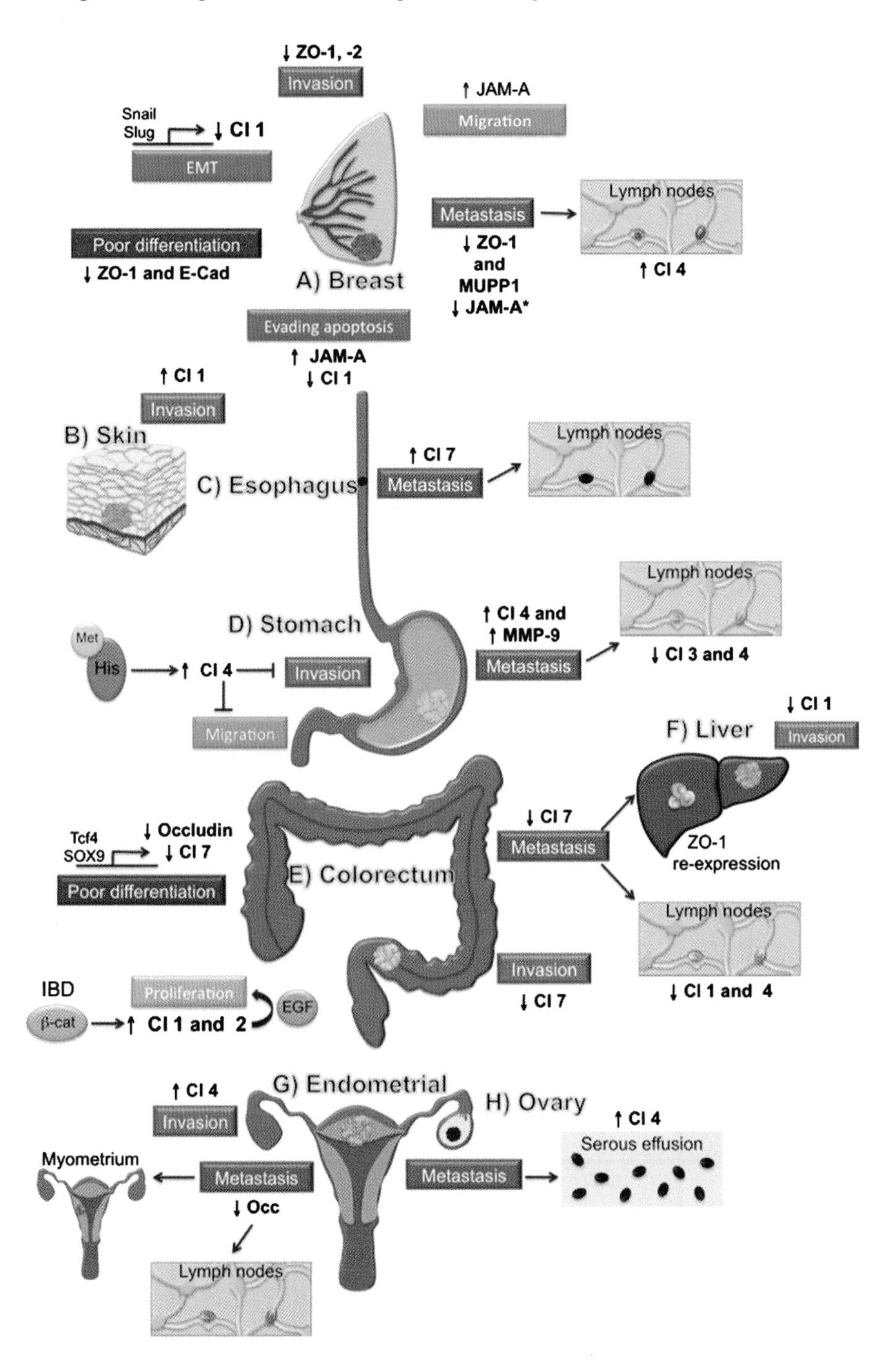

Fig. 9.1 The expression of certain TJ proteins is indicative of the differentiation state of the tumor and of its capacity to proliferate, invade and metastasize

Surprisingly, it has also been observed that the frequency of breast tumors, which are positive to claudins 1 and 4, is significantly higher in ER− than in ER+ tumors. Furthermore, claudins 1 and 4 positivity is associated with the basal like subtype of breast cancers, which display the worst prognosis and reduced patient survival (Blanchard et al. 2009; Kulka et al. 2009). Therefore the expression of claudins 1 and 4 could be employed in the future as a diagnostic tool for basal-like ER− breast cancer subset. With regards to claudin 4, it has additionally been observed that its increased expression in breast cancer and lymph node metastasis is associated with poor prognosis and high tumor grade (Lanigan et al. 2009) (Szasz et al. 2011) (Fig. 9.1a and Table 9.2A).

It is difficult to understand how a breast tumor can pass from a low to a high claudin-1 expression status, and how claudin-1 can switch from having a tumor suppressor role to become instead a tumor enhancer. In accordance with the nonlinear model of breast cancer, it has been speculated that within a breast tumor some cells do not loose claudin-1 and are therefore predetermined to become ER- basal-like breast cancers. Following instead the linear model of breast cancer it is proposed that as cell progress from ER+ to ER−, a different set of transcription factors is expressed that allow the reexpression of claudin-1. The difference in behavior of claudin-1 might depend on the localization on the protein. Thus, when claudin-1 is not present at the TJ and accumulates in the cytosol or is randomly distributed in the plasma membrane, it could exert a tumor-facilitating role by promoting for instance the activation of metalloproteinases (Oku et al. 2006).

9.2.2 Colorectal Cancer

Colorectal cancer (CRC) is the third most common cancer in men and the second in women. Although it is highly preventable through early detection by colonoscopy screening, around 608,00 deaths related to CRC still occur each year worldwide (Ferlay 2008). The adenomatous polyposis coli (APC)/β-catenin pathway has long been recognized as the seminal route that regulates CRC, and in recent times it has become clear that several TJ proteins are indeed among the main targets of this pathway.

9.2.2.1 Peripheral TJ Proteins: ZO-1 and Cingulin

In a study done with samples of stage II colon cancer only 44% of the tumors exhibited normal to elevated expression levels of ZO-1. In addition, the low expression of ZO-1 was associated with a higher tumor grade (Resnick et al. 2005). It is noteworthy that although ZO-1 expression is reduced in primary CRC it is re-expressed in liver metastasized CRC (Kaihara et al. 2003) (Fig. 9.1e).

Instead, the expression of cingulin in primary and metastatic CRC is maintained or upregulated (Citi et al. 1991).

9.2.2.2 Integral TJ Proteins: Occludin and Claudins

Occludin

A decreased expression of occludin in poorly differentiated CRC has been found (Kimura et al. 1997) and a strong inverse trend has been detected between levels of occludin expression and tumor grade (Resnick et al. 2005) (Fig. 9.1e).

Claudins

In CRC an elevated expression of claudins 1 (Grone et al. 2007; Kinugasa et al. 2010), 3 and 4 is observed (Mees et al. 2009; de Oliveira et al. 2005). The increased expression of claudin-1 is accompanied by the cytoplasmic (Grone et al. 2007) and nuclear (Dhawan et al. 2005) accumulation of the protein and the loss of cell polarity.

Claudin-1 up-regulation results from the activation of the β-catenin pathway as claudin-1 transcription is positively regulated by two Tcf4 binding elements in the promoter region of the gene, thus explaining why claudin-1 expression decreases upon β-catenin reduction due to APC transfer into APC deficient colon cancer cells (Miwa et al. 2001). In addition, in CRC the expression of claudin-1 is inversely regulated by the expression of the tumor suppressor Smad4, a signal transduction component of the transforming growth factor-β (TGF-β) family of cytokines. Interestingly, Smad-4 inhibition of claudin-1 expression is independent of TGF-β signaling as TGF-β receptor inhibitor LY364947 does not prevent Smad4 suppression of claudin-1 expression (Shiou et al. 2007). In contrast, the overexpression of Smad7 enhances the formation of colon metastasis in the liver and correlates with an increased expression of claudins 1 and 4 (Halder et al. 2008).

Despite all this evidence, the expression of claudins 1 and 4 has also been found to be significantly lower in cases of lymph node metastasis and in mucinous carcinoma cases (Ersoz et al. 2011), and in another study of CRC, high claudin-1 expression was surprisingly found to lead to a better prognosis (Nakagawa et al. 2011). Moreover, a tissue microarray study, demonstrated by immunohistochemistry analysis that a low expression level of claudin-1 protein is a strong predictor of recurrence and poor patient survival in stage II colon cancer (Resnick et al. 2005). These discrepancies could be due to the employment of different cohorts in the various studies, and could pose a situation similar to the one described above for breast cancer, where in a particular cancer subset, the expression of a claudin like number 1, exerts the opposite prognostic value than in the rest of the carcinomas.

In CRC, there is no agreement on the expression of claudin-2 and its potential as a diagnostic tool. Some studies show decrease immunoreactivity to claudin-2 in tumors cells in comparison to normal tissue, and no correlation of this expression to tumor differentiation or invasive pattern (Hahn-Stromberg et al. 2009). Other reports indicate that CRCs exhibit a high level of claudin-2 mRNA but display claudin-2 protein expression in only a 25.3% of CRCs (Aung et al. 2006), without an obvious correlation between claudin-2 expression and clinicopathological parameters. More recently a cohort with 309 patient CRC samples demonstrated an increase in claudin-2 expression that correlates with cancer progression and a similar increase in claudin-2 expression was found in inflammatory bowel disease (IBD) associated CRC. The role of claudin-2 in CRC appears to be mediated by EGFR as silencing of claudin-2 expression in a colon cancer cell line, reverts the EGF induced increase in cell proliferation (Dhawan et al. 2011).

In addition, it has been observed that in IBD, β-catenin transcriptional activity is elevated, contributing to an increased expression of claudins 1 and 2 that could be involved in the early stages of transformation of IBD into CRC (Weber et al. 2008).

Even though claudin-4 is overexpressed in CRC (de Oliveira et al. 2005; Mees et al. 2009), a correlation has been found between lower levels of claudin-4 expression and more advanced stages of CRC (Resnick et al. 2005).

With regards to claudin-7, a defective distribution of the protein along the crypt surface axis has been observed in CRC. In healthy human colon, the expression of claudin-7 is strong in differentiated postmitotic cells of the surface and low in the crypts due to the inhibitory effect on gene expression exerted by the transcription factors Tcf-4 and Sox-9. Instead, in CRC claudin-7 is overexpressed throughout (Darido et al. 2008). An RT-PCR study revealed increased expression of claudin-7 in CRC (Hewitt et al. 2006) and another analysis showed the upregulation of a complex of proteins containing claudin-7, EpCAM, tetraspanin C0-029 and CD44v6, which is inversely correlated with disease-free survival (Kuhn et al. 2007). In contrast other studies showed that in CRC the reduced expression of claudin-7 mRNA is an early event in the pathology (Bornholdt et al. 2011) and that it correlates with venous invasion and liver metastasis (Oshima et al. 2008). It should be mentioned that hypermethylation of claudin-7 promoter has been detected in 20% of CRC with low claudin-7 expression (Nakayama et al. 2008) (Fig. 9.1e).

Recently it has been found that the expression of claudin-18 correlates with poor survival in patients with CRC and is associated with the gastric phenotype (Matsuda et al. 2010).

9.2.3 Lung Cancer

Lung cancer is the most common cause of death from cancer worldwide (Ferlay 2008). Smoking is responsible for over 90% of cases of lung cancer, but exposure to asbestos, radon or heavy metals also contributes. There are two main groups of lung cancer: small (SCC) and non-small (NSCC) cell carcinomas. The latter has two main histological types, the squamous cell carcinoma characterized for cell keratinization, and adenocarcinoma distinguished for the production of mucin, and a rare type named large cell carcinoma.

The expression pattern of TJ proteins varies in dissimilar ways among the different types of lung cancer and could therefore give rise to a differential diagnosis of lung tumors (Tobioka et al. 2004b; Paschoud et al. 2007; Moldvay et al. 2007;

Sormunen et al. 2007; Merikallio et al. 2011). Thus while occludin remains undetectable in SCC, squamous and large cell carcinomas, ZO-1 and claudins 2, 4 and 7 are positive in squamous cell carcinomas and adenocarcinoma. Instead, in squamous cell carcinoma, claudin-1 is positive, and cingulin and claudins 3 and 5 expression is negative, whereas in adenocarcinomas claudin-1 is negative and claudins 3 and 5 are positive. Large cell carcinomas are positive for claudin-4 and negative for claudins 1 and 7. Pulmonary epithelial metastasis are positive for claudins 1, 2, 3 and 7 and negative for claudin-5 (Table 9.2B).

With regards to prognosis, patients with lung adenocarcinoma that have low claudin-1 expression have a shorter survival rate (Chao et al. 2009)

9.2.4 Prostate Cancer

Prostate cancer is the second most frequent cancer of men and the sixth leading cause of death from cancer in men (Ferlay 2008)

In prostate adenocarcinomas, immunohistochemical analysis revealed the down regulation of claudin-4, accompanied by the over-expression of claudins 2, 3 and 5 in comparison to benign prostatic hyperplasia (Coutinho-Camillo et al. 2011). With regards to prognosis, one study with 141 sections from prostatic adenocarcinomas showed that the decreased expression of claudin-1 correlates with high tumor grade and biochemical disease recurrence, whereas a decrease in claudin-7 correlates with high tumor grade. In contrast expression of claudin-3 correlates with advanced stage tumors and recurrence and expression of claudin-4 correlates with advanced stage (Sheehan et al. 2007). Another study revealed that claudin-3 is overexpressed in prostatic adenocarcinoma and metastatic prostatic adenocarcinoma in comparison to benign prostatic hyperplasia and normal tissue adjacent to prostatic adenocarcinoma, thus suggesting that claudin-3 could be a biomarker for primary and metastatic prostate cancer (Bartholow et al. 2011).

9.2.5 Pancreatic Cancer

Pancreatic cancer is the fourth most common cause of cancer related deaths worldwide. This cancer has a poor prognosis. Thus the 5-year survival rate is 6% and the median survival for locally advanced and metastatic disease that together represent over 80% of individuals, is around 10 and 6 months respectively.

The vast majority of cancers of the pancreas are exocrine in type, and 90% of these are adenocarcinomas, meaning that they arise in the cells lining the ducts of the pancreas. The second most common type of exocrine pancreas cancer is mucinous whose prognosis is slightly better. Rarer types of exocrine pancreatic cancers are solid pseudopapillary tumor (SPT), pancreatoblastoma (PB), undifferentiated carcinoma with osteoclastic-like giant cells and acinar cell carcinomas (ACC). Only a minority of pancreatic tumors, start in the endocrine pancreas, where insulin and other hormones are made. Most of these tumors are benign and are named after the hormone they produce.

Pancreatic adenocarcinoma arises from progressive tissue changes including a lesion named intraductal papillary mucinous neoplasm (IPMN) where an increased expression of claudin-4 has been observed in comparison with adenomas. In addition, a higher expression of claudin-4 is observed in the intestinal type IPMN than in the non-intestinal type, suggesting that claudin-4 expression is associated with neoplastic progression especially with the pathway that leads to intestinal differentiation (Tsutsumi et al. 2011).

Pancreatic endocrine tumors and exocrine cancers exhibit a different expression profile for claudins. Thus endocrine tumors have a high expression of claudins 3 and 7 but are negative for claudins 1, 2 and 4, whereas ductal adenocarcinomas are positive for claudins 1, 4 and 7 and half of them express claudin-2 (Borka et al. 2007). SPT have intense membrane claudin-5 and cytoplasmic claudin-2 staining and lack claudins 3 and 4. Conversely, ACC and PB show a strong membrane expression of claudin-7 like endocrine tumors (Comper et al. 2009) (Table 9.2C).

9.2.6 Stomach Cancer

Stomach cancer is the fourth most common malignancy in the world and the second leading cause of cancer death in both sexes worldwide (Ferlay 2008). More than 95% of stomach cancers are adenocarcinomas, meaning that they originate from the glandular epithelium of the gastric mucosa. There are two types of gastric adenocarcinomas: intestinal and diffuse type. In the former the cells form tubular structures and the tumors are pluristratified and have multiple lumens. Instead in the diffuse type adenocarcinomas, the cells are discohesive and produce large pools of mucus.

In gastric tissue the overexpression of claudins 2 and 7 correlates with gastric carcinogenesis. Thus the percentage of immunochemical claudin-2 positive cases found in a study done with human biopsies were 0% for chronic gastritis, 0% for intestinal type metaplasia, 36% for dysplasia and 74% for gastric intestinal type adenocarcinoma (Song et al. 2008). Claudin-7 was significantly more expressed in intestinal metaplasia, dysplasia and cancer than in normal gastric epithelium. In addition, claudin-7 was more often expressed in intestinal type gastric cancer than in the diffuse type (Park et al. 2007; Johnson et al. 2005).

Instead, in gastric cancer, the expression of claudins 1, 3 and 4 diminishes (Jung et al. 2011) and claudin-11 is silenced via hypermethylation of its promoter region (Agarwal et al. 2009). Claudin-1 transcription is upregulated by the gastric tumor suppressor and transcription factor RUNX3, and in gastric cells from RUNX3 knockout mice the tumorigenicity diminishes by restoring claudin-1 expression (Chang et al. 2010). The lack or diminished expression of claudins 3 and 4 is associated to worse malignancy grades, positive lymphatic invasion and potential metastatic ability (Matsuda et al. 2007). Decreased claudin-4 expression is usually found in diffuse type tumors and poorly differentiated adenocarcinomas (Lee et al. 2005b; Kuo et al. 2006; Lee et al. 2008) and is associated with significantly decreased overall survival rate (Jung et al. 2011; Ohtani et al. 2009). Claudin-4 upregulation in gastric cancer cells suppresses cell invasion and migration, and loss of repressive histone methylations and gain of active histone modifications are associated with claudin-4 overexpression in gastric cells. Therefore the epigenetic de-repression of claudin-4 has been proposed as a potential strategy for treating gastric cancer (Kwon et al. 2011). This proposal however should be taken with caution as claudin-4 expression is associated with matrix metalloproteinase 9 (MMP-9) expression, and has therefore been proposed to play a role in favoring intestinal type gastric cancer generation of distal metastasis (Lee et al. 2008) (Fig. 9.1d and Table 9.2D).

9.2.7 Gynecological Cancer: Cervical, Endometrial and Ovarian Cancers

9.2.7.1 Cervical Cancer

Cervical cancer is the third most common cancer in women and the seventh overall, with a mortality incidence ratio of 52% (Ferlay 2008). Although Pap test has dramatically decreased the incidence and mortality rates of cervical cancer (Eddy 1990), the morphological basis of this test makes it less than ideal for the management of ambiguous or low-grade lesions. Therefore an attempt has been made to understand the molecular basis of cervical tumorigenesis and this has led to discover TJ markers for cervical cancer screening.

cDNA microarrays revealed that claudin-1 is among the top 62 genes that are overexpressed in cervical tumors and high grade intraepithelial lesions (HSIL) (Chen et al. 2003), and immunohistochemistry in cervical tissue samples has indicated increased expression of claudins 1, 2, 4 and 7 in the early phase of carcinogenesis in intraepithelial lesions, which decreases during progression to invasive disease (Sobel et al. 2005; Lee et al. 2005a). Expression of claudin-1 appears to be the strongest in premalignant states and may thus serve as a good diagnostic marker.

9.2.7.2 Endometrial Carcinoma

Endometrial cancer is the sixth most common cancer in women worldwide. Body fatness increases the risk of endometrial cancer and abdominal fatness is probably a cause (Ferlay 2008). Two distinct types of tumors develop in the endometrium. Type I that account for more than 80% of endometrial carcinomas, have an endometrioid histology, favorable prognosis, are estrogen responsive and are preceded by endometrial hyperplasia. Type II have non-endometrioid histology, are serous, clear cell, mucinous, squamous, transitional cell, mesonephric and undifferentiated. These tumors have a poor prognosis, are non-estrogen related and lack association with endometrial hyperplasia.

grow to form a tumor. There are five main types of RCCs: (1) clear cell which is the most common and characterized for the very pale or clear appearance of the cells; (2) papillar, where the cancerous tissue forms finger-like projections or papillae; (3) chromophobe where the cells are pale but large; (4) collecting duct, where the cells form irregular tubes, and (5) unclassified, where the appearance of the cells does not fit any of the other categories.

Renal tumor classification is important because the different subtypes are associated to distinct clinical behavior. For example, clear cell RCC has a worse prognosis compared to papillary or chromophobe cell RCCs, and more importantly, the relatively common benign tumor named renal oncocytoma, cannot be easily distinguished from chromophobe RCC based on histological features. This highlights the importance that molecular markers have for the accurate classification of renal tumors and the subsequent management of patients and prognosis.

Claudins are a useful tool for renal tumor profiling (Schuetz et al. 2005). Thus, clear cell carcinomas are mostly negative for claudin-1, while papillary tumors are positive. Interestingly, claudin-1 positivity in clear cell carcinomas is a prognostic of shortened patient survival (Fritzsche et al. 2008). With regards to claudin-7 expression, chromophobe RCC is positive in 100% cases, papillary RCC in 90%, clear cell RCC in 7% and the benign tumor oncocytoma in 45% (Li et al. 2008). In the nephron, claudins 1 (Reyes et al. 2002) and 7 (Gonzalez-Mariscal et al. 2006) are markers of the distal nephron, therefore these results in addition suggest that chromophobe and papillary RCCs are more closely related to the distal than to the proximal nephron, whereas clear cell RCCs probably derive from the proximal nephron.

The simultaneous analysis of claudins 7 and 8 also serves to identify renal tumors. Thus, membranous claudin-7 and negative claudin-8 is seen in chromophobe RCC and not in oncocytomas, while negative claudin-7 and cytoplasmic claudin-8 is observed in oncocytomas and not in chromophobe RCC (Osunkoya et al. 2009) (Table 9.2H).

9.2.11 Bladder Cancer and Urothelial Carcinoma of the Upper Urinary Tract

Bladder cancer is the 11th most common cancer worldwide (Ferlay 2008). Cigarette smoking and workplace exposure to industrial chemicals involved in rubber, leather, textiles and paint products are linked with bladder cancer (Society 2012). Ninety-five percent of bladder cancer is urothelial/transitional cell carcinoma that is classified as low and high grade. The latter has a worse outcome and exhibits cells that look more abnormal and grow more rapidly. The same type of cancers that grow in the bladder can also grow in other places of the urinary tract such as the lining of the kidney, the ureters and the urethra (Kaufman et al. 2009).

The expression profile of claudins changes during bladder transformation. Thus urothelial papillomas (UPs), papillary urothelial neoplasms of low malignant potential (PUNLMPs) and low-grade urothelial cell carcinomas (LG-UCCs) have significantly decreased claudin-1 expression in comparison to inverted UPs (IUPs). This observation is important because IUPs almost always have a benign behavior, and diagnosing IUP is difficult because it shares morphological characteristics with urothelial carcinoma with inverted growth pattern and might mimic LG-UCC as well. With regards to prognosis it has been observed that LG-UCCs that overexpress claudin-4 are associated to a shorter recurrence free survival, whereas-PUNLMPs that overexpress claudin-1 have longer recurrence free survival (Szekely et al. 2011).

It is noteworthy that the overexpression of claudin-4 in LG-UCCs, is followed in invasive high grade urothelial tumors by a strong downregulation. This effect appears to be due to hypermethylation of a CpG island present within the coding sequence of claudin-4 gene as treatment with a methyl-transferase inhibitor restores the expression of claudin-4 in primary cultures derived from high grade human bladder tumors (Boireau et al. 2007).

Claudin-11 is strongly expressed in the non-invasive bladder cancer cell line RT112/84 in comparison to invasive cell lines T24/83 and EJ138, and in conformity, benign bladder tissue has a more intense staining of claudin-11 than bladder malignant tissue. Interestingly, forced expression of claudin-11 reduces invasion in T24/83 cells (Awsare et al. 2011).

In urothelial carcinoma of the upper urinary tract the increased expression of claudins 3 and 4 correlates with advanced stage and increased expression of claudin-3 is associated with poor overall survival (Nakanishi et al. 2008) (Table 9.2I).

9.2.12 Esophageal Cancer

Esophageal cancer is the eighth most common cancer worldwide and the sixth most common cause of death from cancer (Ferlay 2008). There are two main types of esophageal cancer named according to how cells look: (1) adenocarcinoma that is usually found in the lower part of the esophagus near the stomach, and (2) squamous cell carcinoma, found in the upper part of the esophagus. Around the world the later is the most common type of esophageal cancer (Enzinger and Mayer 2003). Barrett's esophagus is the replacement of the normal squamous epithelium of the lower esophagus by metaplastic columnar epithelium. It appears as a consequence of prolonged gastro-esophageal reflux disease. It is a precancerous condition that predisposes to the development of esophageal adenocarcinoma (Flejou and Svrcek 2007).

Immunohistochemistry analysis of esophagus samples (Gyorffy et al. 2005) (Montgomery et al. 2006) reveals that claudin-1 expression is increased in squamous cell carcinoma in comparison to normal squamous epithelium, that claudin-2 increases in adenocarcinoma in comparison to Barrett's esophagus, and that claudins 3 and 4 are elevated in Barrett's esophagus, high grade dysplasia, adenocarcinoma and metastases in comparison to normal esophagus squamous tissue and in follicular, medullary and poorly differentiated carcinomas. A correlation was found between loss of claudin-1 expression and worse disease free survival (Tzelepi et al. 2008). Other study has revealed that in papillary carcinomas and papillary microcarcinomas primary tumors and lymph node metastases, the expression of claudin-1 is conspicuous (Nemeth et al. 2010) (Table 9.2L).

9.2.17 Malignant Teratoid/Rabdoid and Other Pediatric Tumors of the Central Nervous System

Atypical teratoid/rhabdoid tumors (AT/RTs) are highly aggressive malignant central nervous system (CNS) tumors of children usually less than 2 years old. Their prognosis is very poor having a median survival ranging from 6 to 17 months (Parwani et al. 2005).

Differential diagnosis of AT/RTs from other CNS tumors has been very difficult based on histopathology alone. Fortunately it was observed that in AT/RTs chromosome 22 is lost and in consequence INI1 gene located in chromosome 22 is missing. Since then, the loss of protein INI1 has been used for the diagnosis of AT/RTs. However, there is a need of more molecular markers for AT/RTs as some tumors that display classic AT/RT features retain INI1 staining. Thus, a gene microarray was utilized to compare AT/RTs to other CNS tumors. This array followed my immunohistochemical staining revealed that claudin-6 is the most distinctive molecular marker of AT/RTs in comparison to other brain tumors (Birks et al. 2010). Further studies have however revealed that claudin-6 is also expressed in germinomas and choroid plexus tumors and in around half of Wilms tumors and CNS primitive neuroectodermal tumors (Sullivan et al. 2012).

The presence of claudin-6 in pediatric tumors is worth noting, as the expression of this claudin appears to be regulated developmentally since it is absent or barely detectable in adult tissue (Morita et al. 1999; Morita et al. 2002) and is expressed in embryonic and fetal life (Turksen and Troy 2001, 2002) (Table 9.2M).

9.3 Mesenchymal Neoplasias of the Gastrointestinal Tract

Submucosal tumors (SMTs) are mesenchymal tumors of diverse origin. In the gastrointestinal tract among the benign submucosal tumors are leiomyomas, and the vascular hemangiomas, while the malignant submucosal tumors include leiomyosarcomas, gastrointestinal stromal tumors and the vascular angiosarcomas. Distinguishing between benign and malignant SMTs is complex. Therefore it is of great value to identify novel proteins like claudins that can help in their classification. Accordingly, mesenchymal neoplasias express the following claudins: (1) gastrointestinal stromal tumors (GIST) exhibit claudins 2, 3, 4, 5 and 7; (2) angiosarcomas have claudins 2 and 5; (3) hemangiomas and leiomyomas only express claudin-2, and (4) Leiomyosarcomas exhibit claudins 1–5 and 7 (Gyorffy 2009) (Table 9.2N).

9.3.1 Other Tumors

9.3.1.1 Meningiomas

Meningiomas are tumors arising in the meninges and attached to the dura. Meningiomas are graded as malignancy grade I, II and III, although the vast majority of them are malignancy grade I. Distinguishing between meningiomas and other tumors arising in the meninges is difficult and important because other tumors have a much higher propensity to recur and metastasize than meningiomas. A study showed that 53% of meningiomas are immunoreactive for claudin-1 whereas, fibrous tumors of the meninges, meningeal hemangiopericytomas and vestibular schwannomas are not (Hahn et al. 2006) (Table 9.2O).

9.3.1.2 Sarcomas

A sarcoma is a cancer that arises from transformed cells of mesenchymal origin. The presence of claudin-5 has been explored in a variety of human mesenchymal tumors finding a high expression in angiosarcomas, biphasic synovial sarcomas and Kaposi sarcomas (Miettinen et al. 2011).

Low-grade fibromyxoid sarcoma is a soft tissue sarcoma with recurrent and low metastatic potential that appears morphological similar to perineurioma, a peripheral nerve sheath neoplasm which is usually benign. Distinguishing between the two entities is important for appropriate treatment. However when the expression of claudin-1 was tested it was found expressed in both low-grade fibromyxoid sarcoma and perineurioma (Thway et al. 2009) (Table 9.2P).

9.4 TJ Proteins as Targets for Therapeutic Intervention

During EMT, several TJ proteins, particularly claudins are overexpressed. Therefore strategies have been designed to eliminate cancerous cells by targeting the overexpressed TJ proteins either with antibodies or specific molecules like *Clostridium perfringens* enterotoxin (CPE). The tumor suppressor activity of certain TJ proteins like occludin, has on the other hand led to the idea of reactivating the expression of these proteins in order to induce mesenchymal to epithelia transformation (MET). In addition, molecules and peptides that open the TJ can be used to enhance the paracellular delivery of chemotherapeutic agents to cancerous tissue (Fig. 9.3).

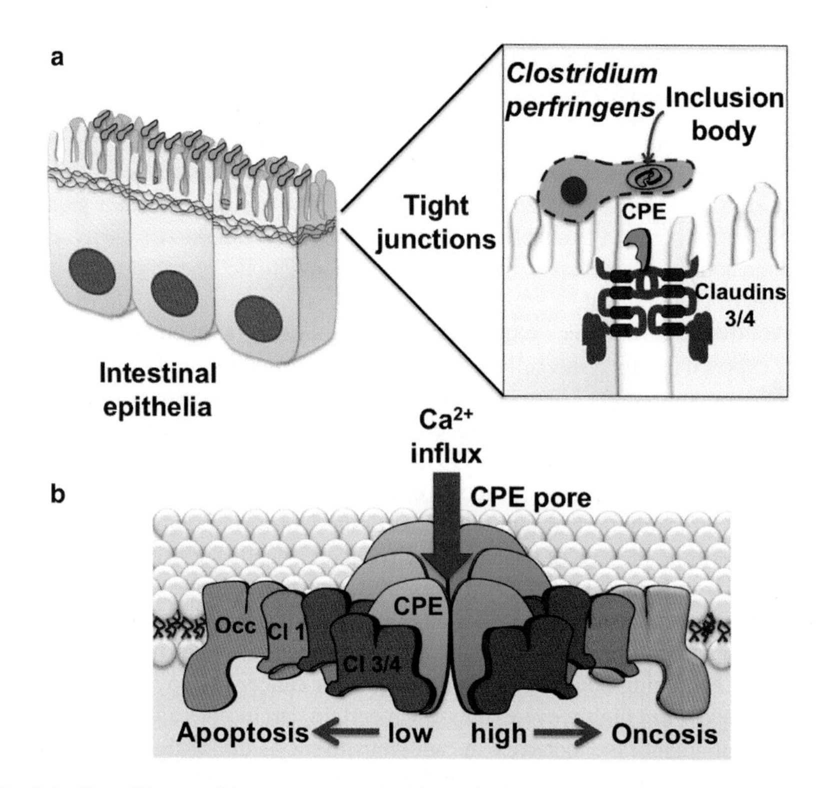

Fig. 9.4 *Clostridium perfringens* enterotoxin (*CPE*) binds to claudins in the plasma membrane and triggers cell death. (**a**) CPE accumulates at intracellular inclusion bodies of bacteria. Upon sporulation, the toxin is released at the intestinal lumen, where it binds to claudins present in the plasma membrane of epithelial cells. (**b**) At the plasma membrane hexamers of CPE form a pore that allows the influx of calcium triggering apoptosis or oncosis in a dose dependent manner. The final CPE complex contains claudins that bind to CPE like claudins 3 and 4, and non CPE binding claudins like claudin-1, as well as occludin

9.4.3.1 The Second Extracellular Loop of Claudins as CPE Receptor

The cellular receptor for CPE was cloned in 1997 from a cDNA expression library of CPE sensitive Vero cells that was transfected into non CPE sensitive L fibroblasts, that were then analyzed by flow cytometric analysis with a biotinylated CPE carboxyl terminal fragment as a probe (Katahira et al. 1997). The CPE receptor (CPR) thus identified was surprisingly found to be similar to the rat withdrawal apoptosis protein (RVP1), whose gene is expressed in the ventral prostate after androgen withdrawal or castration, and the mouse oligodendrocyte specific protein. Since the role of both of these proteins at the time remained unknown, the physiological function of CPR remained unsolved. It was not until 1999, when Dr. Tsukita's group in Japan identified claudins 1 and 2 from an isolated junction fraction from chicken liver (Furuse et al. 1998) and discovered the existence of a claudin gene family, that it became clear that CPR is in fact a claudin, thereafter identified as number 3 (Morita et al. 1999). This identification explains why CPE was able to exert a deleterious effect on HeLa and Vero cells in the presence but not in the absence of Ca^{2+} (Matsuda and Sugimoto 1979). Thus, claudin-3 like all TJ proteins is present at the plasma membrane only when the cells are incubated in Ca^{2+} containing media, since the cation is needed for the formation of E-cadherin based adherens junction that assemble before TJs are formed (Gonzalez-Mariscal et al. 1985).

CPE does not bind to all claudins. Thus, claudins 3, 4, 6, 7, and 9 are high affinity receptors for CPE, while claudins, 1, 2, 5, and 14 are low affinity receptors and claudins 10–13 and 15–24 do not bind CPE at all (Winkler et al. 2009). Interestingly, all the high CPE affinity claudins belong to the group of classic claudins (Claudins 1–10, 14, 15, 17 and 19) that are characterized for having high homology among each other and for sharing a common helix-turn-helix structure in their second extracellular loop (ECL2) (Krause et al. 2008). The latter is the segment in claudins involved in binding CPE (Fujita et al. 2000; Robertson et al. 2010). By substitution mapping, each position in a peptide homologous to the ECL2 of claudin-3 has been substituted by every other amino acid except cysteine, revealing that motif 148NPLVP152 is essential for CPE binding to claudin-3. In agreement, it has been observed that all claudins that bind to CPE have the consensus sequence NP(L/M)(V/L/T)(P/A) (Winkler et al. 2009).

9.4.3.2 CPE Binding Domain for Claudins

Interestingly, the analysis of the residues in CPE involved in binding to claudins reveals that the 30 carboxyl terminal amino acids of CPE are essential for binding to claudins, and that Y306 is the pivotal residue, although Y310, Y312 (Harada et al. 2007) and L315 (Takahashi et al. 2008) also contribute to the interaction. Since Y306K, but not Y306F and Y306W mutants interfered with claudin binding (Ebihara et al. 2007), it appears that the presence of aromatic rings and hydrophobic properties are essential for interaction of CPE with claudins. In fact, the crystal structure of $CPE_{194-319}$ has revealed a nine strand β sandwich similar to the receptor binding domain of other toxins of spore forming bacteria like Colg of *Clostridium histolyticum* and Cry of *Bacillus thuringiensis*, where the critical binding residues form a hydrophobic pit (Van Itallie et al. 2008).

9.4.3.3 CPE/Claudin Interaction

The ECL2 of claudin-5 is essential for the *trans* interaction of the protein and a similar behavior for all classic claudins has been suggested (Piontek et al. 2008). By systematic single mutations that were then transfected to TJ free cells, it became evident that aromatic residues F147, Y148 and Y158 of claudin-5, form an aromatic core among the ECL2 of neighboring claudins that is essential for the *trans* interaction.

Since *trans* interaction among classic claudins and CPE binding is mediated by aromatic residues, it was logical to propose that the association between claudins and CPE relied on the interaction among these critical aromatic residues. Therefore it was a surprise to find that the removal of the aromatic residues in the ECL2 of claudins 2, 3 and 5 increased CPE binding capacity, and that the aromatic residues in the ECL2 do not fit sterically with the aromatic residues in CPE when analyzed in a molecular model (Winkler et al. 2009). In fact, it has been observed that the carboxyl terminal region of CPE binds claudins in the plasma membrane that have not yet become incorporated to TJs (Winkler et al. 2009) and that the residues in claudin that interact with CPE are those present in the C-terminal flank of the loop and in the following trans membrane 4 region, and not the aromatic residues in the flanking helices (Winkler et al. 2009).

9.4.3.4 CPE Activity on Epithelial Cells

After CPE binds to claudins present on the plasma membrane of epithelial cells, a small SDS sensitive complex of around 90 kDa is formed (Wieckowski et al. 1994). This complex contains both claudins that act as CPE receptors as well as other associated claudins, like nonclassical ones, that cannot bind CPE by themselves. Gel shift analysis has revealed that this complex then oligomerizes into an SDS resistant one of around 450 kDa, named the CPE hexamer-1 (CH-1) for containing claudins and six CPE molecules per complex (Robertson et al. 2007). The CH-1 complex assembles in the plasma membrane and forms with amino acids 81–106 of CPE a pore (Smedley et al. 2007) that results in calcium influx that at low doses triggers death by apoptosis, whereas at high concentration kills the cells by oncosis, a process characterized, as the greek word *ónkos* indicates, by cell swelling (Chakrabarti and McClane 2005; Chakrabarti et al. 2003). During this process a second larger complex of around 650 kDa named CH-2 is formed and in contrast to the other, this one also contains occludin (Singh et al. 2000). The formation of CH-2 complex completely disrupts TJs and leads to the internalization of occludin and claudins (Fig. 9.4b).

9.4.3.5 CPE for Cancer Therapy

Claudins 3 and 4 are abundantly expressed in breast (Kominsky et al. 2004), uterine (Santin et al. 2007b), prostate (Maeda et al. 2012; Long et al. 2001), pancreatic (Michl et al. 2001) and ovarian (Gao et al. 2011) cancers. As claudins 3 and 4 are specific receptors for CPE, the idea emerged of using this toxin as a novel claudin directed therapy for malignant tumors. However since claudins 3 and 4 are also expressed in other normal tissues such as urinary bladder (Acharya et al. 2004), kidney (Reyes et al. 2002), lung (Wang et al. 2003), cornea (Yoshida et al. 2009), cochlea (Florian et al. 2003) and the gastrointestinal tract (Rahner et al. 2001), systemic administration is not a viable approach, and instead the local administration of CPE has been employed to treat these cancers.

In the case of breast cancer, CPE induces cytolysis exclusively in the cell lines expressing claudins 3 and 4, and when applied intratumorally, CPE significantly reduces tumor volume. Necrotic reactions are also observed when CPE is administered to freshly resected primary breast carcinoma samples (Kominsky et al. 2004). Furthermore, in a murine model of breast cancer metastasis to the brain, intracranial CPE treatment inhibited tumor growth, without generating any toxicity, apparently due to the fact that claudins 3 and 4 are absent in normal brain tissue (Kominsky et al. 2007). In a similar manner, bone metastatic androgen independent prostate adenocarcinoma cells are lysed upon treatment with CPE (Long et al. 2001).

With regards to ovarian cancer, in vitro exposure of primary tumors to CPE, produces cell death, regardless of cell's resistance to chemotherapy (Santin et al. 2005), and CPE treatment sensitizes ovarian cancer cells to clinically relevant chemotherapeutic drugs like taxol and carboplatin (Gao et al. 2011). Furthermore, in an *in vivo* mouse xenograft model with chemotherapy resistant human ovarian cancer cells OVA-1, CPE produced a significant inhibitory effect on tumor progression (Santin et al. 2005).

In uterine serous papillary carcinoma (USPC) cell lines, CPE administration produces a dose dependent cytotoxic effect and in vivo intratumoral injections induce the disappearance of large subcutaneous mouse xenografts (Santin et al. 2007b). Similarly in pancreatic cancer, intratumoral injections of CPE to xenografts of the Panc-1 cell line, produce areas of tumor cell necrosis and a significant reduction in the tumor growth rate (Michl et al. 2001). PKC α inhibitors enhance the sensitivity of pancreatic cancer cells to CPE. This occurs as PKC α activation, frequently observed in pancreatic cancer, delocalizes claudin-4 from the cell borders (Kyuno et al. 2011).

Recently, CPE has been successfully employed as suicide gene therapy for the selective treatment of cancer cells and tumors that overexpress claudins 3 and 4 (Walther et al. 2012). Thus the expression of vectors bearing wild type CPE cDNA or translation optimized CPE, produces a rapid and intense cytotoxic effect in tumor cell lines that express claudins 3 and 4, heightened by CPE release by bystander effect. Additionally, the nonviral intratumoral in vivo gene transfer of translation optimized CPE leads to reduced tumor growth.

9.4.3.6 Use of the C-Terminal Domain of CPE for Cancer Treatment

The C-terminal half of CPE (C-CPE) has claudin binding activity and is non cytoxic, as it lacks the regions of the amino domain necessary for insertion into the plasma membrane and pore formation (Hanna et al. 1989). These properties allow C-CPE to be used in cancer for two specific purposes: As a specific carrier for cytotoxic drugs to cancerous cells that overexpress claudins 3 and 4, and as a tool to enhance the delivery of chemotherapeutic agents through the paracellular pathway.

C-CPE/Toxins Fusion Proteins

Conventional cancer therapies have their limitations. For example, chemotherapy targets uncontrolled cell division that constitutes the hallmark of transformed cells. However due to tumor progression, the response to chemotherapy becomes modest as the growth rate of the cells in a tumor decreases. Radiation therapy also has its drawbacks as it affects not only the tumor but also the surrounding normal tissue, and surgery is applicable only at the initial stages of solid tumors. This situation shows the need of novel therapies such as fusion proteins/immunotoxins, which are targeted therapies aimed at cancerous tissues and not to normal tissue. These fusion proteins contain toxins selected for their action as protein synthesis inhibitors, derived from plants (e.g. ricin, gelonin), bacteria (diphtheria toxin, pseudomonas exotoxin), fungus (e.g. restrictocin) or animals (e.g. hemolytic toxin from sea anemones), which are fused genetically or by chemical ligation to factors with the ability to target the conjugated toxin to cancerous cells. Cancer cells have specific antigens overexpressed in their surface, therefore cytokines, antibodies and growth factors that selectively associate to them can be used as targeting molecules [for review see (Potala et al. 2008)].

C-CPE has been employed as a target factor for tumor therapy. Thus, C-CPE fused to tumor necrosis factor ($CPE_{290-319}$-TNF) (Yuan et al. 2009) exerts a cytotoxic effect in human ovarian cancer cells 2008, and fused to the *diphtheria* toxin fragment A (DTA-C-CPE) and to the protein synthesis inhibitory factor derived from *Pseudomonas aeuruginosa* exotoxin (C-CPE-PSIF), produces cytotoxicity only in claudin-4 expressing L fibroblasts (Kakutani et al. 2010b; Saeki et al. 2009). In addition, the intratumoral injection of C-CPE-PSIF triggers a significant reduction in the tumor growth rate of mice xenografts of the breast cancer cell line 4T1 (Saeki et al. 2009).

C-CPE as a Drug Delivery Enhancer

The capacity of C-CPE to remove specific claudins from the plasma membrane (Sonoda et al. 1999) can be used to modulate the degree of paracellular sealing of epithelial cells, and in cancer it could be used to increase the delivery of chemotherapeutic agents to the interior of tumors expressing claudins 3 and 4. As proof of principle of this approach, it has been demonstrated that molecules that specifically interact with the second extracellular loop of claudins such as homologous peptides, open the paracellular pathway and have been hence claimed as novel strategies to enhance drug delivery (Quay et al. 2007).

To examine if C-CPE could be used to enhance drug delivery, the absorption of dextran was assessed in an in situ loop assay in rat jejunum. C-CPE proved to be 400 fold more potent at enhancing dextran absorption than the clinically used enhancer capric acid and this effect is not accompanied by injury of the intestinal mucosa as assessed by leakage of lactose dehydrogenase (Kondoh et al. 2005). This observation lead the way for studying if by targeting claudin-4 in the nasal epithelium, C-CPE could be employed for mucosal vaccination. While parenteral vaccination only induces systemic immune responses and requires injections, which are invasive and painful for patients, mucosal vaccines elicit both systemic and mucosal responses, and provide humoral and cell-mediated protection not only at the original mucosal surface, but throughout the body. In addition, mucosal vaccination is needle-free. However, the problem with mucosal vaccination is that in order to elicit an immune response, antigens need to reach the mucosa associated lymphoid tissues (MALT) that lie below the epithelia, usually monolayers that cover the cavities and ducts of the body. In order to deliver antigens to MALT, several strategies have been attempted, including the use of microparticles, liposomes, saponins and chitosans (Ryan et al. 2001), and more recently of C-CPE. Nasal vaccination has been a prime objective, as it produces greater systemic responses than other mucosal immunizations and induces mucosal IgA antibody responses in the salivary glands, respiratory, genital and intestinal tracts (Staats et al. 1997; Imaoka et al. 1998). Thus when a fusion protein of the model antigen ovalbumin (OVA) and C-CPE is used for nasal immunization, it induces the production of OVA-specific IgG serum and nasal, vaginal and fecal IgA, and provokes anti tumor activity in mice inoculated with OVA expressing thymoma cells (Kakutani et al. 2010a; Suzuki et al. 2010).

C-CPE also has the potential to enhance the absorption of biologics. For instance, C-CPE enhances nasal absorption of human parathyroid hormone derivative $hPTH_{1-34}$ and when 10 amino acids of N-terminal of C-CPE are deleted its solubility increases by 30 fold, making it capable of enhancing jejunal and pulmonary absorption of $hPTH_{1-34}$ (Uchida et al. 2010).

In order to optimize C-CPE function as a paracellular drug delivery system, several deletions and mutations have been done in the original C-CPE comprising amino acids 184–319. For example, C-CPE opening of the paracellular pathway is lost by deletion of amino acids 184–219 (Masuyama et al. 2005) or 290–319 (Kondoh et al. 2005), as with the alanine substitution of tyrosines 306, 310, 312, and leucine 315 (Harada et al. 2007; Takahashi et al. 2008). Instead, paracellular delivery is enhanced with the deletion of amino acids 184–204 coupled to the alanine substitution of asparagine 309 and serine 313 (Takahashi et al. 2011) or with a mutant that starts at amino acid 194 and contains the latter alanine substitutions (Suzuki et al. 2012) (Fig. 9.5).

9.4.4 TJ Proteins as Targets for the Enhancement of Drug Delivery

Although chemotherapy is often received by infusion therapy, a rising number of oral chemotherapies are being developed to give cancer patients a less invasive treatment option, to avoid multiple office visits and to give them a sense of control over their own care. However, patients on regimen that combine oral and parenteral therapy find it more convenient to receive their entire regimen by infusion therapy (Weingart et al. 2008). This situation highlights the convenience of having an exclusively oral regime of chemotherapy for cancer patients. The latter has not been easy to develop, as hydrophilic

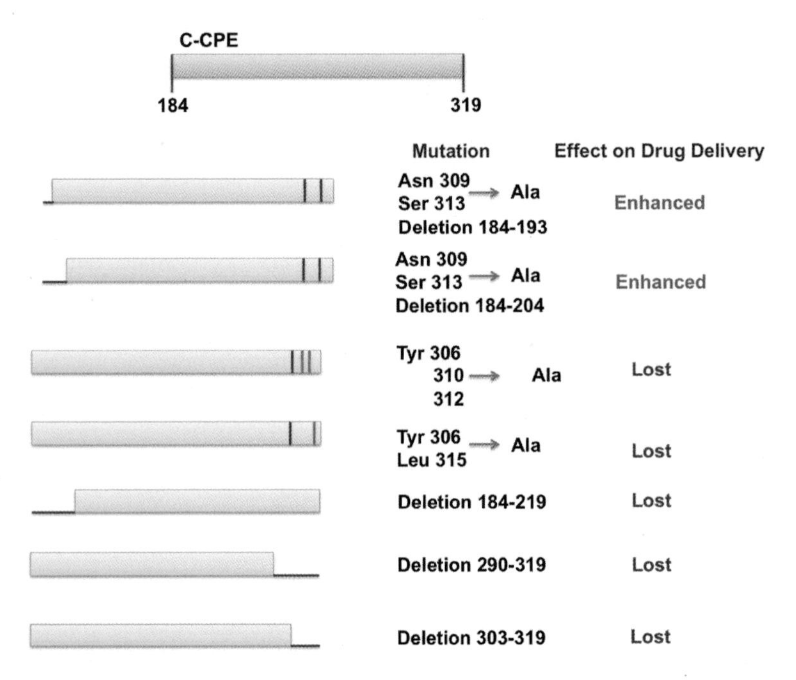

Fig. 9.5 Optimization of C-CPE$_{184-319}$ for paracellular drug delivery. Alanine substitutions at Asn 309 and Ser313 together with deletion of 9–20 residues in the amino terminal segment of C-CPE enhance drug delivery. Instead opening of the paracellular pathway is lost by deletion of 35 or 16 residues from the respective amino or carboxyl terminals of C-CPE, as with the alanine substitution of Tyr 306 coupled with alanine substitutions at Tyr 310 and 312 or Leu 315

drugs cannot cross epithelial membranes and transit through the transcellular route involves transport mechanisms such as channels, carriers and pumps, which are tightly regulated. In addition, drugs that are going to be moved by these transport systems, need to be modified in order to be recognized as transport targets.

The paracellular route instead allows the passage of hydrophilic molecules across epithelia and endothelia. Hence drug delivery strategies have started to focus on ways to modulate TJ sealing. Although no reports have yet appeared indicating the use of the paracellular route for the transit of chemotherapeutic agents through epithelia, here we will give a brief description of the strategies so far developed to enhance drug delivery by targeting TJ proteins.

9.4.4.1 Peptides Homologous to Sequences in the Extracellular Loops of Occludin, Claudins and JAMs

The first indication suggesting that TJ adhesion could be modulated by the addition of extracellular peptides arose after observing that synthetic peptides homologous to the first extracellular loop of occludin inhibited adhesion induced in fibroblasts by exogenous occludin (Van Itallie and Anderson 1997). Since then several synthetic peptides homologous to the first and second loops of occludin (Lacaz-Vieira et al. 1999; Tavelin et al. 2003; Wong and Gumbiner 1997; Nusrat et al. 2005) and claudins (Quay et al. 2007) or that correspond to segments present in the membrane distal Ig loop of the extracellular domain of human JAM-A have been reported to reduce TER and to increase paracellular permeability .

In cell adhesion proteins the minimal sequences required for the establishment of cell contacts are named *c*ell *a*dhesion *r*ecognition (CAR) sequences, and are composed of at least three amino acids. In the case of occludin the sequence LYHY present in the second extracellular loop constitutes a CAR sequence (Blaschuk et al. 2002b) that has been claimed in patents for the purpose of drug delivery (Blaschuk et al. 2001). A similar situation is encountered with JAM-A and claudins where a CAR motif that corresponds to the sequence DPK of JAM-A (Blaschuk et al. 2003) or to the amino terminal end of the first extracellular loop of claudins 1–9 (Blaschuk et al. 2004, 2002a), has been claimed to enhance the passage of drugs through the paracellular pathway.

9.4.4.2 TJ Proteins Silencing with siRNA and Antisense Oligonucleotides

siRNAs and antisense oligonuleotides have been designed to silence the expression of genes for TJ proteins. For the transient opening of TJs in the human respiratory epithelium siRNA for JAM-A, occludin and claudins 1, 3, 4, 9, 12 and 20 have been developed (Cui and Dutzar 2005). Nasal drug delivery appears to be very promising for the use of siRNAs as this route is non-invasive, involves no first pass metabolism, and has the potential for direct delivery to the central nervous system. Antisense ZO-1 siRNAs that have been chemically modified with phosphorothioate linkages, to improve binding and selectivity, have successfully been employed to decrease flow resistance induced by steroids in the human eye Schlemm's canal and trabecular network (Underwood et al. 1999). In addition several patents have been granted claiming the use of antisense oligonucleotides against JAM-C (Dobie 2003) and occludin (O'Mahony 2000) for enhancing drug delivery.

9.4.4.3 Other Protein and Peptides that Open the Paracellular Pathway

ZOT and Zonulin/Prehaptoglobin-2

The zonula occludens toxin (ZOT) originally identified as the factor responsible for residual diarrhea generated with *Vibrio cholera* vaccines of cholera toxin, opens the paracellular pathway by triggering a chain reaction that starts with activation of PKCα, induction of actin polymerization, contraction of the perijunctional actinomiosin ring and the subsequent membrane displacement of TJ proteins that ultimately leads to opening of the paracellular pathway. A 12 kDa carboxyl fragment of ZOT named Delta G retains the ability to increase the paracellular transit of low bioavailability agents, and in this segment a sequence named AT1002 consisting of amino acids FCIGRL constitutes the minimal ZOT receptor binding domain and is capable of enhancing the transport of several therapeutic agents after nasal, intraduodenal and intratracheal administration (Song and Eddington 2012).

Interestingly, an endogenous protein named Zonulin opens the TJ after binding to the same receptor activated by ZOT: the EGF receptor (EGFR) via proteinase activated receptor 2 (PAR_2) activation. Proteomic analysis revealed that zonulin is the precursor for haptoglobulin 2 (Tripathi et al. 2009), a protein that in mature form scavenges free hemoglobin to inhibit its oxidative activity. Although no function had been previously ascribed to zonulin/prehaptoglobin-2, it now constitutes the only known physiological modulator of TJ sealing and as such appears to be up-regulated in autoimmune, inflammatory and neoplastic diseases (Fasano 2011). Thus, in glioma, oral squamous cell carcinoma, lung, pancreatic and hepatocellular cancer zonulin/prehaptoglobin-2 is overexpressed.

VP8

Rotaviruses are a major cause of diarrhea in young children. Rotaviruses use integrins as cell receptors in epithelia cells. However, when rotaviruses reach the intestinal lumen, their integrin receptors located on the basolateral membrane, are hidden below the TJ. This observation led to an effort to explore if rotavirus outer capsid proteins had the capacity to open TJs. Thus it was found that VP8, a protein that arises from the proteolitic cleavage of VP4, a protein that forms spikes that project from the outermost layer of proteins of the rotavirus, opens the TJ in a dose dependent and reversible manner. The capacity of VP8 to open the TJ and enhance drug delivery was demonstrated in vivo, when VP8 allowed insulin administered orally to reduce blood glucose concentration (Nava et al. 2004).

PN159

With the aim of translocating polar compounds such as antisense oligonucleotides and peptides across the plasma membrane in a model independent of endocytic uptake, amphipatic peptides were designed containing a high relative abundance of positively charged amino acids such as lysine or arginine, and non-polar, hydrophobic amino acids (Hallbrink et al. 2001). Unexpectedly, one of these amphipatic peptides named PN159 with sequence KLALKLALKALKAALKLA, decreases TER and increases paracellular permeability. Therefore this peptide has been claimed to enhance drug delivery of therapeutic agents by the paracellular route (Cui et al. 2006)

References

Acharya P, Beckel J, Ruiz WG, Wang E, Rojas R, Birder L, Apodaca G (2004) Distribution of the tight junction proteins ZO-1, occludin, and claudin-4, -8, and -12 in bladder epithelium. Am J Physiol Renal Physiol 287(2):F305–F318. doi:10.1152/ajprenal.00341.2003

Agarwal R, Mori Y, Cheng Y, Jin Z, Olaru AV, Hamilton JP, David S, Selaru FM, Yang J, Abraham JM, Montgomery E, Morin PJ, Meltzer SJ (2009) Silencing of claudin-11 is associated with increased invasiveness of gastric cancer cells. PLoS One 4(11):e8002. doi:10.1371/journal.pone.0008002

Apostolopoulos V, McKenzie IF (1994) Cellular mucins: targets for immunotherapy. Crit Rev Immunol 14(3–4):293–309

Aro K, Rosa LE, Bello IO, Soini Y, Makitie AA, Salo T, Leivo I (2011) Expression pattern of claudins 1 and 3-an auxiliary tool in predicting behavior of mucoepidermoid carcinoma of salivary gland origin. Virchows Arch Int J Pathol 458(3):341–348. doi:10.1007/s00428-010-1026-1

Aung PP, Mitani Y, Sanada Y, Nakayama H, Matsusaki K, Yasui W (2006) Differential expression of claudin-2 in normal human tissues and gastrointestinal carcinomas. Virchows Arch Int J Pathol 448(4):428–434. doi:10.1007/s00428-005-0120-2

Awsare NS, Martin TA, Haynes MD, Matthews PN, Jiang WG (2011) Claudin-11 decreases the invasiveness of bladder cancer cells. Oncol Rep 25(6):1503–1509. doi:10.3892/or.2011.1244

Bartholow TL, Chandran UR, Becich MJ, Parwani AV (2011) Immunohistochemical profiles of claudin-3 in primary and metastatic prostatic adenocarcinoma. Diagn Pathol 6:12. doi:10.1186/1746-1596-6-12

Bartlett JG (2002) Clinical practice. Antibiotic-associated diarrhea. N Engl J Med 346(5):334–339. doi:10.1056/NEJMcp011603

Bello IO, Vilen ST, Niinimaa A, Kantola S, Soini Y, Salo T (2008) Expression of claudins 1, 4, 5, and 7 and occludin, and relationship with prognosis in squamous cell carcinoma of the tongue. Hum Pathol 39(8):1212–1220. doi:10.1016/j.humpath.2007.12.015

Bignotti E, Tassi RA, Calza S, Ravaggi A, Romani C, Rossi E, Falchetti M, Odicino FE, Pecorelli S, Santin AD (2006) Differential gene expression profiles between tumor biopsies and short-term primary cultures of ovarian serous carcinomas: identification of novel molecular biomarkers for early diagnosis and therapy. Gynecol Oncol 103(2):405–416. doi:10.1016/j.ygyno.2006.03.056

Birks DK, Kleinschmidt-DeMasters BK, Donson AM, Barton VN, McNatt SA, Foreman NK, Handler MH (2010) Claudin 6 is a positive marker for atypical teratoid/rhabdoid tumors. Brain Pathol 20(1):140–150. doi:10.1111/j.1750-3639.2008.00255.x

Blanchard AA, Skliris GP, Watson PH, Murphy LC, Penner C, Tomes L, Young TL, Leygue E, Myal Y (2009) Claudins 1, 3, and 4 protein expression in ER negative breast cancer correlates with markers of the basal phenotype. Virchows Arch Int J Pathol 454(6):647–656. doi:10.1007/s00428-009-0770-6

Blaschuk OW, Gour BJ, Symonds JM (2001) Compounds and methods and modulating tissue permeability US Patent US628864B1

Blaschuk OW, Gourd BJ, Symonds JM (2002a) Compounds and methods for modulating claudin-mediated functions US Patent US2002193294

Blaschuk OW, Oshima T, Gour BJ, Symonds JM, Park JH, Kevil CG, Trocha SD, Michaud S, Okayama N, Elrod JW, Alexander JS, Sasaki M (2002b) Identification of an occludin cell adhesion recognition sequence. Inflammation 26(4):193–198

Blaschuk OW, Gourd BJ, Symonds JM (2003) Compounds and methods for modulating junctional adhesion molecule-mediated functions US Patent US2003027761

Blaschuk OW, Gourd BJ, Symonds JM (2004) Compounds and methods for modulating claudin-mediated functions US Patent US20046756356

Boireau S, Buchert M, Samuel MS, Pannequin J, Ryan JL, Choquet A, Chapuis H, Rebillard X, Avances C, Ernst M, Joubert D, Mottet N, Hollande F (2007) DNA-methylation-dependent alterations of claudin-4 expression in human bladder carcinoma. Carcinogenesis 28(2):246–258. doi:10.1093/carcin/bgl120

Borka K, Kaliszky P, Szabo E, Lotz G, Kupcsulik P, Schaff Z, Kiss A (2007) Claudin expression in pancreatic endocrine tumors as compared with ductal adenocarcinomas. Virchows Arch Int J Pathol 450(5):549–557. doi:10.1007/s00428-007-0406-7

Bornholdt J, Friis S, Godiksen S, Poulsen SS, Santoni-Rugiu E, Bisgaard HC, Lothe IM, Ikdahl T, Tveit KM, Johnson E, Kure EH, Vogel LK (2011) The level of claudin-7 is reduced as an early event in colorectal carcinogenesis. BMC Cancer 11:65. doi:10.1186/1471-2407-11-65

Chakrabarti G, McClane BA (2005) The importance of calcium influx, calpain and calmodulin for the activation of CaCo-2 cell death pathways by Clostridium perfringens enterotoxin. Cell Microbiol 7(1):129–146. doi:10.1111/j.1462-5822.2004.00442.x

Chakrabarti G, Zhou X, McClane BA (2003) Death pathways activated in CaCo-2 cells by Clostridium perfringens enterotoxin. Infect Immun 71(8):4260–4270

Chang TL, Ito K, Ko TK, Liu Q, Salto-Tellez M, Yeoh KG, Fukamachi H, Ito Y (2010) Claudin-1 has tumor suppressive activity and is a direct target of RUNX3 in gastric epithelial cells. Gastroenterology 138(1):255–265 e251–253. doi:10.1053/j.gastro.2009.08.044

Chao YC, Pan SH, Yang SC, Yu SL, Che TF, Lin CW, Tsai MS, Chang GC, Wu CH, Wu YY, Lee YC, Hong TM, Yang PC (2009) Claudin-1 is a metastasis suppressor and correlates with clinical outcome in lung adenocarcinoma. Am J Respir Crit Care Med 179(2):123–133. doi:10.1164/rccm.200803-456OC

Chen Y, Miller C, Mosher R, Zhao X, Deeds J, Morrissey M, Bryant B, Yang D, Meyer R, Cronin F, Gostout BS, Smith-McCune K, Schlegel R (2003) Identification of cervical cancer markers by cDNA and tissue microarrays. Cancer Res 63(8):1927–1935

Choi YL, Kim J, Kwon MJ, Choi JS, Kim TJ, Bae DS, Koh SS, In YH, Park YW, Kim SH, Ahn G, Shin YK (2007) Expression profile of tight junction protein claudin 3 and claudin 4 in ovarian serous adenocarcinoma with prognostic correlation. Histol Histopathol 22(11):1185–1195

Chung YE, Park MS, Park YN, Lee HJ, Seok JY, Yu JS, Kim MJ (2009) Hepatocellular carcinoma variants: radiologic-pathologic correlation. AJR Am J Roentgenol 193(1):W7–W13. doi:10.2214/AJR.07.3947

Citi S, Amorosi A, Franconi F, Giotti A, Zampi G (1991) Cingulin, a specific protein component of tight junctions, is expressed in normal and neoplastic human epithelial tissues. Am J Pathol 138(4):781–789

Comper F, Antonello D, Beghelli S, Gobbo S, Montagna L, Pederzoli P, Chilosi M, Scarpa A (2009) Expression pattern of claudins 5 and 7 distinguishes solid-pseudopapillary from pancreatoblastoma, acinar cell and endocrine tumors of the pancreas. Am J Surg Pathol 33(5):768–774. doi:10.1097/PAS.0b013e3181957bc4

Coutinho-Camillo CM, Lourenco SV, da Fonseca FP, Soares FA (2011) Claudin expression is dysregulated in prostate adenocarcinomas but does not correlate with main clinicopathological parameters. Pathology 43(2):143–148. doi:10.1097/PAT.0b013e3283428099

Cui K, Dutzar B (2005) Method for opening thight junctions US Patent US2005129679

Cui K, Chen SC, Houston ME, Quay SC (2006) Tight junction modulator peptide PN159 for enhanced mucosal delivery of therapeutic compounds US Patent US20060062758

Darido C, Buchert M, Pannequin J, Bastide P, Zalzali H, Mantamadiotis T, Bourgaux JF, Garambois V, Jay P, Blache P, Joubert D, Hollande F (2008) Defective claudin-7 regulation by Tcf-4 and Sox-9 disrupts the polarity and increases the tumorigenicity of colorectal cancer cells. Cancer Res 68(11):4258–4268. doi:10.1158/0008-5472.CAN-07-5805

de Oliveira SS, de Oliveira IM, De Souza W, Morgado-Diaz JA (2005) Claudins upregulation in human colorectal cancer. FEBS Lett 579(27):6179–6185. doi:10.1016/j.febslet.2005.09.091

Dhawan P, Singh AB, Deane NG, No Y, Shiou SR, Schmidt C, Neff J, Washington MK, Beauchamp RD (2005) Claudin-1 regulates cellular transformation and metastatic behavior in colon cancer. J Clin Invest 115(7):1765–1776. doi:10.1172/JCI24543

Dhawan P, Ahmad R, Chaturvedi R, Smith JJ, Midha R, Mittal MK, Krishnan M, Chen X, Eschrich S, Yeatman TJ, Harris RC, Washington MK, Wilson KT, Beauchamp RD, Singh AB (2011) Claudin-2 expression increases tumorigenicity of colon cancer cells: role of epidermal growth factor receptor activation. Oncogene 30(29):3234–3247. doi:10.1038/onc.2011.43

Dobie KW (2003) US Patent US2003232034

Ebihara C, Kondoh M, Harada M, Fujii M, Mizuguchi H, Tsunoda S, Horiguchi Y, Yagi K, Watanabe Y (2007) Role of Tyr306 in the C-terminal fragment of Clostridium perfringens enterotoxin for modulation of tight junction. Biochem Pharmacol 73(6):824–830. doi:10.1016/j.bcp. 2006.11.013

Eddy DM (1990) Screening for cervical cancer. Ann Intern Med 113(3):214–226

Enzinger PC, Mayer RJ (2003) Esophageal cancer. N Engl J Med 349(23):2241–2252. doi:10.1056/NEJMra035010
Ersoz S, Mungan S, Cobanoglu U, Turgutalp H, Ozoran Y (2011) Prognostic importance of Claudin-1 and Claudin-4 expression in colon carcinomas. Pathol Res Pract 207(5):285–289. doi:10.1016/j.prp. 2011.01.011
Farquhar MG, Palade GE (1963) Junctional complexes in various epithelia. J Cell Biol 17:375–412
Fasano A (2011) Zonulin and its regulation of intestinal barrier function: the biological door to inflammation, autoimmunity, and cancer. Physiol Rev 91(1):151–175. doi:10.1152/physrev.00003.2008
Ferlay J, Shin HR, Bray F, Forman D, Mathers C, Parkin DM (2010) GLOBOCAN 2008, cancer incidence and mortality worldwide: IARC CancerBase No. 10 [Internet]. Int J Cancer. http://globocan.iarc.fr [cited 11/06/12], 17 June 2010
Flejou JF, Svrcek M (2007) Barrett's oesophagus–a pathologist's view. Histopathology 50(1):3–14. doi:10.1111/j.1365-2559.2006.02569.x
Florian P, Amasheh S, Lessidrensky M, Todt I, Bloedow A, Ernst A, Fromm M, Gitter AH (2003) Claudins in the tight junctions of stria vascularis marginal cells. Biochem Biophys Res Commun 304(1):5–10
Fritzsche FR, Oelrich B, Johannsen M, Kristiansen I, Moch H, Jung K, Kristiansen G (2008) Claudin-1 protein expression is a prognostic marker of patient survival in renal cell carcinomas. Clin Cancer Res Off J Am Assoc Cancer Res 14(21):7035–7042. doi:10.1158/1078-0432.CCR-08-0855
Fujita K, Katahira J, Horiguchi Y, Sonoda N, Furuse M, Tsukita S (2000) Clostridium perfringens enterotoxin binds to the second extracellular loop of claudin-3, a tight junction integral membrane protein. FEBS Lett 476(3):258–261
Furuse M, Fujita K, Hiiragi T, Fujimoto K, Tsukita S (1998) Claudin-1 and -2: novel integral membrane proteins localizing at tight junctions with no sequence similarity to occludin. J Cell Biol 141(7):1539–1550
Furuse M, Sasaki H, Tsukita S (1999) Manner of interaction of heterogeneous claudin species within and between tight junction strands. J Cell Biol 147(4):891–903
Gao Z, Xu X, McClane B, Zeng Q, Litkouhi B, Welch WR, Berkowitz RS, Mok SC, Garner EI (2011) C terminus of Clostridium perfringens enterotoxin downregulates CLDN4 and sensitizes ovarian cancer cells to Taxol and Carboplatin. Clin Cancer Res Off J Am Assoc Cancer Res 17(5):1065–1074. doi:10.1158/1078-0432.CCR-10-1644
Goldenberg MM (1999) Trastuzumab, a recombinant DNA-derived humanized monoclonal antibody, a novel agent for the treatment of metastatic breast cancer. Clin Ther 21(2):309–318. doi:10.1016/S0149-2918(00)88288-0
Gonzalez-Mariscal L, Chavez de Ramirez B, Cereijido M (1985) Tight junction formation in cultured epithelial cells (MDCK). J Membr Biol 86(2):113–125
Gonzalez-Mariscal L, Betanzos A, Nava P, Jaramillo BE (2003) Tight junction proteins. Prog Biophys Mol Biol 81(1):1–44
Gonzalez-Mariscal L, Namorado Mdel C, Martin D, Sierra G, Reyes JL (2006) The tight junction proteins claudin-7 and -8 display a different subcellular localization at Henle's loops and collecting ducts of rabbit kidney. Nephrol Dial Transplant Off Publ Eur Dial Transplant Assoc Eur Ren Assoc 21(9):2391–2398. doi:10.1093/ndt/gfl255
Grone J, Weber B, Staub E, Heinze M, Klaman I, Pilarsky C, Hermann K, Castanos-Velez E, Ropcke S, Mann B, Rosenthal A, Buhr HJ (2007) Differential expression of genes encoding tight junction proteins in colorectal cancer: frequent dysregulation of claudin-1, -8 and -12. Int J Colorectal Dis 22(6):651–659. doi:10.1007/s00384-006-0197-3
Gyorffy H (2009) Study of claudins and prognostic factors in some gastrointestinal diseases. Magy Onkol 53(4):377–383. doi:10.1556/MOnkol.53.2009.4.7
Gyorffy H, Holczbauer A, Nagy P, Szabo Z, Kupcsulik P, Paska C, Papp J, Schaff Z, Kiss A (2005) Claudin expression in Barrett's esophagus and adenocarcinoma. Virchows Arch Int J Pathol 447(6):961–968. doi:10.1007/s00428-005-0045-9
Hahn HP, Bundock EA, Hornick JL (2006) Immunohistochemical staining for claudin-1 can help distinguish meningiomas from histologic mimics. Am J Clin Pathol 125(2):203–208. doi:10.1309/G659-FVVB-MG7U-4RPQ

Hahn-Stromberg V, Edvardsson H, Bodin L, Franzen L (2009) Tumor volume of colon carcinoma is related to the invasive pattern but not to the expression of cell adhesion proteins. APMIS Acta Pathol Microbiol Immunol Scand 117(3):205–211. doi:10.1111/j.1600-0463.2008.00011.x

Halder SK, Rachakonda G, Deane NG, Datta PK (2008) Smad7 induces hepatic metastasis in colorectal cancer. Br J Cancer 99(6):957–965. doi:10.1038/sj.bjc.6604562

Hallbrink M, Floren A, Elmquist A, Pooga M, Bartfai T, Langel U (2001) Cargo delivery kinetics of cell-penetrating peptides. Biochim Biophys Acta 1515(2):101–109

Hanna PC, Wnek AP, McClane BA (1989) Molecular cloning of the 3′ half of the Clostridium perfringens enterotoxin gene and demonstration that this region encodes receptor-binding activity. J Bacteriol 171(12):6815–6820

Harada M, Kondoh M, Ebihara C, Takahashi A, Komiya E, Fujii M, Mizuguchi H, Tsunoda S, Horiguchi Y, Yagi K, Watanabe Y (2007) Role of tyrosine residues in modulation of claudin-4 by the C-terminal fragment of Clostridium perfringens enterotoxin. Biochem Pharmacol 73(2):206–214. doi:10.1016/j.bcp. 2006.10.002

Harris RC (1991) Potential physiologic roles for epidermal growth factor in the kidney. Am J Kidney Dis Off J Natl Kidney Found 17(6):627–630

Hewitt KJ, Agarwal R, Morin PJ (2006) The claudin gene family: expression in normal and neoplastic tissues. BMC Cancer 6:186. doi:10.1186/1471-2407-6-186

Higashi Y, Suzuki S, Sakaguchi T, Nakamura T, Baba S, Reinecker HC, Nakamura S, Konno H (2007) Loss of claudin-1 expression correlates with malignancy of hepatocellular carcinoma. J Surg Res 139(1):68–76. doi:10.1016/j.jss.2006.08.038

Hoevel T, Macek R, Swisshelm K, Kubbies M (2004) Reexpression of the TJ protein CLDN1 induces apoptosis in breast tumor spheroids. Int J Cancer J Int Cancer 108(3):374–383. doi:10.1002/ijc.11571

Hoover KB, Liao SY, Bryant PJ (1998) Loss of the tight junction MAGUK ZO-1 in breast cancer: relationship to glandular differentiation and loss of heterozygosity. Am J Pathol 153(6):1767–1773. doi:10.1016/S0002-9440(10)65691-X

Imaoka K, Miller CJ, Kubota M, McChesney MB, Lohman B, Yamamoto M, Fujihashi K, Someya K, Honda M, McGhee JR, Kiyono H (1998) Nasal immunization of nonhuman primates with simian immunodeficiency virus p55gag and cholera toxin adjuvant induces Th1/Th2 help for virus-specific immune responses in reproductive tissues. J Immunol 161(11):5952–5958

Jemal A, Siegel R, Ward E, Hao Y, Xu J, Thun MJ (2009) Cancer statistics, 2009. CA Cancer J Clin 59(4):225–249. doi:10.3322/caac.20006

Jerant AF, Johnson JT, Sheridan CD, Caffrey TJ (2000) Early detection and treatment of skin cancer. Am Fam Physician 62(2):357–368, 375–356, 381–352

Johnson RG, Sheridan JD (1971) Junctions between cancer cells in culture: ultrastructure and permeability. Science 174(4010):717–719

Johnson AH, Frierson HF, Zaika A, Powell SM, Roche J, Crowe S, Moskaluk CA, El-Rifai W (2005) Expression of tight-junction protein claudin-7 is an early event in gastric tumorigenesis. Am J Pathol 167(2):577–584. doi:10.1016/S0002-9440(10)62999-9

Jung H, Jun KH, Jung JH, Chin HM, Park WB (2011) The expression of claudin-1, claudin-2, claudin-3, and claudin-4 in gastric cancer tissue. J Surg Res 167(2):e185–e191. doi:10.1016/j.jss.2010.02.010

Kaihara T, Kawamata H, Imura J, Fujii S, Kitajima K, Omotehara F, Maeda N, Nakamura T, Fujimori T (2003) Redifferentiation and ZO-1 reexpression in liver-metastasized colorectal cancer: possible association with epidermal growth factor receptor-induced tyrosine phosphorylation of ZO-1. Cancer Sci 94(2):166–172

Kakutani H, Kondoh M, Fukasaka M, Suzuki H, Hamakubo T, Yagi K (2010a) Mucosal vaccination using claudin-4-targeting. Biomaterials 31(20):5463–5471. doi:10.1016/j.biomaterials.2010.03.047

Kakutani H, Kondoh M, Saeki R, Fujii M, Watanabe Y, Mizuguchi H, Yagi K (2010b) Claudin-4-targeting of diphtheria toxin fragment A using a C-terminal fragment of Clostridium perfringens enterotoxin. Eur J Pharm Biopharm Off J Arbeitsgemeinschaft fur Pharmazeutische Verfahrenstechnik eV 75(2):213–217. doi:10.1016/j.ejpb.2010.03.003

Katahira J, Inoue N, Horiguchi Y, Matsuda M, Sugimoto N (1997) Molecular cloning and functional characterization of the receptor for Clostridium perfringens enterotoxin. J Cell Biol 136(6):1239–1247

Kaufman DS, Shipley WU, Feldman AS (2009) Bladder cancer. Lancet 374(9685):239–249. doi:10.1016/S0140-6736(09)60491-8

Kim TH, Huh JH, Lee S, Kang H, Kim GI, An HJ (2008) Down-regulation of claudin-2 in breast carcinomas is associated with advanced disease. Histopathology 53(1):48–55. doi:10.1111/j.1365-2559.2008.03052.x

Kimura Y, Shiozaki H, Hirao M, Maeno Y, Doki Y, Inoue M, Monden T, Ando-Akatsuka Y, Furuse M, Tsukita S, Monden M (1997) Expression of occludin, tight-junction-associated protein, in human digestive tract. Am J Pathol 151(1):45–54

Kinugasa T, Akagi Y, Yoshida T, Ryu Y, Shiratuchi I, Ishibashi N, Shirouzu K (2010) Increased claudin-1 protein expression contributes to tumorigenesis in ulcerative colitis-associated colorectal cancer. Anticancer Res 30(8):3181–3186

Kleinberg L, Holth A, Trope CG, Reich R, Davidson B (2008) Claudin upregulation in ovarian carcinoma effusions is associated with poor survival. Hum Pathol 39(5):747–757. doi:10.1016/j.humpath.2007.10.002

Kominsky SL, Argani P, Korz D, Evron E, Raman V, Garrett E, Rein A, Sauter G, Kallioniemi OP, Sukumar S (2003) Loss of the tight junction protein claudin-7 correlates with histological grade in both ductal carcinoma in situ and invasive ductal carcinoma of the breast. Oncogene 22(13):2021–2033. doi:10.1038/sj.onc.1206199

Kominsky SL, Vali M, Korz D, Gabig TG, Weitzman SA, Argani P, Sukumar S (2004) Clostridium perfringens enterotoxin elicits rapid and specific cytolysis of breast carcinoma cells mediated through tight junction proteins claudin 3 and 4. Am J Pathol 164(5):1627–1633. doi:10.1016/S0002-9440(10)63721-2

Kominsky SL, Tyler B, Sosnowski J, Brady K, Doucet M, Nell D, Smedley JG 3rd, McClane B, Brem H, Sukumar S (2007) Clostridium perfringens enterotoxin as a novel-targeted therapeutic for brain metastasis. Cancer Res 67(17):7977–7982. doi:10.1158/0008-5472.CAN-07-1314

Kondoh M, Masuyama A, Takahashi A, Asano N, Mizuguchi H, Koizumi N, Fujii M, Hayakawa T, Horiguchi Y, Watanbe Y (2005) A novel strategy for the enhancement of drug absorption using a claudin modulator. Mol Pharmacol 67(3):749–756. doi:10.1124/mol.104.008375

Konecny GE, Agarwal R, Keeney GA, Winterhoff B, Jones MB, Mariani A, Riehle D, Neuper C, Dowdy SC, Wang HJ, Morin PJ, Podratz KC (2008) Claudin-3 and claudin-4 expression in serous papillary, clear-cell, and endometrioid endometrial cancer. Gynecol Oncol 109(2):263–269. doi:10.1016/j.ygyno.2008.01.024

Kramer F, White K, Kubbies M, Swisshelm K, Weber BH (2000) Genomic organization of claudin-1 and its assessment in hereditary and sporadic breast cancer. Hum Genet 107(3):249–256

Krause G, Winkler L, Mueller SL, Haseloff RF, Piontek J, Blasig IE (2008) Structure and function of claudins. Biochim Biophys Acta 1778(3):631–645. doi:10.1016/j.bbamem.2007.10.018

Kuhn S, Koch M, Nubel T, Ladwein M, Antolovic D, Klingbeil P, Hildebrand D, Moldenhauer G, Langbein L, Franke WW, Weitz J, Zoller M (2007) A complex of EpCAM, claudin-7, CD44 variant isoforms, and tetraspanins promotes colorectal cancer progression. Mol Can Res MCR 5(6):553–567. doi:10.1158/1541-7786.MCR-06-0384

Kulka J, Szasz AM, Nemeth Z, Madaras L, Schaff Z, Molnar IA, Tokes AM (2009) Expression of tight junction protein claudin-4 in basal-like breast carcinomas. Pathol Oncol Res POR 15(1):59–64. doi:10.1007/s12253-008-9089-x

Kuo WL, Lee LY, Wu CM, Wang CC, Yu JS, Liang Y, Lo CH, Huang KH, Hwang TL (2006) Differential expression of claudin-4 between intestinal and diffuse-type gastric cancer. Oncol Rep 16(4):729–734

Kuo SJ, Chien SY, Lin C, Chan SE, Tsai HT, Chen DR (2010) Significant elevation of CLDN16 and HAPLN3 gene expression in human breast cancer. Oncol Rep 24(3):759–766

Kwon MJ, Kim SH, Jeong HM, Jung HS, Kim SS, Lee JE, Gye MC, Erkin OC, Koh SS, Choi YL, Park CK, Shin YK (2011) Claudin-4 overexpression is associated with epigenetic derepression

Michl P, Buchholz M, Rolke M, Kunsch S, Lohr M, McClane B, Tsukita S, Leder G, Adler G, Gress TM (2001) Claudin-4: a new target for pancreatic cancer treatment using Clostridium perfringens enterotoxin. Gastroenterology 121(3):678–684

Miettinen M, Sarlomo-Rikala M, Wang ZF (2011) Claudin-5 as an immunohistochemical marker for angiosarcoma and hemangioendotheliomas. Am J Surg Pathol 35(12):1848–1856. doi:10.1097/PAS.0b013e318229a401

Miwa N, Furuse M, Tsukita S, Niikawa N, Nakamura Y, Furukawa Y (2001) Involvement of claudin-1 in the beta-catenin/Tcf signaling pathway and its frequent upregulation in human colorectal cancers. Oncol Res 12(11–12):469–476

Miyamori H, Takino T, Kobayashi Y, Tokai H, Itoh Y, Seiki M, Sato H (2001) Claudin promotes activation of pro-matrix metalloproteinase-2 mediated by membrane-type matrix metalloproteinases. J Biol Chem 276(30):28204–28211. doi:10.1074/jbc.M103083200

Miyamoto K, Kusumi T, Sato F, Kawasaki H, Shibata S, Ohashi M, Hakamada K, Sasaki M, Kijima H (2008) Decreased expression of claudin-1 is correlated with recurrence status in esophageal squamous cell carcinoma. Biomed Res 29(2):71–76

Moldvay J, Jackel M, Paska C, Soltesz I, Schaff Z, Kiss A (2007) Distinct claudin expression profile in histologic subtypes of lung cancer. Lung Cancer 57(2):159–167. doi:10.1016/j.lungcan.2007.02.018

Montgomery E, Mamelak AJ, Gibson M, Maitra A, Sheikh S, Amr SS, Yang S, Brock M, Forastiere A, Zhang S, Murphy KM, Berg KD (2006) Overexpression of claudin proteins in esophageal adenocarcinoma and its precursor lesions. Appl Immunohistochem Mol Morphol AIMM/Off Publ Soc Appl Immunohistochem 14(1):24–30. doi:10.1097/01.pai.0000151933.04800.1c

Moreno-Bueno G, Salvador F, Martin A, Floristan A, Cuevas EP, Santos V, Montes A, Morales S, Castilla MA, Rojo-Sebastian A, Martinez A, Hardisson D, Csiszar K, Portillo F, Peinado H, Palacios J, Cano A (2011) Lysyl oxidase-like 2 (LOXL2), a new regulator of cell polarity required for metastatic dissemination of basal-like breast carcinomas. EMBO Mol Med 3(9):528–544. doi:10.1002/emmm.201100156

Morita K, Furuse M, Fujimoto K, Tsukita S (1999) Claudin multigene family encoding four-transmembrane domain protein components of tight junction strands. Proc Natl Acad Sci U S A 96(2):511–516

Morita K, Furuse M, Yoshida Y, Itoh M, Sasaki H, Tsukita S, Miyachi Y (2002) Molecular architecture of tight junctions of periderm differs from that of the maculae occludentes of epidermis. J Invest Dermatol 118(6):1073–1079. doi:10.1046/j.1523-1747.2002.01774.x

Morohashi S, Kusumi T, Sato F, Odagiri H, Chiba H, Yoshihara S, Hakamada K, Sasaki M, Kijima H (2007) Decreased expression of claudin-1 correlates with recurrence status in breast cancer. Int J Mol Med 20(2):139–143

Murakami M, Giampietro C, Giannotta M, Corada M, Torselli I, Orsenigo F, Cocito A, d'Ario G, Mazzarol G G, Confalonieri S, Di Fiore PP, Dejana E (2011) Abrogation of junctional adhesion molecule-A expression induces cell apoptosis and reduces breast cancer progression. PLoS One 6(6):e21242. doi:10.1371/journal.pone.0021242

Myal Y, Leygue E, Blanchard AA (2010) Claudin 1 in breast tumorigenesis: revelation of a possible novel "claudin high" subset of breast cancers. J Biomed Biotechnol 2010:956897. doi:10.1155/2010/956897

Naik MU, Naik TU, Suckow AT, Duncan MK, Naik UP (2008) Attenuation of junctional adhesion molecule-A is a contributing factor for breast cancer cell invasion. Cancer Res 68(7):2194–2203. doi:10.1158/0008-5472.CAN-07-3057

Nakagawa S, Miyoshi N, Ishii H, Mimori K, Tanaka F, Sekimoto M, Doki Y, Mori M (2011) Expression of CLDN1 in colorectal cancer: a novel marker for prognosis. Int J Oncol 39(4):791–796. doi:10.3892/ijo.2011.1102

Nakanishi K, Ogata S, Hiroi S, Tominaga S, Aida S, Kawai T (2008) Expression of occludin and claudins 1, 3, 4, and 7 in urothelial carcinoma of the upper urinary tract. Am J Clin Pathol 130(1):43–49. doi:10.1309/U77A6BTEXVCA5D0E

Nakayama F, Semba S, Usami Y, Chiba H, Sawada N, Yokozaki H (2008) Hypermethylation-modulated downregulation of claudin-7 expression promotes the progression of colorectal carcinoma. Pathobiol J Immunopathol Mol Cell Biol 75(3):177–185. doi:10.1159/000124978

Nava P, Lopez S, Arias CF, Islas S, Gonzalez-Mariscal L (2004) The rotavirus surface protein VP8 modulates the gate and fence function of tight junctions in epithelial cells. J Cell Sci 117(Pt 23):5509–5519. doi:10.1242/jcs.01425

Nemeth Z, Szasz AM, Somoracz A, Tatrai P, Nemeth J, Gyorffy H, Szijarto A, Kupcsulik P, Kiss A, Schaff Z (2009a) Zonula occludens-1, occludin, and E-cadherin protein expression in biliary tract cancers. Pathol Oncol Res POR 15(3):533–539. doi:10.1007/s12253-009-9150-4

Nemeth Z, Szasz AM, Tatrai P, Nemeth J, Gyorffy H, Somoracz A, Szijarto A, Kupcsulik P, Kiss A, Schaff Z (2009b) Claudin-1, -2, -3, -4, -7, -8, and -10 protein expression in biliary tract cancers. J Histochem Cytochem Off J Histochem Soc 57(2):113–121. doi:10.1369/jhc.2008.952291

Nemeth J, Nemeth Z, Tatrai P, Peter I, Somoracz A, Szasz AM, Kiss A, Schaff Z (2010) High expression of claudin-1 protein in papillary thyroid tumor and its regional lymph node metastasis. Pathol Oncol Res POR 16(1):19–27. doi:10.1007/s12253-009-9182-9

Nishino R, Honda M, Yamashita T, Takatori H, Minato H, Zen Y, Sasaki M, Takamura H, Horimoto K, Ohta T, Nakanuma Y, Kaneko S (2008) Identification of novel candidate tumour marker genes for intrahepatic cholangiocarcinoma. J Hepatol 49(2):207–216. doi:10.1016/j.jhep.2008.03.025

Nusrat A, Brown GT, Tom J, Drake A, Bui TT, Quan C, Mrsny RJ (2005) Multiple protein interactions involving proposed extracellular loop domains of the tight junction protein occludin. Mol Biol Cell 16(4):1725–1734. doi:10.1091/mbc.E04-06-0465

O'Mahony DJCG (2000) US Patent US20006346613

Offner S, Hekele A, Teichmann U, Weinberger S, Gross S, Kufer P, Itin C, Baeuerle PA, Kohleisen B (2005) Epithelial tight junction proteins as potential antibody targets for pancarcinoma therapy. Cancer Immunol Immunother CII 54(5):431–445. doi:10.1007/s00262-004-0613-x

Ohtani S, Terashima M, Satoh J, Soeta N, Saze Z, Kashimura S, Ohsuka F, Hoshino Y, Kogure M, Gotoh M (2009) Expression of tight-junction-associated proteins in human gastric cancer: downregulation of claudin-4 correlates with tumor aggressiveness and survival. Gastric Cancer Off J Int Gastric Cancer Assoc Japanese Gastric Cancer Assoc 12(1):43–51. doi:10.1007/s10120-008-0497-0

Oku N, Sasabe E, Ueta E, Yamamoto T, Osaki T (2006) Tight junction protein claudin-1 enhances the invasive activity of oral squamous cell carcinoma cells by promoting cleavage of laminin-5 gamma2 chain via matrix metalloproteinase (MMP)-2 and membrane-type MMP-1. Cancer Res 66(10):5251–5257. doi:10.1158/0008-5472.CAN-05-4478

Orban E, Szabo E, Lotz G, Kupcsulik P, Paska C, Schaff Z, Kiss A (2008) Different expression of occludin and ZO-1 in primary and metastatic liver tumors. Pathol Oncol Res POR 14(3):299–306. doi:10.1007/s12253-008-9031-2

Osanai M, Murata M, Nishikiori N, Chiba H, Kojima T, Sawada N (2006) Epigenetic silencing of occludin promotes tumorigenic and metastatic properties of cancer cells via modulations of unique sets of apoptosis-associated genes. Cancer Res 66(18):9125–9133. doi:10.1158/0008-5472.CAN-06-1864

Osanai M, Murata M, Nishikiori N, Chiba H, Kojima T, Sawada N (2007) Occludin-mediated premature senescence is a fail-safe mechanism against tumorigenesis in breast carcinoma cells. Cancer Sci 98(7):1027–1034. doi:10.1111/j.1349-7006.2007.00494.x

Oshima T, Kunisaki C, Yoshihara K, Yamada R, Yamamoto N, Sato T, Makino H, Yamagishi S, Nagano Y, Fujii S, Shiozawa M, Akaike M, Wada N, Rino Y, Masuda M, Tanaka K, Imada T (2008) Reduced expression of the claudin-7 gene correlates with venous invasion and liver metastasis in colorectal cancer. Oncol Rep 19(4):953–959

Osunkoya AO, Cohen C, Lawson D, Picken MM, Amin MB, Young AN (2009) Claudin-7 and claudin-8: immunohistochemical markers for the differential diagnosis of chromophobe renal cell carcinoma and renal oncocytoma. Hum Pathol 40(2):206–210. doi:10.1016/j.humpath.2008.07.002

Palekar MS, Sirsat SM (1975) Lanthanum staining of cell surface and junctional complexes in normal and malignant human oral mucosa. J Oral Pathol 4(5):231–243

Pan XY, Wang B, Che YC, Weng ZP, Dai HY, Peng W (2007) Expression of claudin-3 and claudin-4 in normal, hyperplastic, and malignant endometrial tissue. Int J Gynecol Cancer Off J Int Gynecol Cancer Society 17(1):233–241. doi:10.1111/j.1525-1438.2006.00748.x

Park JY, Park KH, Oh TY, Hong SP, Jeon TJ, Kim CH, Park SW, Chung JB, Song SY, Bang S (2007) Up-regulated claudin 7 expression in intestinal-type gastric carcinoma. Oncol Rep 18(2):377–382

Parwani AV, Stelow EB, Pambuccian SE, Burger PC, Ali SZ (2005) Atypical teratoid/rhabdoid tumor of the brain: cytopathologic characteristics and differential diagnosis. Cancer 105(2):65–70. doi:10.1002/cncr.20872

Paschoud S, Bongiovanni M, Pache JC, Citi S (2007) Claudin-1 and claudin-5 expression patterns differentiate lung squamous cell carcinomas from adenocarcinomas. Mod Patho Off J U S Can Acad Pathol Inc 20(9):947–954. doi:10.1038/modpathol.3800835

Patonai A, Erdelyi-Belle B, Korompay A, Somoracz A, Straub BK, Schirmacher P, Kovalszky I, Lotz G, Kiss A, Schaff Z (2011) Claudins and tricellulin in fibrolamellar hepatocellular carcinoma. Virchows Arch Int J Pathol 458(6):679–688. doi:10.1007/s00428-011-1077-y

Piontek J, Winkler L, Wolburg H, Muller SL, Zuleger N, Piehl C, Wiesner B, Krause G, Blasig IE (2008) Formation of tight junction: determinants of homophilic interaction between classic claudins. FASEB J Off Publ Fed Am Soc Exp Biol 22(1):146–158. doi:10.1096/fj.07-8319com

Potala S, Sahoo SK, Verma RS (2008) Targeted therapy of cancer using diphtheria toxin-derived immunotoxins. Drug Discov Today 13(17–18):807–815. doi:10.1016/j.drudis.2008.06.017

Quay SC, Quay S, Lamharzi NFKT (2007) Method of enhancing transmucosal delivery of therapeutic compounds US Patent US2007077283

Rahner C, Mitic LL, Anderson JM (2001) Heterogeneity in expression and subcellular localization of claudins 2, 3, 4, and 5 in the rat liver, pancreas, and gut. Gastroenterology 120(2):411–422

Randi G, Malvezzi M, Levi F, Ferlay J, Negri E, Franceschi S, La Vecchia C (2009) Epidemiology of biliary tract cancers: an update. Ann Oncol Off J Eur Soc Med Oncol/ESMO 20(1):146–159. doi:10.1093/annonc/mdn533

Resnick MB, Konkin T, Routhier J, Sabo E, Pricolo VE (2005) Claudin-1 is a strong prognostic indicator in stage II colonic cancer: a tissue microarray study. Mod Patho Off J U S Can Acad Pathol Inc 18(4):511–518. doi:10.1038/modpathol.3800301

Reyes JL, Lamas M, Martin D, del Carmen NM, Islas S, Luna J, Tauc M, Gonzalez-Mariscal L (2002) The renal segmental distribution of claudins changes with development. Kidney Int 62(2):476–487. doi:10.1046/j.1523-1755.2002.00479.x

Robertson SL, Smedley JG 3rd, Singh U, Chakrabarti G, Van Itallie CM, Anderson JM, McClane BA (2007) Compositional and stoichiometric analysis of Clostridium perfringens enterotoxin complexes in Caco-2 cells and claudin 4 fibroblast transfectants. Cell Microbiol 9(11):2734–2755. doi:10.1111/j.1462-5822.2007.00994.x

Robertson SL, Smedley JG 3rd, McClane BA (2010) Identification of a claudin-4 residue important for mediating the host cell binding and action of Clostridium perfringens enterotoxin. Infect Immun 78(1):505–517. doi:10.1128/IAI.00778-09

Ryan EJ, Daly LM, Mills KH (2001) Immunomodulators and delivery systems for vaccination by mucosal routes. Trends Biotechnol 19(8):293–304

Saeki R, Kondoh M, Kakutani H, Tsunoda S, Mochizuki Y, Hamakubo T, Tsutsumi Y, Horiguchi Y, Yagi K (2009) A novel tumor-targeted therapy using a claudin-4-targeting molecule. Mol Pharmacol 76(4):918–926. doi:10.1124/mol.109.058412

Salido EC, Fisher DA, Barajas L (1986) Immunoelectron microscopy of epidermal growth factor in mouse kidney. J Ultrastruct Mol Struct Res 96(1–3):105–113

Santin AD, Cane S, Bellone S, Palmieri M, Siegel ER, Thomas M, Roman JJ, Burnett A, Cannon MJ, Pecorelli S (2005) Treatment of chemotherapy-resistant human ovarian cancer xenografts in C.B-17/SCID mice by intraperitoneal administration of Clostridium perfringens enterotoxin. Cancer Res 65(10):4334–4342. doi:10.1158/0008-5472.CAN-04-3472

Santin AD, Bellone S, Marizzoni M, Palmieri M, Siegel ER, McKenney JK, Hennings L, Comper F, Bandiera E, Pecorelli S (2007a) Overexpression of claudin-3 and claudin-4 receptors in uterine serous papillary carcinoma: novel targets for a type-specific therapy using Clostridium perfringens enterotoxin (CPE). Cancer 109(7):1312–1322. doi:10.1002/cncr.22536

Santin AD, Bellone S, Siegel ER, McKenney JK, Thomas M, Roman JJ, Burnett A, Tognon G, Bandiera E, Pecorelli S (2007b) Overexpression of Clostridium perfringens enterotoxin receptors

claudin-3 and claudin-4 in uterine carcinosarcomas. Clin Cancer Res Off J Am Assoc Cancer Res 13(11):3339–3346. doi:10.1158/1078-0432.CCR-06-3037
Sauer T, Pedersen MK, Ebeltoft K, Naess O (2005) Reduced expression of Claudin-7 in fine needle aspirates from breast carcinomas correlate with grading and metastatic disease. Cytopathol Off J Br Soc Clin Cytol 16(4):193–198. doi:10.1111/j.1365-2303.2005.00257.x
Schuetz AN, Yin-Goen Q, Amin MB, Moreno CS, Cohen C, Hornsby CD, Yang WL, Petros JA, Issa MM, Pattaras JG, Ogan K, Marshall FF, Young AN (2005) Molecular classification of renal tumors by gene expression profiling. J Mol Diagn JMD 7(2):206–218. doi:10.1016/S1525-1578(10)60547-8
Sheehan GM, Kallakury BV, Sheehan CE, Fisher HA, Kaufman RP Jr, Ross JS (2007) Loss of claudins-1 and -7 and expression of claudins-3 and -4 correlate with prognostic variables in prostatic adenocarcinomas. Hum Pathol 38(4):564–569. doi:10.1016/j.humpath.2006.11.007
Sherman S, Klein E, McClane BA (1994) Clostridium perfringens type A enterotoxin induces tissue damage and fluid accumulation in rabbit ileum. J Diarrhoeal Dis Res 12(3):200–207
Shiou SR, Singh AB, Moorthy K, Datta PK, Washington MK, Beauchamp RD, Dhawan P (2007) Smad4 regulates claudin-1 expression in a transforming growth factor-beta-independent manner in colon cancer cells. Cancer Res 67(4):1571–1579. doi:10.1158/0008-5472.CAN-06-1680
Singh U, Van Itallie CM, Mitic LL, Anderson JM, McClane BA (2000) CaCo-2 cells treated with Clostridium perfringens enterotoxin form multiple large complex species, one of which contains the tight junction protein occludin. J Biol Chem 275(24):18407–18417. doi:10.1074/jbc.M001530200
Singh U, Mitic LL, Wieckowski EU, Anderson JM, McClane BA (2001) Comparative biochemical and immunocytochemical studies reveal differences in the effects of Clostridium perfringens enterotoxin on polarized CaCo-2 cells versus Vero cells. J Biol Chem 276(36):33402–33412. doi:10.1074/jbc.M104200200
Smalley KS, Brafford P, Haass NK, Brandner JM, Brown E, Herlyn M (2005) Up-regulated expression of zonula occludens protein-1 in human melanoma associates with N-cadherin and contributes to invasion and adhesion. Am J Pathol 166(5):1541–1554. doi:10.1016/S0002-9440(10)62370-X
Smedley JG 3rd, Uzal FA, McClane BA (2007) Identification of a prepore large-complex stage in the mechanism of action of Clostridium perfringens enterotoxin. Infect Immun 75(5):2381–2390. doi:10.1128/IAI.01737-06
Sobel G, Paska C, Szabo I, Kiss A, Kadar A, Schaff Z (2005) Increased expression of claudins in cervical squamous intraepithelial neoplasia and invasive carcinoma. Hum Pathol 36(2):162–169. doi:10.1016/j.humpath.2004.12.001
Sobel G, Nemeth J, Kiss A, Lotz G, Szabo I, Udvarhelyi N, Schaff Z, Paska C (2006) Claudin 1 differentiates endometrioid and serous papillary endometrial adenocarcinoma. Gynecol Oncol 103(2):591–598. doi:10.1016/j.ygyno.2006.04.005
Society AC (2012) Cancer facts and figures 2012. Am Cancer Soc
Soini Y (2005) Expression of claudins 1, 2, 3, 4, 5 and 7 in various types of tumours. Histopathology 46(5):551–560. doi:10.1111/j.1365-2559.2005.02127.x
Song KH, Eddington ND (2012) The impact of AT1002 on the delivery of ritonavir in the presence of bioadhesive polymer, carrageenan. Arch Pharm Res 35(5):937–943. doi:10.1007/s12272-012-0520-1
Song X, Li X, Tang Y, Chen H, Wong B, Wang J, Chen M (2008) Expression of claudin-2 in the multistage process of gastric carcinogenesis. Histol Histopathol 23(6):673–682
Sonoda N, Furuse M, Sasaki H, Yonemura S, Katahira J, Horiguchi Y, Tsukita S (1999) Clostridium perfringens enterotoxin fragment removes specific claudins from tight junction strands: evidence for direct involvement of claudins in tight junction barrier. J Cell Biol 147(1):195–204
Sormunen R, Paakko P, Kaarteenaho-Wiik R, Soini Y (2007) Differential expression of adhesion molecules in lung tumours. Histopathology 50(2):282–284. doi:10.1111/j.1365-2559.2007.02574.x
Staats HF, Montgomery SP, Palker TJ (1997) Intranasal immunization is superior to vaginal, gastric, or rectal immunization for the induction of systemic and mucosal anti-HIV antibody responses. AIDS Res Hum Retroviruses 13(11):945–952

Sullivan LM, Yankovich T, Le P, Martinez D, Santi M, Biegel JA, Pawel BR, Judkins AR (2012) Claudin-6 is a nonspecific marker for malignant rhabdoid and other pediatric tumors. Am J Surg Pathol 36(1):73–80. doi:10.1097/PAS.0b013e31822cfa7e

Sung CO, Han SY, Kim SH (2011) Low expression of claudin-4 is associated with poor prognosis in esophageal squamous cell carcinoma. Ann Surg Oncol 18(1):273–281. doi:10.1245/s10434-010-1289-4

Suzuki H, Kakutani H, Kondoh M, Watari A, Yagi K (2010) The safety of a mucosal vaccine using the C-terminal fragment of Clostridium perfringens enterotoxin. Die Pharmazie 65(10):766–769

Suzuki H, Kondoh M, Kakutani H, Yamane S, Uchida H, Hamakubo T, Yagi K (2012) The application of an alanine-substituted mutant of the C-terminal fragment of Clostridium perfringens enterotoxin as a mucosal vaccine in mice. Biomaterials 33(1):317–324. doi:10.1016/j.biomaterials.2011.09.048

Szasz AM, Tokes AM, Micsinai M, Krenacs T, Jakab C, Lukacs L, Nemeth Z, Baranyai Z, Dede K, Madaras L, Kulka J (2011) Prognostic significance of claudin expression changes in breast cancer with regional lymph node metastasis. Clin Exp Metastasis 28(1):55–63. doi:10.1007/s10585-010-9357-5

Szekely E, Torzsok P, Riesz P, Korompay A, Fintha A, Szekely T, Lotz G, Nyirady P, Romics I, Timar J, Schaff Z, Kiss A (2011) Expression of claudins and their prognostic significance in noninvasive urothelial neoplasms of the human urinary bladder. J Histochem Cytochem Off J Histochem Soc 59(10):932–941. doi:10.1369/0022155411418829

Takahashi A, Komiya E, Kakutani H, Yoshida T, Fujii M, Horiguchi Y, Mizuguchi H, Tsutsumi Y, Tsunoda S, Koizumi N, Isoda K, Yagi K, Watanabe Y, Kondoh M (2008) Domain mapping of a claudin-4 modulator, the C-terminal region of C-terminal fragment of Clostridium perfringens enterotoxin, by site-directed mutagenesis. Biochem Pharmacol 75(8):1639–1648. doi:10.1016/j.bcp. 2007.12.016

Takahashi A, Kondoh M, Uchida H, Kakamu Y, Hamakubo T, Yagi K (2011) Mutated C-terminal fragments of Clostridium perfringens enterotoxin have increased affinity to claudin-4 and reversibly modulate tight junctions in vitro. Biochem Biophys Res Commun 410(3):466–470. doi:10.1016/j.bbrc.2011.05.161

Tassi RA, Bignotti E, Falchetti M, Ravanini M, Calza S, Ravaggi A, Bandiera E, Facchetti F, Pecorelli S, Santin AD (2008) Claudin-7 expression in human epithelial ovarian cancer. Int J Gynecol Cancer Off J Int Gynecol Cancer Society 18(6):1262–1271. doi:10.1111/j.1525-1438.2008.01194.x

Tavelin S, Hashimoto K, Malkinson J, Lazorova L, Toth I, Artursson P (2003) A new principle for tight junction modulation based on occludin peptides. Mol Pharmacol 64(6):1530–1540. doi:10.1124/mol.64.6.1530

Thway K, Fisher C, Debiec-Rychter M, Calonje E (2009) Claudin-1 is expressed in perineurioma-like low-grade fibromyxoid sarcoma. Hum Pathol 40(11):1586–1590. doi:10.1016/j.humpath.2009.04.003

Tobioka H, Isomura H, Kokai Y, Tokunaga Y, Yamaguchi J, Sawada N (2004a) Occludin expression decreases with the progression of human endometrial carcinoma. Hum Pathol 35(2):159–164

Tobioka H, Tokunaga Y, Isomura H, Kokai Y, Yamaguchi J, Sawada N (2004b) Expression of occludin, a tight-junction-associated protein, in human lung carcinomas. Virchows Arch Int J Pathol 445(5):472–476. doi:10.1007/s00428-004-1054-9

Tokes AM, Szasz AM, Juhasz E, Schaff Z, Harsanyi L, Molnar IA, Baranyai Z, Besznyak I Jr, Zarand A, Salamon F, Kulka J (2012) Expression of tight junction molecules in breast carcinomas analysed by array PCR and immunohistochemistry. Pathol Oncol Res POR 18(3):593–606. doi:10.1007/s12253-011-9481-9

Tripathi A, Lammers KM, Goldblum S, Shea-Donohue T, Netzel-Arnett S, Buzza MS, Antalis TM, Vogel SN, Zhao A, Yang S, Arrietta MC, Meddings JB, Fasano A (2009) Identification of human zonulin, a physiological modulator of tight junctions, as prehaptoglobin-2. Proc Natl Acad Sci U S A 106(39):16799–16804. doi:10.1073/pnas.0906773106

Tsutsumi K, Sato N, Cui L, Mizumoto K, Sadakari Y, Fujita H, Ohuchida K, Ohtsuka T, Takahata S, Tanaka M (2011) Expression of claudin-4 (CLDN4) mRNA in intraductal papillary mucinous neoplasms of the pancreas. Mod Patho Off J U S Can Acad Pathol Inc 24(4):533–541. doi:10.1038/modpathol.2010.218

Turksen K, Troy TC (2001) Claudin-6: a novel tight junction molecule is developmentally regulated in mouse embryonic epithelium. Develop Dyn Off Publ Am Assoc Anat 222(2):292–300. doi:10.1002/dvdy.1174

Turksen K, Troy TC (2002) Permeability barrier dysfunction in transgenic mice overexpressing claudin 6. Development 129(7):1775–1784

Tzelepi VN, Tsamandas AC, Vlotinou HD, Vagianos CE, Scopa CD (2008) Tight junctions in thyroid carcinogenesis: diverse expression of claudin-1, claudin-4, claudin-7 and occludin in thyroid neoplasms. Mod Patho Off J U S Can Acad Pathol Inc 21(1):22–30. doi:10.1038/modpathol.3800959

Uchida H, Kondoh M, Hanada T, Takahashi A, Hamakubo T, Yagi K (2010) A claudin-4 modulator enhances the mucosal absorption of a biologically active peptide. Biochem Pharmacol 79(10):1437–1444. doi:10.1016/j.bcp.2010.01.010

Ueda J, Semba S, Chiba H, Sawada N, Seo Y, Kasuga M, Yokozaki H (2007) Heterogeneous expression of claudin-4 in human colorectal cancer: decreased claudin-4 expression at the invasive front correlates cancer invasion and metastasis. Pathobiol J Immunopathol Mol Cell Biol 74(1):32–41. doi:10.1159/000101049

Underwood JL, Murphy CG, Chen J, Franse-Carman L, Wood I, Epstein DL, Alvarado JA (1999) Glucocorticoids regulate transendothelial fluid flow resistance and formation of intercellular junctions. Am J Physiol 277(2 Pt 1):C330–C342

Usami Y, Chiba H, Nakayama F, Ueda J, Matsuda Y, Sawada N, Komori T, Ito A, Yokozaki H (2006) Reduced expression of claudin-7 correlates with invasion and metastasis in squamous cell carcinoma of the esophagus. Hum Pathol 37(5):569–577. doi:10.1016/j.humpath.2005.12.018

Van Itallie CM, Anderson JM (1997) Occludin confers adhesiveness when expressed in fibroblasts. J Cell Sci 110(Pt 9):1113–1121

Van Itallie CM, Betts L, Smedley JG 3rd, McClane BA, Anderson JM (2008) Structure of the claudin-binding domain of Clostridium perfringens enterotoxin. J Biol Chem 283(1):268–274. doi:10.1074/jbc.M708066200

Walther W, Petkov S, Kuvardina ON, Aumann J, Kobelt D, Fichtner I, Lemm M, Piontek J, Blasig IE, Stein U, Schlag PM (2012) Novel Clostridium perfringens enterotoxin suicide gene therapy for selective treatment of claudin-3- and -4-overexpressing tumors. Gene Ther 19(5):494–503. doi:10.1038/gt.2011.136

Wang F, Daugherty B, Keise LL, Wei Z, Foley JP, Savani RC, Koval M (2003) Heterogeneity of claudin expression by alveolar epithelial cells. Am J Respir Cell Mol Biol 29(1):62–70. doi:10.1165/rcmb.2002-0180OC

Wartofsky L (2010) Increasing world incidence of thyroid cancer: increased detection or higher radiation exposure? Hormones (Athens) 9(2):103–108

Weber CR, Nalle SC, Tretiakova M, Rubin DT, Turner JR (2008) Claudin-1 and claudin-2 expression is elevated in inflammatory bowel disease and may contribute to early neoplastic transformation. Lab Inves J Tech Method Pathol 88(10):1110–1120. doi:10.1038/labinvest.2008.78

Weeraratna AT, Becker D, Carr KM, Duray PH, Rosenblatt KP, Yang S, Chen Y, Bittner M, Strausberg RL, Riggins GJ, Wagner U, Kallioniemi OP, Trent JM, Morin PJ, Meltzer PS (2004) Generation and analysis of melanoma SAGE libraries: SAGE advice on the melanoma transcriptome. Oncogene 23(12):2264–2274. doi:10.1038/sj.onc.1207337

Weingart SN, Brown E, Bach PB, Eng K, Johnson SA, Kuzel TM, Langbaum TS, Leedy RD, Muller RJ, Newcomer LN, O'Brien S, Reinke D, Rubino M, Saltz L, Walters RS (2008) NCCN Task Force report: oral chemotherapy. J Natl Compr Cancer Netw JNCCN 6(Suppl 3):S1–S14

Wieckowski EU, Wnek AP, McClane BA (1994) Evidence that an approximately 50-kDa mammalian plasma membrane protein with receptor-like properties mediates the amphiphilicity of specifically bound Clostridium perfringens enterotoxin. J Biol Chem 269(14):10838–10848

Williams HK (2000) Molecular pathogenesis of oral squamous carcinoma. Mol Pathol MP 53(4):165–172

Winkler L, Gehring C, Wenzel A, Muller SL, Piehl C, Krause G, Blasig IE, Piontek J (2009) Molecular determinants of the interaction between Clostridium perfringens enterotoxin fragments and claudin-3. J Biol Chem 284(28):18863–18872. doi:10.1074/jbc.M109.008623

Wong V, Gumbiner BM (1997) A synthetic peptide corresponding to the extracellular domain of occludin perturbs the tight junction permeability barrier. J Cell Biol 136(2):399–409

Wu Q, Liu Y, Ren Y, Xu X, Yu L, Li Y, Quan C (2010) Tight junction protein, claudin-6, downregulates the malignant phenotype of breast carcinoma. Eur J Cancer Prev 19(3):186–194. doi:10.1097/CEJ.0b013e328337210e

Yoshida Y, Ban Y, Kinoshita S (2009) Tight junction transmembrane protein claudin subtype expression and distribution in human corneal and conjunctival epithelium. Invest Ophthalmol Vis Sci 50(5):2103–2108. doi:10.1167/iovs.08-3046

Yuan X, Lin X, Manorek G, Kanatani I, Cheung LH, Rosenblum MG, Howell SB (2009) Recombinant CPE fused to tumor necrosis factor targets human ovarian cancer cells expressing the claudin-3 and claudin-4 receptors. Mol Cancer Ther 8(7):1906–1915. doi:10.1158/1535-7163.MCT-09-0106

Zhu Y, Brannstrom M, Janson PO, Sundfeldt K (2006) Differences in expression patterns of the tight junction proteins, claudin 1, 3, 4 and 5, in human ovarian surface epithelium as compared to epithelia in inclusion cysts and epithelial ovarian tumours. Int J Cancer J Int Cancer 118(8):1884–1891. doi:10.1002/ijc.21506

Chapter 10
VEGF-Mediated Effects on Brain Microvascular Endothelial Tight Junctions and Transmigration of Breast Cancer Cells Across the Blood-Brain Barrier

Shalom Avraham, Shuxian Jiang, Lili Wang, Yigong Fu, and Hava Karsenty Avraham

Abstract Breast cancer spreads to the bones, lungs, brain and liver. While the mechanisms of breast metastasis to bones and lungs have been studied and characterized, there is little information about the molecular basis for metastasis to the brain. The disruption of the blood brain barrier (BBB) by brain metastases of triple-negative and basal-type breast cancers, but not of Her2/neu positive breast cancer, was reported recently. To metastasize to the brain, breast cancer cells must attach to brain microvascular endothelial cells (BMECs) and invade the BBB. Structurally, the BBB is a barrier comprised mainly of BMECs. BBB function is maintained by the tight junctions protein complexes (TJs) between adjacent BMECs which is comprised mainly of zonula occudens-1 and -2 (ZO-1 and -2), occludin, claudins, junctional adhesion molecules, actin and cingulin. Very little is known about extravasation of breast cancer cells across the BBB and their effects on TJs expression and function as well as their interactions with BMECs.

Here, we focused on VEGF as a mediator for breast cancer cell extravasation across the BBB and the function of VEGF on TJs expression and activation of BMECs. Our results showed that breast cells with high homing capacity to the brain secrete high levels of VEGF but no Angiopoietin-2. VEGF induced significant changes in the localization and distribution of ZO-1 and claudin-5 in BMECs as well as alterations in BMEC permeability and integrity, resulting in transmigration of tumor cells across the BBB. Further, VEGF acts as an important biological determinant in remodeling and cytoskeletal reorganization in BMECs, leading to formation of brain microvasculature niche for colonization of breast cancer in brain. Thus, therapies targeted to TJs in the BBB in conjunction with anti-antiangiogenic therapies may hold significant promise for the treatment and prevention of breast cancer metastasis to the brain.

S. Avraham • S. Jiang • L. Wang • Y. Fu • H.K. Avraham (✉)
The Division of Experimental Medicine, Beth Israel Deaconess Medical Center and Harvard Medical School, Research North, 99 Brookline Avenue, Room 330F, Boston, MA 02215, USA
e-mail: havraham@bidmc.harvard.edu

T.A. Martin and W.G. Jiang (eds.), *Tight Junctions in Cancer Metastasis*, Cancer Metastasis - Biology and Treatment 19, DOI 10.1007/978-94-007-6028-8_10,

Keywords Breast cancer • Metastasis • Brain microvascular endothelial cells • Blood-brain barrier • Adhesion • Transmigration • Colonization • Tight junctions • ZO-1 • Claudin-5

Abbreviations

BBB	Blood-brain barrier
BMEC	Brain microvascular endothelial cells
CNS	Central nervous system
EC	Endothelial cells
FAs	Focal adhesion sites
GAPDH	Glyceraldehyde-3-phosphate dehydrogenase
HBMECs	Human brain microvascular endothelial cells
HUVECs	Human umbilical vein endothelial cells
IHC	Immunohistochemistry
IF	Immunostaining
TEER	Trans-endothelial electrical resistance
TJs	Tight junctions
VEGF	Vascular endothelial growth factor
VEGFR	Vascular endothelial growth factor receptor
WB	Western blotting

10.1 Introduction

Brain metastasis entails several biological functions that collectively enable cancerous cells to disseminate from a primary site and colonize in brain (Sharma and Abraham 2007; Cheng and Hung 2007; Eichler and Loeffler 2007; Kaal and Vecht 2007). The sequential steps of brain metastasis include local invasion, intravasation, survival in the circulation following extravasation and colonization of breast tumor cells in brain. The BBB, with its tight layer of endothelial cells and astrocyte foot processes around the brain parenchyma, is infiltrated by circulating cancer cells that possess highly specialized functions, many of such mechanisms remain to be uncharacterized (Palmieri et al. 2006; Aragon-Ching and Zujewski 2007). Breast cancer cells extravasate through the BBB capillaries TJs and are lodged in brain capillaries as single cells, suggesting that brain metastases result from the ability of these cells to breach the BBB (Palmieri et al. 2006; Aragon-Ching and Zujewski 2007; Hawkins and Davis 2005).

An important step in metastasis to the brain is the interaction of brain microvascular endothelial cells (BMECs) with cancer cells and the transmigration of tumor cells across BMECs. The BMECs are the major constituent of the BBB (Hawkins and Davis 2005). For metastatic cells to reach the brain parenchyma, they need to first cross the blood-brain barrier (BBB), a highly selective BMEC barrier that

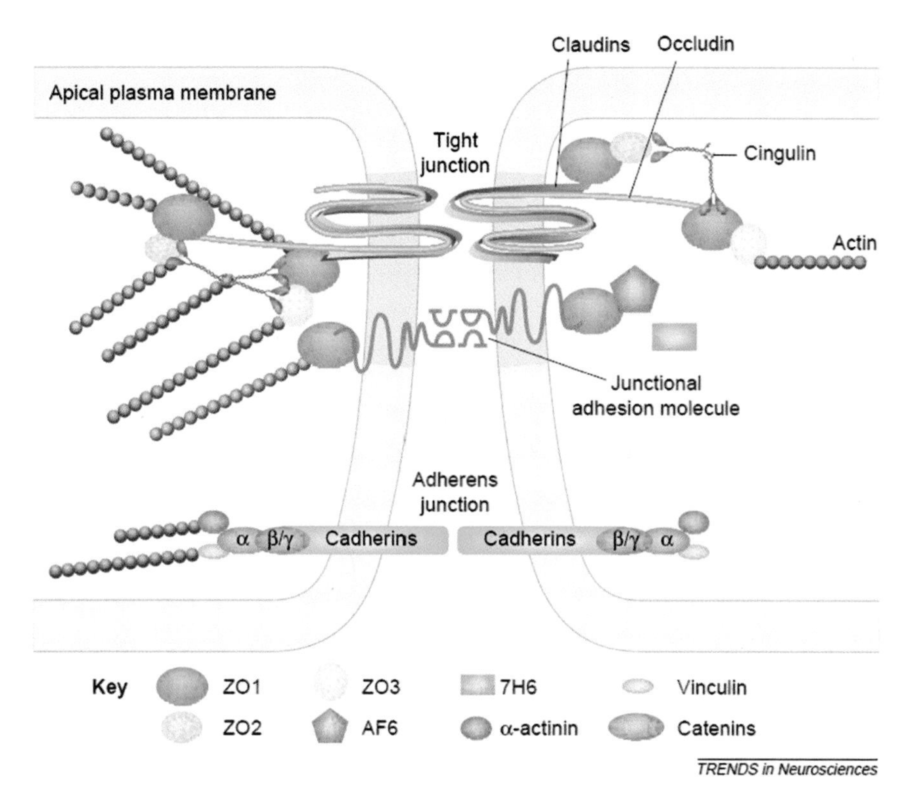

Fig. 10.1 Proposed interactions of the major proteins associated with tight junctions (TJs) at the blood-brain barrier (BBB). Tight junctions and adherens junctions exist in the blood-brain barrier (*BBB*). Three transmembrane proteins, Claudins, Occludin, and Junctional adhesion molecules (*JAMs*), form integral tight junctions between adjacent endothelial cells. They provide the primary "seal" and paracellular permeability regulation of the BBB. Other accessory proteins, such as Zonula occludens (*ZO-1* and *ZO-2*), AF6, 7H6, and Cingulin play as structure support, regulation, location recognition, and signal transduction proteins for the tight junctions. Adherens junctions consist of one transmembrane protein, VE-Cadherin, and three sturcutre support proteins, Catenins (α, β, γ), α-actinin, and Vinculin, that link to major cytoskeleton proteins, actin. For more details, please see Huber et al. (2001) (This figure was modified based on Fig. 10.2 shown in Huber et al. (2001) with permission by Dr. Davis (Huber et al. 2001))

separates the central nervous system (CNS) from the peripheral circulation in the brain. In terms of their structures and properties (see Hawkins and Davis 2005; Harhaj and Antonetti 2004; Wolburg and Lippoldt 2002 for review), BMECs are different from endothelium in other tissues. In the brain, the "neurovascular unit" is comprised of the cerebral microvascular endothelium, neurons, astrocytes, pericytes and the extracellular matrix (Hawkins and Davis 2005; Harhaj and Antonetti 2004; Wolburg and Lippoldt 2002). The permeability properties of the BBB are determined mainly by the expression and regulation of TJs complexes and adherens junctions (AJs) located between the adjacent endothelial cells (Fig. 10.1). These inter-endocellular junctional proteins contribute to the dynamic regulation of passage of ions, nutrients, cells as well as potential neurotoxins across the brain (Wolburg and Lippoldt 2002). It is commonly accepted that cohesive assembly and interactions of the TJ molecules (e.g. claudins, occludin, ZO-1) primarily give the barrier its permeability properties for maintaining cerebral homeostasis.

Disruption of intercellular protein structures and increased BBB permeability are frequently observed in the development and progression of diseases that affect the CNS (Harhaj and Antonetti 2004). For instance, impairment of the BBB leads to an increase in permeability and formation of edema in the brain. In the event of inflammation, mediators such as histamine, bradykinin, and Substance P have been shown to cause rapid formation of endothelial "gaps" (Rameshwar 2007; Chappa et al. 2006), which in turn induced changes in BBB permeability in vivo. While changes in BBB permeability are often seen as the cause of many pathological conditions (Hawkins and Davis 2005), in some circumstances the failure of BBB is identified as the outcome (Hawkins and Davis 2005). As metastatic tumor cells adhere to BBB endothelium, the generally accepted assumption is that they induce a retraction of the endothelium and penetrate through the exposed vascular basement membrane beneath. Many studies have demonstrated that after binding to specific components in the vascular membrane, tumor cells initiate extravasation, colonization and growth at secondary organ sites. Most tumor metastasis studies were focused on the effects of degrading enzymes (e.g. metalloproteinase) or adhesion molecules (e.g. alpha-v-beta3) on tumor cells. Little is known about the role of TJs of the microvascular endothelium integrity during tumor cell transmigration across the BBB.

10.1.1 Tight Junctions and Blood-Brain Barrier Integrity

The junctional complexes, which include adherens junctions (AJs), tight junctions (TJs) and gap junctions (Pardridge 1983; Morita et al. 1999; Musch et al. 2006; Schneeberger and Lynch 2004; Harhaj and Antonetti 2004; Umeda et al. 2006; Bradbury 1993; Bradbury and Deane 1993) (See Fig. 10.1), are focused in the interendothelial space of the BMECs. The gap junctions mediate intercellular communication between endothelial cells while AJs and TJs act to restrict the permeability across the endothelium. Composed of VE-cadherin, *AJs* are important structures for vascular adhesion of endothelial cells, catenins (α, β, γ), vinculin, and α-actinin (Fig. 10.1), focal adhesions, contact inhibition during vascular growth, remodeling, initiation of cell polarity and partial regulation of paracellular permeability (Harhaj and Antonetti 2004). The *TJs* are composed of a dynamic complex of claudins, occludin and the junctional adhesion molecules (JAM-1 and JAM-2) (JAMs), and of belt-like zonula occluden proteins (ZO-1 and ZO-2) (Fig. 10.1, Morita et al. 1999; Musch et al. 2006; Schneeberger and Lynch 2004; Harhaj and Antonetti 2004; Umeda et al. 2006; Bradbury 1993). In the brain microenvironment The TJs confer low paracellular permeability and high electrical resistance (Harhaj and Antonetti 2004).

The ZO proteins, in particular, are involved in clustering of protein complexes to the cell membrane, and link transmembrane proteins of the TJ to the actin cytoskeleton (Umeda et al. 2006).

10.1.2 *Identification of Genes that Facilitate Breast Cancer Metastasis to the Brain*

The disruption of the BBB by brain metastasis was observed in triple-negative and basal-type breast cancers (TNBCs) but not in Her2/neu-positive breast cancer (Yonemori et al. 2010).

Examination of human brain tissue revealed decreased expression of TJs in BMECs in pathological diseases (Harhaj and Antonetti 2004; Wolburg and Lippoldt 2002; Schneeberger and Lynch 2004). During brain development, BMECs gradually acquire the ability to form a highly selective barrier (Wolburg and Lippoldt 2002) and astrocytes play an inductive role in this process (Hawkins and Davis 2005). BMECs are joined together by intercellular TJs that are responsible for acquisition of highly selective permeability. This dynamic property is unique to BBB TJs and probably dependent on the higher density of structural organization of proteins such as occludin, claudins, JAM-1, JAM-2 and ZO-1 in the BBB TJs (Hawkins and Davis 2005; Harhaj and Antonetti 2004; Wolburg and Lippoldt 2002). ZO-1 binds to actin filaments and to the C-terminus of occludin and claudins which couples the structural and dynamic properties of perijunctional actin to the paracellular barrier. Most of the studies on TJs were performed on epithelial cells and less information is available on TJs in BMECs. Therefore, to prevent breast cancer metastasis to the brain, it is essential to focus on the role of HBMEC-TJs in human BMECs (HBMECs) function.

Several genes were shown to facilitate the development of brain metastasis and they include the cyclo-oxygenase COX-2, the EGFR ligand HBEGF, and α 2, 6-sialyltransferase ST6GALNAC5 (Bos et al. 2009; Nguyen et al. 2009). In this study, COX-2 and EGFR ligands were examined for their roles as mediators for BBB transmigration using BBB in vitro model that includes HUVEC cells (Bos et al. 2009), of which are quite different from HBMECs and therefore may not represent the BMECs.

Reactive astrocytes were identified as the most active cells that immediately localize to individual tumor cells invading the brain after their extravasation across the BBB (Lorger and Felding-Habermann 2010; Fitzgerald et al. 2008). Further, the vascular basement was found to promote co-option and colonization of tumor cells in the brain (Carbonell et al. 2009).

10.1.3 *VEGF and Brain Metastasis*

Vascular endothelial growth factor (VEGF) is primarily responsible for angiogenesis, vasculogenesis and vascular permeability (Machein and Plate 2000). VEGF is a member of a family of growth factors that include PIGF-1 and-2 (Placenta growth factor), VEGF-B, VEGF-C, VEGF-D and VEGF-E (Maglione et al. 1991; Olofsson et al. 1996; Jaukov et al. 1996; Grimmond et al. 1996; Yamada et al. 1997; Meyer et al. 1999). VEGF binds to endothelial cells via interactions with high-affinity tyrosine kinase receptors Flt-1 (VEGFR-1) and Flk-2 (VEGFR-2) (Shibuya et al. 1990; De Vries et al. 1992; Terman et al. 1992; Millauer et al. 1993). VEGF is regulated by hypoxia at both the transcriptional and post-transcriptional levels (Ikeda et al. 1995; Damert et al. 1997). Transcriptional regulation of HIF-1 (hypoxia inducible factor-1), which accumulates under hypoxia and binds to a HIF-1 consensus sequence, is located in the 5′ flanking region of VEGF gene (Forsythe et al. 1996).

Marked neovascularization is a hallmark of brain tumor (Machein and Plate 2000). The expression patterns of VEGF and its receptors in brain tumors indicate that VEGF plays a major role in brain tumor angiogenesis and formation (Machein and Plate 2000). VEGF is essential for angiogenesis and BBB functioning. Our previous studies (Radisavljevic et al. 2000; Lee et al. 2002, 2003, 2004, 2005, 2006; Price et al. 2001, 2004; Avraham et al. 2003) showed that VEGF upregulated ICAM-1 via phosphatidylinositol 3 OH-kinase/AKT/Nitric oxide pathway and modulated migration of HBMECs. Previous studies by other groups also showed that VEGF increased BMEC monolayer permeability by affecting occludin expression and TJ assembly (Wang et al. 2001; Mayhan 1999; Zhang et al. 2000). Using human cytokine cDNA array, we found that VEGF induced a significant increase in expression of monocytes chemoattractant protein-1, the chemokine receptor CXCR4 as well as IL-8 in HBMECs (Lee et al. 2002). VEGF increased IL-8 production in HBMECs by activating nuclear factor-κB via calcium and phosphatidylinositol 3-kinase pathways (Lee et al. 2002, 2003). We also showed that VEGF secreted from breast cancer cells significantly increased the adhesion and penetration of these cells across the HBMECs monolayer, via observation of changes in VE-cadherin induced by SU-1498 inhibitor for VEGFR-2 and calcium chelator (Lee et al. 2003). VEGF also regulated focal adhesion assembly in HBMECs through activation of FAK and RAFTK/Pyk2 (Avraham et al. 2003). These focal adhesions are complexes comprised of scaffolding and signaling proteins organized by adhesion to the extracellular matrix (ECM). Further, VEGF upregulated the expression of α6 integrin and α6β1 integrin in HBMECs, both were important for VEGF-induced adhesion, migration, in vivo angiogenesis and tumor angiogenesis (Lee et al. 2006).

Thus, understanding how VEGF axis modulates the BBB integrity and permeability should contribute to better understanding of the mechanisms involved in BBB impairment and breast cancer metastasis to the brain (Fig. 10.2) and may lead to better therapeutic strategies for treatment modalities targeted to the BMECs using specific inhibitors for VEGF/VEGFR-2 pathways.

10.2 Results

10.2.1 VEGF and Angiopoietin-2 Secretion by Breast Cancer Cells

Using VEGF and Angiopoietin-2 ELISA assays, breast cancer cells secreted high levels of VEGF (Table 10.1). As shown in Table 10.1, breast cancer cells that "home"

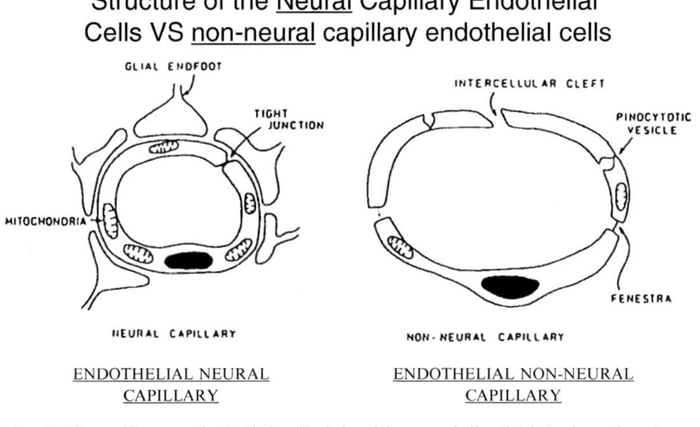

1. In CNS capillary endothelial cells joined by specialized 'tight junctions'.
2. Astrocytes involved in regulating ionic microenvironment (particularly $Ca^{2+}$and K^{+} ions).

- Functional differences between brain microvascular endothelial cells and peripheral endothelial cells:
  - Highly selective permeability to most substances.
  - Increased expression of transport and carrier proteins: receptor mediated endocytosis.
  - Limited paracellular and transcellular transport.

Fig. 10.2 Schematic presentation of tumor cell penetration across the BBB

Table 10.1 Secretion of VEGF and Angiopoietin 2 by breast cancer cells. Quantification of VEGF and Angiopoietin 2 levels in breast cancer cells as indicated. Cells (1×10^6) were grown in serum free media for 24 h. The results are presented as the mean of VEGF levels and Ang2 levels respectively (pg/ml; $n = 4$)

	Cell lines	VEGF (ng/ml)	Angiopoietin-2 (pg/ml)
1.	GFAP-MDA-MB-231	3.11 ± 0.027*	N.D.
2.	GFAP-MDA-MB-231-BrM2	3.086 ± 0.15*	N.D.
3.	GFAP-CN34- BrM2	3.036 ± 0.06*	N.D.
4.	MCF-10A	N.D.	N.D.

N.D. not detected
*$p < 0.05$ versus MCF-10A non-tumorogenic epithelial cells

to the brain have high levels of VEGF but do not secrete Ang2. These cells also express VEGFR-1/VEGFR-2 as well as Tie2 (data not shown) which may result in constitutive activation of VEGF receptors on tumor cells.

10.2.2 Cocultures of HBMEC and Human Astrocytes as an In-Vitro 3D Human Blood-Brain Barrier Model

Cocultures were established and initiated by transferring inserts containing astrocytes directly into wells containing confluent HBMECs as detailed previously (Avraham et al. 2008). The cocultures were tested for the maintenance of a tight endothelial permeability barrier by determination of the paracellular permeability of the BBB monolayers for ^{3}H-inulin and ^{14}C-sucrose (Avraham et al. 2008). HBMEC monolayers were always monitored after various treatments to ensure that the monolayers are intact. Further, we continuously monitored the human astrocytes used in the 3D cocultures to ensure that these cells are not altered or overgrown by other cell types.

10.2.3 Effects of Breast Tumor Cells on Adhesion, Transmigration and Permeability Changes in HBMECs

We observed that in comparison to breast tumor cells that do not secrete VEGF (MDA-MB-231/sh VEGF), cell adhesion (Table 10.2a) and transmigration (Table 10.2b) were significantly increased in breast tumor cells that do secrete VEGF (MDA-MB-231) in the presence of tumor cells. The increased adhesion is mediated by binding of VEGF secreted from tumor cells to HBMECs expressing VEGFR-2 receptors. Further, the permeability changes induced by adhesion of breast tumor cells on HBMEC were measured by Trans Endothelial Electrical Resistance (TEER). Significant changes in HBMEC permeability were observed following adhesion of tumor cells secreting VEGF, as compared to tumor cells lacking expression of VEGF (Table 10.2c).

10.2.4 Tight Junction Expression in HBMECs Following Adhesion of Tumor Cells

We examined TJ expression following adhesion of tumor cells to HBMEC monolayer. Within 30 min, TJs of the HBMEC monolayer were disrupted and by 24 h, redistribution of ZO-1 (Fig. 10.3a) and claudin-5 (Fig. 10.3b) to the cytosol were significant. Of note, we previously showed that VEGF disrupts HBMEC endothelial cell-cell junctions and increased the permeability of brain microvascular capillaries (Lee et al. 2003).

10.2.5 Effects of VEGF on KDR/VEGFR-2 and Tie2 Receptors on HBMECs

Exposure of HBMEC to VEGF induced VEGFR-2 translocation from the cell surface to the nucleus in a time-dependent manner, as shown in Fig. 10.4a. Further, VEGF

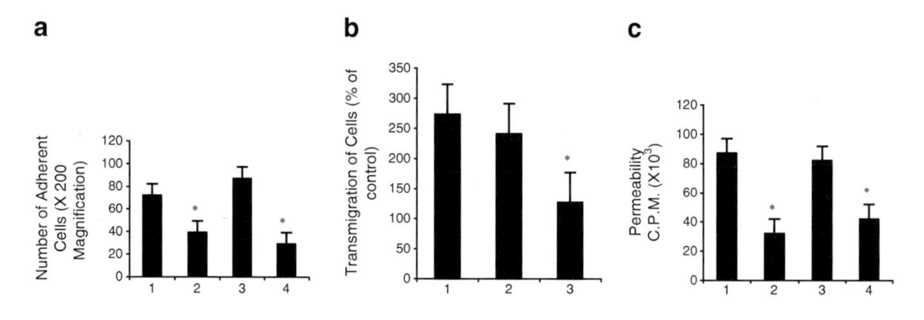

Table 10.2 (**a**) Adhesion of breast tumor cells to HBMEC co-cultures: About 100,000 HBMECs were added to fibronectin-coated 24 well Transculture inserts with pore sizes of 0.4 μm and grown for 5 days in 5% CO_2 at 37 °C in 3D-cocultures with human astrocytes. 100,000 Di I labeled tumor cells in 500 μl were added to each well. After incubation for 2 h, the cells were fixed and washed extensively with PBS to remove floating tumor cells. Attached labeled tumor cells were observed by a fluorescent microscope and counted from 10 random fields of ×200 magnification. The results are mean ±S.D. of triplicate samples. *Lane 1*: MDA-MB-231, *Lane 2*: MDA-MB-231-shVEGF, *Lane 3*: MDA-MB-231-shControl, and *Lane 4*: MDA-MB-231-VEGF inhibitor treated with SU6668. *$p < 0.05$ as compared to parental control cells. (**b**) Transmigration of breast tumor cells across HBMEC co-cultures: The tumor cells were added to HBMECs co-cultures. DiI labeled tumor cells were added to the apical chamber for 6 h. After incubation, the apical chamber was fixed and washed extensively with PBS. Only the migrating labeled tumor cells were observed by a fluorescent microscope and counted from 10 random fields. The results are presented as the mean ±S.D. of triplicate samples. *Lane 1*: MDA-MB-231, *Lane 2*: MDA-MB-231/shControl, *Lane 3*: MDA-MB-231/shVEGF. *$p < 0.05$ as compared to MDA-MB-231/sh control. (**c**) Effects of breast tumor cells on permeability of HBMEC co-cultures: Approximately 100,000 HMBECs were added to factor-coated 24 well Transculture inserts with pore sizes of 0.4 μm and grown for 5 days in 5% CO_2 at 37° C, in a 3D co-cultures with human astrocytes. 100,000 Di I labeled tumor cells were added to each well for 2 h. Cells were washed extensively, and 0.4 ml of the fresh medium containing {^{3}H} inulin (1 μ Ci) was added to the apical chamber. The basolateral chamber was filled with 0.6 ml of the same medium without {^{3}H} inulin. After 2 h of incubation, 30 μl medium was collected from the basolateral chamber and the amount of {^{3}H} inulin across the monolayers was determined by scintilation counting. The data represent the mean values of total cpm of three separate experiments. *Lane 1*: MDA-MB-231/shControl, *Lane 2*: MDA-MB-231/shVEGF, *Lane 3*: CN34-BrM2/shControl, *Lane 4*: CN34- BrM2/shVEGF. *$p < 0.05$ as compared to parental control cells

induced Tie-2 redistribution and translocation within 15 min of VEGF exposure and by 30 min, Tie-2 was translocated back to the cell surface as shown in Fig. 10.4, indicating the "cross-talk" between VEGF and Tie-2 receptors to be a short-term "cross-talk"-mediated effects.

10.2.6 VEGF Induces Cytoskeletal Remodeling of HBMECs

VEGF also induced cytoskeletal reorganization as shown by redistribution of the focal adhesion kinase (FAK) (Fig. 10.5) in a time-dependent manner. Cytoskeletal reorganization is required for endothelial cell migration. As shown in Fig. 10.5, following 30 min of VEGF exposure of HBMECs, FAK was translocated from focal adhesion sites to the cytosol.

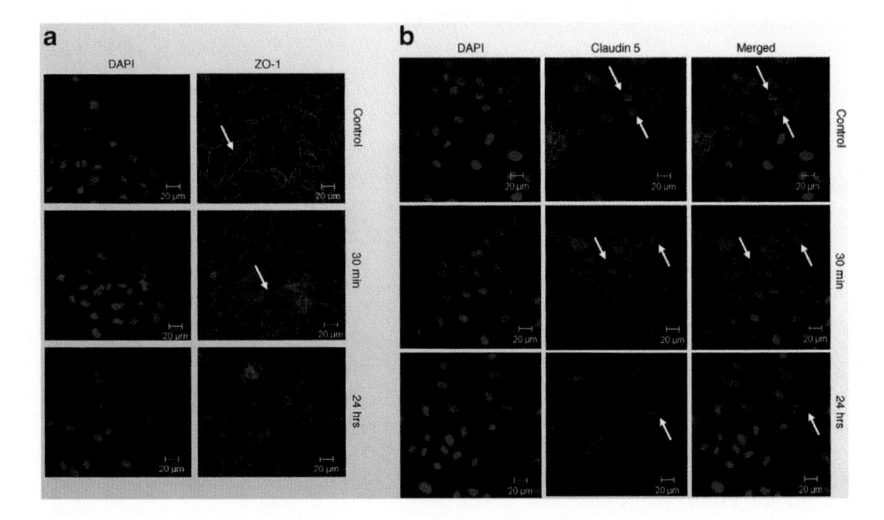

Fig. 10.3 Effects of VEGF on ZO-1 (*Panel A*) and claudin-5 (*Panel B*) localization in HBMECs. HBMECs were exposed to VEGF (100 ng/ml) for the indicated times and immunostained with ZO-1 or claudin-5 (*red*). Nuclei were counterstained with DAPI staining (*blue*) in HBMECs. Images were capture using a confocal microscopy

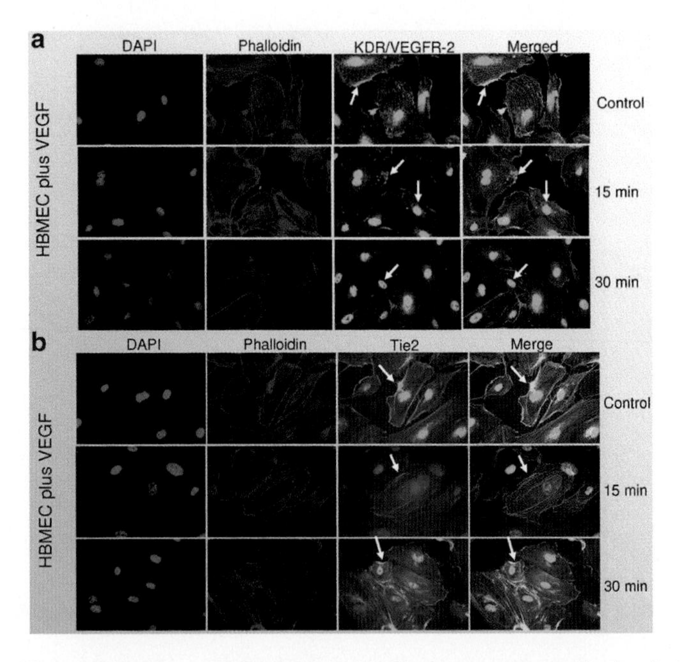

Fig. 10.4 Effects of VEGF on KDR/VEGF-2 and Tie-2 receptors localization in HBMECs. HBMECs were exposed to VEGF (100 ng/ml) for the indicated times and immunostained with KDR/VEGFR-2 (*green color*) and Phalloidin, which specifically binds to F-actin (*red color*). Nuclei were counterstained with DAPI staining (*blue*) in HBMECs. Images were captured using a confocal microscopy

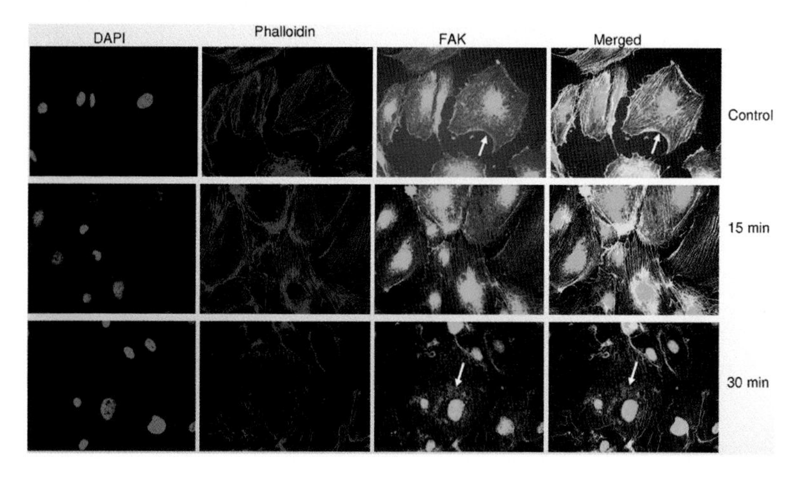

Fig. 10.5 Effects of VEGF on FAK expression and cytoskeletal reorganization. HBMECs were exposed to VEGF (100 ng/ml) for the indicated times and immunostained with FAK (*green color*) and Phalloidin (*red color*). Phalloidin binds to F-actin. Nuclei were counterstained with DAPI staining (*blue*). Images were captured using a confocal microscopy

10.3 Discussion

Breast cancer cell metastasis is a complex clinical and biological problem and entails several biological functions that collectively enable cancerous cells to disseminate from a primary site and colonize in brain (Bos et al. 2009; Nguyen et al. 2009). The sequential steps of brain metastasis include local invasion, intravasation, and survival in the circulation following extravasation and colonization of breast tumor cells in brain. The BBB, with its tight layer of microvascular endothelial cells and astrocytes foot processes, resists the infiltration of circulating breast cancer cells into the brain parenchyma (Cheng and Hung 2007; Abbott et al. 2006; Dejana 2004). Breast cancer cells extravasate through capillaries with highly selective TJs and are lodged in brain capillaries as single cells, suggesting that brain breast tumor cell metastases result from the ability of these cells to breach the BBB (Carbonell et al. 2009).

Our results showed that breast tumor cells secrete high levels of VEGF but not Angiopoietin-2 (Table 10.1). The tumor cells also express VEGFR-1 and VEGFR-2, suggesting an autocrine loop signaling pathway mediated by VEGF in tumor cells (Lee et al. 2007). VEGF induced significant changes in the localization and distribution of ZO-1 and claudin-5 in BMECs as well as changes in BMEC permeability and integrity (Table 10.2, Fig. 10.3). Further, VEGF acts as important biological determinant in adhesion of breast tumor cells to BMECs (Table 10.2), which is the first crucial step before breast tumor cell transmigration cross the BMECs. VEGF also disrupted the integrity of the BMECs by modulating the localization and distribution of TJs in BMECs (Fig. 10.3) and remodeling of BMECs, as shown by actin cytoskeletal reorganization (Fig. 10.5). Taken together, VEGF is an important modulator of TJs in BMECs and can enhance the transmigration of tumor cells across the BBB.

Disruption of the BBB-BMEC-TJs was observed in various diseases can lead to impaired BBB function and compromise homeostasis of the CNS. Failure of the BBB is a critical event in the development and progression of several diseases that affect the CNS, including metabolic brain tumors and metastasis development. Therefore, therapies targeted to the BBB-BMEC-TJs should hold significant promise for the treatment and prevention of breast cancer cell metastasis to the brain.

The majority of both in vivo experimental models of brain metastasis and analysis of human clinical specimens showed that over 95% of early breast tumor cell micrometastases demonstrated vascular co-option with little evidence for isolated neurotropic growth (Carbonell et al. 2009). The neurovasculature was identified as the critical partner for brain tumor growth (Carbonell et al. 2009), while cancer survival was recognized by the ability of cancer cells to specifically interact with components of the microvessel endothelial cells (Carbonell et al. 2009; Gevertz and Torquato 2006). Thus, the existing microvasculature is a key niche for malignant progression and therapies targeted to the BBB should hold significant promise for the treatment of breast tumor cells metastases in the brain.

Loss of the quiescent endothelial phenotype is a common feature of tumor angiogenesis (Saharinen et al. 2004; Lohela et al. 2009; Huber et al. 2001; Mayhan 1999; Zhang et al. 2000). The functions of Tie receptors and Angiopoietins in BMECs following contact with tumor cells in quiescent stable brain endothelium, as compared to activated/angiogenic brain endothelium, are unknown. Understanding and elucidation of the specific roles of VEGF in seeding of primary tumor cells in brain and in the transition of HBMECs from quiescent to angiogenic endothelium, may lead to significant advance in delineating the signaling pathways of VEGF/Ang2 axis in BMECs that lead to vascular remodeling, co-option of tumor cells with BMECs, and colonization in brain.

Blocking of the VEGF-VEGFR2 pathway is accepted as one of the first anti-angiogenic approaches. However, since tumors often develop evasive resistance to this therapy, the development of new anti-angiogenic approaches is necessary for successful anti-angiogenic therapy. This can be achieved by better understanding of the receptors and pathways involved in vascular remodeling in the brain.

In conclusion, this study provides additional findings and insights into the molecular mechanisms of VEGF axis in HBMEC integrity and as mediator that facilitates tumor cell metastasis to the brain.

References

Abbott NJ, Rönnbäck L, Hansson E (2006) Astrocyte-endothelial interactions at the blood-brain barrier. Nat Rev Neurosci 7(1):41–53

Aragon-Ching JB, Zujewski JA (2007) CNS metastasis: an old problem in a new guise. Clin Cancer Res 13:1644–1647

Avraham HK, Lee TH, Koh Y, Kim TA, Jiang S, Sussman M, Samarel AM, Avraham S (2003) Vascular endothelial growth factor regulates focal adhesion assembly in human brain microvascular endothelial cells through activation of the focal adhesion kinase and related adhesion focal tyrosine kinase. J Biol Chem 278(38):36661–36668

Avraham HK*, Lu TS*, Seng S, Tachado SD, Koziel H, Makriyannis A, Avraham S (2008) Cannabinoids inhibit HIV-1 Gp120-mediated insults in brain microvascular endothelial cells. J Immunol 181(9):6406–6416 (The first two authors contributed equally to this study)
Bos PD, Zhang XH, Nadal C, Shu W, Gomis RR, Nguyen DX, Minn AJ, van de Vijver MJ, Gerald WL, Foekens JA, Massagué J (2009) Genes that mediate breast cancer metastasis to the brain. Nature 459(7249):1005–1009
Bradbury MW (1993) The blood-brain barrier. Exp Physiol 78:453–472
Bradbury MW, Deane R (1993) Permeability of the blood-brain barrier to lead. Neurotoxicology 14:131–136
Carbonell WS, Ansorge O, Sibson N, Muschel R (2009) The vascular basement membrane as "soil" in brain metastasis. PLoS One 4(6):e5857
Chappa AK, Audus KL, Lunte SM (2006) Characteristics of substance P transport across the blood-brain barrier. Pharm Res 23:1201–1208
Cheng X, Hung MC (2007) Breast cancer brain metastases. Cancer Metastasis Rev 26:635–643
Damert A, Machein M, Breier G, Fujita MQ, Hanahan D, Risau W, Plate KH (1997) Up-regulation of vascular endothelial growth factor expression in a rat glioma is conferred by two distinct hypoxia-driven mechanisms. Cancer Res 57:3860–3864
De Vries C, Escobedo JA, Ueno H, Houck K, Ferrara N, Williams LT (1992) The fms-like tyrosine kinase, a receptor for vascular endothelial growth factor. Science 255:989–991
Dejana E (2004) Endothelial cell-cell junctions: happy together. Nat Rev Mol Cell Biol 5(4):261–270
Eichler AF, Loeffler JS (2007) Multidisciplinary management of brain metastases. Oncologist 12:884–898
Fitzgerald DP, Palmieri D, Hua E, Hargrave E, Herring JM, Qian Y, Vega-Valle E, Weil RJ, Stark AM, Vortmeyer AO, Steeg PS (2008) Reactive glia are recruited by highly proliferative brain metastases of breast cancer and promote tumor cell colonization. Clin Exp Metastasis 25(7):799–810
Forsythe JA, Jiang BH, Iyer NV, Agani F, Leung SW, Koos RD, Semenza GL (1996) Activation of vascular endothelial growth factor gene transcription by hypoxia-inducible factor-1. Mol Cell Biol 16:4604–4613
Gevertz JL, Torquato S (2006) Modeling the effects of vasculature evolution on early brain tumor growth. J Theor Biol 243(4):517–531
Grimmond S, Lagercrantz J, Drinkwater C, Sums G, Townson S, Pollock P, Gotley D, Carson E, Rakar S, Nordenskjold M, Ward I, Hayward N, Weber G (1996) Cloning and characterization of a novel human gene related to vascular endothelial growth factor. Genome Res 6:124–131
Harhaj NS, Antonetti DA (2004) Regulation of tight junctions and loss of barrier function in pathophysiology. Int J Biochem Cell Biol 36:1206–1237
Hawkins BT, Davis TP (2005) The blood-brain barrier/neurovascular unit in health and disease. Pharmacol Rev 57:173–185
Huber JD, Egleton RD, Davis TP (2001) Molecular physiology and pathophysiology of tight junctions in the blood-brain barrier. Trends Neurosci 24(12):719–725
Ikeda E, Achen MG, Breier G (1995) RisauW: hypoxia-induced transcriptional activation and increased mRNA stability of vascular endothelial growth factor inC6glioma cells. J Biol Chem 270:19761–19766
Jaukov V, Pasujola K, Kaipanen A, Chilvo D, Lahtinen I, Kukk E, Saksela O, Kalkkinen N, Alitalo K (1996) A novel vascular endothelial growth factor, VEGF-C, is a ligand for the flt-4 (VEGFR-3) and KDR (VEGFR-2) receptor tyrosine kinases. EMBO J 15:290–298
Kaal EC, Vecht CJ (2007) CNS complications of breast cancer: current and emerging treatment options. CNS Drugs 21:559–579
Lee TH, Avraham H, Lee SH, Avraham S (2002) Vascular endothelial growth factor modulates neutrophil transendothelial migration via up-regulation of interleukin-8 in human brain microvascular endothelial cells. J Biol Chem 277(12):10445–10451
Lee TH, Avraham HK, Jiang S, Avraham S (2003) Vascular endothelial growth factor modulates the transendothelial migration of MDA-MB-231 breast cancer cells through regulation of brain microvascular endothelial cell permeability. J Biol Chem 278(7):5277–5284

Lee BC, Lee TH, Avraham S, Avraham HK (2004) Involvement of the chemokine receptor CXCR4 and its ligand stromal cell-derived factor 1alpha in breast cancer cell migration through human brain microvascular endothelial cells. Mol Cancer Res 2(6):327–338
Lee BC, Lee TH, Zagozdzon R, Avraham S, Usheva A, Avraham HK (2005) Carboxyl-terminal Src kinase homologous kinase negatively regulates the chemokine receptor CXCR4 through YY1 and impairs CXCR4/CXCL12 (SDF-1alpha)-mediated breast cancer cell migration. Cancer Res 65(7):2840–2845
Lee TH, Seng S, Li H, Kennel SJ, Avraham HK, Avraham S (2006) Integrin regulation by vascular endothelial growth factor in human brain microvascular endothelial cells: role of alpha6beta1 integrin in angiogenesis. J Biol Chem 281(52):40450–40460
Lee TH, Seng S, Sekine M, Hinton C, Fu Y, Avraham HK, Avraham S (2007) Vascular endothelial growth factor mediates intracrine survival in human breast carcinoma cells through internally expressed VEGFR1/FLT1. PLoS Med 4(6):e186
Lohela M, Bry M, Tammela T, Alitalo K (2009) VEGFs and receptors involved in angiogenesis versus lymphangiogenesis. Curr Opin Cell Biol 21(2):154–165
Lorger M, Felding-Habermann B (2010) Capturing changes in the brain microenvironment during initial steps of breast cancer brain metastasis. Am J Pathol 176(6):2958–2971
Machein MR, Plate KH (2000) VEGF in brain tumors. J Neurooncol 50:109–120
Maglione D, Guerriero V, Viglietto G, Delli-Bovi P, Persico MG (1991) Isolation of human placental cDNA coding for a protein related to the vascular permeability factor. Proc Natl Acad Sci USA 88:9267–9271
Mayhan WG (1999) VEGF increases permeability of the blood-brain barrier via a nitric oxide synthase/cGMP-dependent pathway. Am J Physiol 276(5 Pt 1):C1148–C1153
Meyer M, Clauss M, Lepple-Wienhues A, Waltenberger J, Augustin HG, Ziche M, Lanz C, Buttner M, Rziha HJ, Dehio C (1999) A novel vascular endothelial growth factor encoded by Orf virus, VEGF-E, mediates angiogenesis via signalling through VEGFR-2 (KDR) but not VEGFR-1 (Flt-1) receptor tyrosine kinases. EMBO J 15:363–374
Millauer B, Wizigmann-Voos S, Schnurch H, Martinez R, Moller NP, Risau W, Ullrich A (1993) High affinity VEGF binding and development expression suggest flk-1 as a major regulator of vasculogenesis and angiogenesis. Cell 72:835–846
Morita K, Sasaki H, Furuse M, Tsukita S (1999) Endothelial claudin: claudin-5/TMVCF constitutes tight junction strands in endothelial cells. J Cell Biol 147:185–194
Musch MW, Walsh-Reitz MM, Chang EB (2006) Roles of ZO-1, occludin, and actin in oxidant-induced barrier disruption. Am J Physiol Gastrointest Liver Physiol 290:G222–G231
Nguyen DX, Bos PD, Massagué J (2009) Metastasis: from dissemination to organ-specific colonization. Nat Rev Cancer 9(4):274–284
Olofsson B, Pajusola K, Kaipanen A, von Euler G, Joukov V, Saksela O, Orpana A, Pettersson RF, Alitalo K, Eriksson U (1996) Vascular endothelial growth factor B, a novel growth factor for endothelial cells. Proc Natl Acad Sci U S A 93:2576–2581
Palmieri D, Smith QR, Lockman PR, Bronder J, Gril B, Chambers AF, Weil RJ, Steeg PS (2006) Brain metastases of breast cancer. Breast Dis 26:139–147
Pardridge WM (1983) Brain metabolism: a perspective from the blood-brain barrier. Physiol Rev 63:1481–1535
Price DJ, Miralem T, Jiang S, Steinberg R, Avraham H (2001) Role of vascular endothelial growth factor in the stimulation of cellular invasion and signaling of breast cancer cells. Cell Growth Differ 12(3):129–135
Price DJ, Avraham S, Jiang S, Fu Y, Avraham HK (2004) Role of the aging vasculature and Erb B-2 signaling in epidermal growth factor-dependent intravasion of breast carcinoma cells. Cancer 101(1):198–205
Radisavljevic Z, Avraham H, Avraham S (2000) Vascular endothelial growth factor up-regulates ICAM-1 expression via the phosphatidylinositol 3 OH-kinase/AKT/nitric oxide pathway and modulates migration of brain microvascular endothelial cells. J Biol Chem 275(27):20770–20774
Rameshwar P (2007) Implication of possible therapies targeted for the tachykinergic system with the biology of neurokinin receptors and emerging related proteins. Recent Patents CNS Drug Discov 2:79–84

Saharinen P, Tammela T, Karkkainen MJ, Alitalo K (2004) Lymphatic vasculature: development, molecular regulation and role in tumor metastasis and inflammation. Trends Immunol 25(7):387–395

Schneeberger EE, Lynch RD (2004) The tight junction: a multifunctional complex. Am J Physiol Cell Physiol 286:C1213–C1228

Sharma M, Abraham J (2007) CNS metastasis in primary breast cancer. Expert Rev Anticancer Ther 7:1561–1566

Shibuya M, Yamaguchi S, Yamane A, Ikeda T, Tojo A, Matsushime H, Sato M (1990) Nucleotide sequence and expression of a novel human receptor-type tyrosine kinase gene (flt) closely related to fms family. Oncogene 5:519–524

Terman BC, Dougher-Vermazen M, Carrion ME, Dimitrov D, Armellino DC, Gospodarowicz D, Boehlen P (1992) Identification of the KDR tyrosine kinase receptor as a receptor for vascular endothelial growth factor. Biochem Biophys Res Commun 187:1579–1586

Umeda K, Ikenouchi J, Katahira-Tayama S, Furuse K, Sasaki H, Nakayama M, Matsui T, Tsukita S, Furuse M (2006) ZO-1 and ZO-2 independently determine where claudins are polymerized in tight-junction strand formation. Cell 126:741–754

Wang W, Dentler WL, Borchardt RT (2001) VEGF increases BMEC monolayer permeability by affecting occludin expression and tight junction assembly. Am J Physiol Heart Circ Physiol 280(1):H434–H440

Wolburg H, Lippoldt A (2002) Tight junctions of the blood-brain barrier: development, composition and regulation. Vascul Pharmacol 38:323–337

Yamada Y, Nezu J, Shimane M, Hirata Y (1997) Molecular cloning of a novel vascular endothelial growth factor, VEGF-D. Genomics 15:483–488

Yonemori K, Tsuta K, Ono M, Shimizu C, Hirakawa A, Hasegawa T, Hatanaka Y, Narita Y, Shibui S, Fujiwara Y (2010) Disruption of the blood brain barrier by brain metastases of triple-negative and basal-type breast cancer but not HER2/neu-positive breast cancer. Cancer 116(2): 302–308

Zhang ZG, Zhang L, Jiang Q, Zhang R, Davies K, Powers C, Bruggen N, Chopp M (2000) VEGF enhances angiogenesis and promotes blood-brain barrier leakage in the ischemic brain. J Clin Invest 106(7):829–838

Chapter 11
Claudin-5 and Cancer Metastasis

Cláudia Malheiros Coutinho-Camillo, Silvia Vanessa Lourenço, and Fernando Augusto Soares

Abstract The claudin family of proteins plays a critical role in the maintenance of epithelial and endothelial tight junctions, maintenance of the cytoskeleton and in cell signaling. While the exact functions of claudins in cancer cells are not fully understood, some studies suggest that claudins are involved in survival and invasion of cancer cells. Claudin-5, forming the backbone of tight junctions in blood vessels and lymphatic endothelial cells, has an important role in maintaining the homeostasis of the tissue microenvironment and tumors showing claudin-5 expression could have some additional affinity for blood vessels. Metastasis is a complex phenomenon that

C.M. Coutinho-Camillo (✉)
Department of Anatomic Pathology, Hospital A.C. Camargo, São Paulo, Brazil

Centro Internacional de Pesquisa e Ensino, Hospital AC Camargo, Rua Taguá, 440 – Primeiro andar, São Paulo 01508-010, Brazil;
e-mail: claumcc@terra.com.br

S.V. Lourenço
Discipline of General Pathology, Dental School, University of São Paulo, São Paulo, Brazil

Av Prof Lineu Prestes, 2227, São Paulo 05508-000, Brazil
e-mail: silvialourenco@usp.br

F.A. Soares
Department of Anatomic Pathology, Hospital A.C. Camargo, São Paulo, Brazil

Discipline of General Pathology, Dental School, University of São Paulo, São Paulo, Brazil
e-mail: soaresfernando@terra.com.br

T.A. Martin and W.G. Jiang (eds.), *Tight Junctions in Cancer Metastasis*, Cancer Metastasis - Biology and Treatment 19, DOI 10.1007/978-94-007-6028-8_11,

requires a number of specific steps such as decreased adhesion, increased motility and invasion, proteolysis, and resistance to apoptosis. Several studies about significance of claudin-5 expression in cancer cells were reported, suggesting that it may play an important role in regulation of metastasis, angiogenesis and tumor growth.

Keywords Claudins • Claudin-5 • Metastasis • Tumorigenesis • Cancer • Tight junction • Cell proliferation • Invasion • Cell survival • Cell-cell adhesion

11.1 Tight Junctions

Tight junctions (TJs) are the most apical cellular adhesion structures, constituting paracellular barriers that regulate flux of ions and proteins in addition to maintaining cell polarity (Anderson 2001). These junctions are composed of several membrane and peripheral proteins – mainly ocludins, zonula ocludens ZO-1, 2 and 3 proteins and claudins (Langbein et al. 2002; Chiba et al. 2003; Gonzalez-Mariscal et al. 2003) (Fig. 11.1). Due to their ability to recruit signalling cascades, tight junction proteins have also been hypothesized to control proliferation, differentiation, and other cellular functions (Anderson 2001; Morin 2005; Singh et al. 2010).

Claudins, discovered in 1998 are the main sealing proteins of TJ and appear to be important in the assembly and stabilization of TJ strands. They are connected to the actin cytoskeleton and participate in intracellular signaling (Furuse et al. 1998, 1999; Tsukita and Furuse 1999).

The claudin family of proteins regulates the formation and function of tight junctions (Furuse et al. 1998). To date, there are at least 27 human claudin family members, which are expressed in epithelial and endothelial cells. Claudins are integral transmembrane proteins that have four membrane-spanning regions and range in molecular weight from 20 to 27 KDa (Kominsky 2006 ; Furuse 2010; Turksen and Troy 2011). The sealing of tight junctions by claudins is likely to be mediated in part by phosphorylation. The cytoplasmic C-terminus of claudins contains not only several phosphorylation sites but also a PDZ-binding motif, to which PDZ domain-containing proteins, such as the tight junction proteins ZO-1, -2, and -3, bind (Kominsky 2006) (Fig. 11.2).

Studies have shown that the epithelial tight junctions are dynamic structures and are subject to modulation during epithelial tissue remodeling, wound repair, inflammation and neoplastic pathogenesis (Singh et al. 2010). Cell-cell adhesion is critical in the establishment and maintenance of cell polarity and cell-cell adhesiveness is generally reduced in various human cancers. Tumor cells are dissociated from the tumor masses, lose their cell polarity and infiltrate the stroma, leading to subsequent metastatic events. This process is a crucial step for tumour progression and the suppression of cell-cell adhesiveness may trigger the release of cancer cells from the primary cancer nests and confer invasive properties on a tumor (Frixen et al. 1991; Hirohashi 1998; Martin and Jiang 2009). Previous reports have indicated a relationship between tumors and alterations in tight junctions' structure (Turksen and Troy 2004; Singh et al. 2010; Ouban and Ahmed 2010; Turksen and Troy 2011).

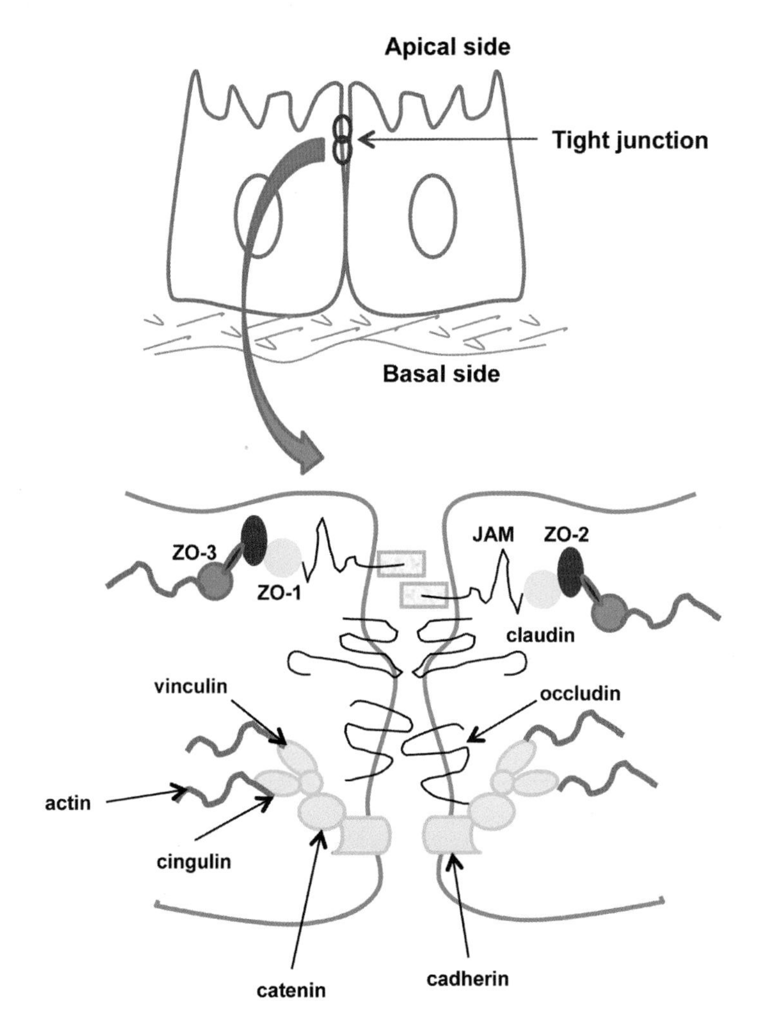

Fig. 11.1 Schematic representation of tight junction structure and major components of the complex. Tight junctions are composed of cytoplasmic proteins, such as zonula occludens (*ZO-1* to *ZO-3*), in addition to three transmembrane proteins: occludin, junctional adhesion molecules (*JAMs*) and claudins

11.2 Claudins and Tumorigenesis

The distribution of claudin subtypes has been examined in several tissues and organs. The patterns of expression of these molecules vary according to the tissue and lesion. Gain or losses of these molecules appear to relate to lesion progression and type (Morin 2005; Soini 2005; Hewitt et al. 2006; Kominsky 2006; Oliveira and Morgado-Díaz 2007; Ouban and Ahmed 2010; Singh et al. 2010).

In tumor tissues, claudins are mainly present in neoplasms originating from epithelia, endothelia and mesothelia (Soini 2005; Soini et al. 2006a). Tumors of

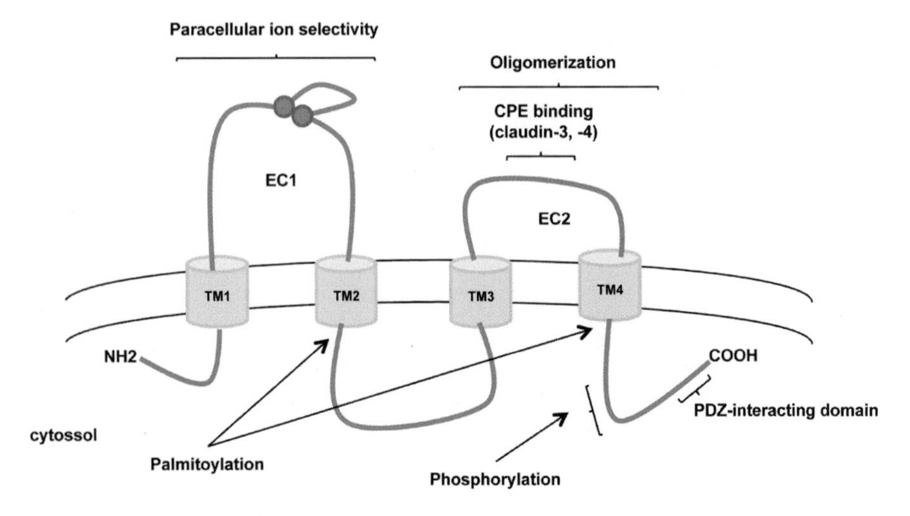

Fig. 11.2 Schematic representation of claudin structure. The claudin protein consists of four transmembrane (*TM1* to *TM4*) and 2 extracellular loops (*EC1* and *EC2*). The C-terminus contains a phosphorylation site and a PDZ-binding motif

hematopoietic origin do not express claudins, and their expression has been found only in a few sarcomas, such as synovial sarcoma, for instance (Billings et al. 2004; Soini 2005). In epithelial tumors, expression of claudins may vary depending on the site and cellular origin. Glandular and squamous epithelia express claudins differently and consequently, claudin expression is also different in tumors originating from squamous epithelial or glandular cells (Väre and Soini 2010).

While the exact functions of claudins in cancer cells are not fully understood, some studies suggest that claudins are involved in survival and invasion of cancer cells (Michl et al. 2003; Agarwal et al. 2005; Morin 2005). Tumorigenesis is accompanied by changes in adhesion, such as tight junction (TJ) disruption and the subsequent loss of cell-cell adhesion. These events can mediate the loss of cell differentiation and uncontrolled cell proliferation, leading to the cell-invasive phenotype in malignant neoplasms. Loss of claudin expression has also been associated with epithelial-mesenchymal transition (EMT), a phenomenon where tumor cells attain properties of mesenchymal cells, making it easier for them to invade and metastasize (Väre and Soini 2010).

Claudin expression could be lost as a direct result of protein down-regulation as well as phosphorylation by mitogen-activated protein kinase (MAPK), protein kinase C, cyclic adenosine-5′-monophosphate (cAMP)-dependent protein kinase and with-no-lysine(K)-4 kinase (WNK-4), leading to junction disruption and cytoplasmic internalization. Regulation of claudin expression is also dependent upon the immediate tissue microenvironment and, in this way, several growth factors and cytokines will have a role, such as epidermal growth factor (EGF), transforming growth factor (TGF)-beta, tumor necrosis factor (TNF)-alpha and interferon-gamma, among others (Oliveira and Morgado-Díaz 2007; Singh et al. 2010).

Dysregulation of adhesion molecules can also manifest with gain in their expression in several types of neoplasms. There are tumors, such as ovarian, cervical, hepatocellular, colorectal and prostate carcinoma (Agarwal et al. 2005; Lee et al. 2005; Cheung et al. 2005; Dhawan et al. 2005; Coutinho-Camillo et al. 2011), in which claudin expression is up-regulated. This may seem paradoxal, given that claudins are known to be involved in sealing epithelial junctions. One possible mechanism is that aberrant claudin expression may contribute to neoplasia by directly altering tight junction structure and function or that claudins may also affect cell signaling pathways (Singh et al. 2010).

Some studies have shown that claudin overexpression activates TCF-LEF/beta-catenin, an important cell cycle control pathway. This complex acts as a transcription factor that induces the expression of oncogenes that affect cell proliferation, survival, and invasion (myc, cyclin D1, MMP-7) (Dhawan et al. 2005). Claudins may also account for the invasive behavior of cancer cells, since there are able to interact with membrane-type matrix metalloproteinases (MT-MMPs) and promote MMP activity (Oliveira and Morgado-Díaz 2007). Colon cancer cells expressing high levels of claudin-1 were able to increase cell migration and MMP-2 and -9 activity (Dhawan et al. 2005). Another study reported that in oral squamous cell carcinoma cell lines, claudin-1 expression also enhances invasion through the increased expression of matrix metalloproteinase (MMP)-2 (Oku et al. 2006).

11.3 Claudin-5 and Cancer Metastasis

Early studies suggested that the diverse barrier functions of different epithelial and endothelial tissues reflect the use of different claudins. Claudin-5 is, for instance, strongly expressed in endothelial cells and takes part in the formation of the blood brain barrier (Väre and Soini 2010). Claudin-5, forming the backbone of tight junctions in blood vessels and lymphatic endothelial cells, has an important role in maintaining the homeostasis of the tissue microenvironment (Chiba et al. 2010). However, the role of claudin-5 expression in cancer mestastasis has not been elucidated.

Metastasis is a complex phenomenon that requires a number of specific steps such as decreased adhesion, increased motility and invasion, proteolysis, and resistance to apoptosis: cells detach from a primary, vascularized tumor, penetrate the surrounding tissue, enter nearby blood vessels (intravasation) and circulate in the vascular system. Some of these cells eventually adhere to blood vessel walls and are able to extravasate and migrate into the local tissue, where they can form a secondary tumor (Wirtz et al. 2011).

Considering that interaction and penetration of the vascular endothelium by dissociated cancer cells is an important step during the metastatic process, tight junctions (TJs) represent the first barrier that cancer cells must overcome in order to metastasize (Martin and Jiang 2009). Martin et al. (2002) have demonstrated that TJ of vascular endothelium in vivo function as a barrier between blood and tissues against metastatic cancer cells. Several studies have demonstrated TJ leakiness in tumoral tissues (Tobioka et al. 1996; Soler et al. 1999; Mullin et al. 2000).

Regulation of vascular permeability is one of the most important functions of endothelial cells, and endothelial cells from different organ sites show different degrees of permeability. Tumor blood vessels are more permeable on macromolecular diffusion than normal tissue vessels. However, the cause and mechanism of hyperpermeability of human vessels had not been clear (Martin and Jiang 2009). Sakaguchi et al. (2008) have speculated that disruption of the integrity of interepithelial tight junctions may be one of the causes of hyperpermeability. To test this hypothesis, they examined the expression status of claudin-5 and found that an attenuated expression of claudin-5 in sinusoidal endothelial cells (SECs) was associated with increased hepatitic or fibrotic grade and lower claudin-5-microvessel density (MVD) was associated with poor differentiation and vasculobiliary invasion. Down-regulated claudin-5 expression in tumor vessels may serve as a useful predictor for poor prognosis in hepatocellular carcinoma.

Dendritic cells are equipped with tight junction proteins that enable them to migrate into the luminal space by passing through the mucosal layer composed of epithelial cells expressing identical tight junction proteins (Rescigno et al. 2001; Sung et al. 2006). Cancer cells with aberrant claudin-5 expression may tend to colonize in the lymphatic sinus because claudin-5-positive cancer cells have an affinity for the sinus-lining endothelial cells of lymph nodes that express the specific molecular architecture of cellular junctions, including claudin-5 (Pfeiffer et al. 2008). Alternatively, aberrant expression of claudin-5 may reflect the malignant potential of cancer cells because claudin-5 regulates cell proliferation in a variety of tumor tissues (Soini et al. 2006b; Takala et al. 2007; Chiba et al. 2010).

Miyamori et al. (2001) have shown that claudin-5 promotes processing of pro-MMP-2 by MT1-MMP. Expression of claudin-5 not only replaced TIMP-2 in pro-MMP-2 activation by MT1-MMP but also promoted activation of pro-MMP-2 mediated by all MT-MMPs. MT1-MMP plays an essential role in angiogenesis. Therefore, they speculated that it is possible that claudins may regulate MT1-MMP activity during angiogenesis and/or that interaction of MT1-MMP with claudins would affect TJ functions such as vascular permeability.

11.4 Claudin-5 Expression in Cancer Tissues

Several studies about significance of claudin-5 expression in cancer cells were reported (Table 11.1). Soini (2005) has investigated claudin expression in a variety of tumors. Lymphomas, naevocytic lesions and soft tissue lesions were negative for claudin-5. Vascular tumors express mainly claudin-5 although its expression has also been reported in several epithelial tumors such as bladder, breast, colon, endometrium, oesophagus, lung, pancreas, prostate, stomach and thyroid whereas renal and hepatocellular carcinomas express hardly any claudin-5.

Table 11.1 Expression of claudin-5 in cancer

Tumor	Expression	Reference
Breast carcinoma	Up	Soini (2004), Szasz et al. (2011)
Ovarian surface epithelial carcinomas	Up	Soini and Talvensaari-Mattila (2006)
Ovarian sex cord stromal tumors	Down	Soini and Talvensaari-Mattila (2006)
Endometrial carcinoma	Up	Sobel et al. (2006)
Gastric carcinoma	Up	Soini et al.(2006b)
Adenocarcinoma of the lung	Up	Paschoud et al. (2007)
Squamous cell carcinoma of the lung	Down	Paschoud et al. (2007)
Oral squamous cell carcinoma	Down	Lourenço et al. (2010)
Ameloblastomas	Down	Bello et al. (2007)
Squamous cell carcinoma of the tongue	Down	Bello et al. (2008)
Esophageal adenocarcinoma	Up	Takala et al. (2007)
Esophageal squamous cell carcinoma	Down	Takala et al. (2007)
Prostatic carcinoma	Up	Coutinho-Camillo et al. (2011)
	Down	Väre et al. (2008), Seo et al. (2010)
Metastatic adenocarcinoma of the pleura	Up	Soini et al. (2006a)
Mesothelioma	Down	Soini et al. (2006a)

Bello et al. (2007) investigating claudin expression in ameloblastomas and developing human teeth found that claudin-5 was negative in most peripheral cells except in ameloblastic carcinomas, in which weak immunostaining was observed.

Bello et al. (2008) also found weak immunoreactivity for claudin-5 in superficial and invasive front areas of squamous cell carcinoma of the tongue. Loss of claudin expression, especially claudin-5 in oral squamous cell carcinoma (OSCC) is also significant, being more striking in poorly differentiated OSCC, but its clinical-pathological value is yet to be determined (Fig. 11.3) (Lourenço et al. 2010).

In lung cancer, claudin-5 expression pattern varies according to histological subtype: adenocarcinomas are positive for claudin-5 whereas squamous cell carcinomas are negative suggesting different pathways in tumor development and progression (Paschoud et al. 2007).

Mullin et al. (2006) have demonstrated that Barrett's metaplasia exhibits a paracellular transepithelial leak to small nonelectrolytes and low levels of claudin-5 when compared to normal esophageal mucosa. Another study by Chiba et al. (2010) demonstrated that in superficial esophageal squamous cell carcinoma, claudin-5 expression may be useful for predicting lymph node metastasis and thereby reducing the number of patients undergoing additional surgery after successful endoscopic resection.

A study by Fedwick et al. (2005) demonstrated regulation of claudin-5 during *Helicobacter pylori* infection. Gastric cancer is associated with *H. pylori* infection and Soini et al. (2006b) observed 50% of the gastric adenocarcinoma cases positive for claudin-5. Expression of claudin-5 seems to be associated with biological markers associated with tumor growth, such as proliferation and apoptosis and seems to be lower in diffuse-type carcinomas.

Chiba H, Gotoh T, Kojima T et al (2003) Hepatocyte nuclear factor (HNF)-4α triggers formation of functional tight junction and establishment of polarized epithelial morphology in F9 embryonal carcinoma cells. Exp Cell Res 286:288–297

Chiba T, Kawachi H, Kawano T, Kumagai J, Kitagaki K, Sekine M, Uchida K, Kobayashi M, Sugihara K, Eishi Y (2010) Independent histological risk factors for lymph node metastasis of superficial esophageal squamous cell carcinoma; implication of claudin-5 immunohistochemistry for expanding the indications of endoscopic resection. Dis Esophagus 23(5):398–407

Coutinho-Camillo CM, Lourenço SV, da Fonseca FP, Soares FA (2011) Claudin expression is dysregulated in prostate adenocarcinomas but does not correlate with main clinicopathological parameters. Pathology 43(2):143–148

De Torres Ramirez I (2007) Prognosis and predictive factors of prostate cancer in the prostatic biopsy. Actas Urol Esp 31(9):1025–1044

Dhawan P, Singh AB, Deane NG et al (2005) Claudin-1 regulates cellular transformation and metastatic behaviour in colon cancer. J Clin Invest 115:1765–1776

Fedwick JP, Lapointe TK, Meddings JB, Sherman PM, Buret AG (2005) Helicobacter pylori activates myosin light-chain kinase to disrupt claudin-4 and claudin-5 and increase epithelial permeability. Infect Immun 73(12):7844–7852

Frixen UH, Behrens J, Sachs M et al (1991) E-cadherin-mediated cell-cell adhesion prevents invasiveness of human carcinoma cells. J Cell Biol 113:173–185

Furuse M (2010) Molecular basis of the core structure of tight junctions. Cold Spring Harb Perspect Biol 2(1):a002907

Furuse M, Fujita K, Hiiragi T, Fujimoto K, Tsukita S (1998) Claudin-1 and -2: novel integral membrane proteins localizing at tight junctions with no sequence similarity to occluding. J Cell Biol 41:1539–1550

Furuse M, Sasaki H, Tsukita S (1999) Manner of interaction of heterogenous claudin species within and between tight junction strands. J Cell Biol 147:891–903

Gonzalez-Mariscal L, Betanzos A, Nava P, Jaramillo BE (2003) Tight junction proteins. Prog Biophys Mol Biol 81:1–44

Hewitt KJ, Agarwal R, Morin PJ (2006) The claudin gene family: expression in normal and neoplastic tissues. BMC Cancer 6:186

Hirohashi S (1998) Inactivation of the E-cadherin-mediated cell adhesion system in human cancers. Am J Pathol 153:333–339

Kominsky SL (2006) Claudins: emerging targets for cancer therapy. Expert Rev Mol Med 8(18):1–11

Langbein L, Grund C, Kuhn C et al (2002) Tight junctions and compositionally related junctional structures in mammalian stratified epithelia and cell cultures derived therefrom. Eur J Cell Biol 81:419–435

Lee JW, Lee SJ, Seo J et al (2005) Increased expressions of claudin-1 and claudin-7 during the progression of cervical neoplasia. Gynecol Oncol 97:53–59

Lourenço SV, Coutinho-Camillo CM, Buim ME, Pereira CM, Carvalho AL, Kowalski LP, Soares FA (2010) Oral squamous cell carcinoma: status of tight junction claudins in the different histopathological patterns and relationship with clinical parameters. A tissue-microarray-based study of 136 cases. J Clin Pathol 63(7):609–614

Marchesi F, Piemonti L, Mantovani A, Allavena P (2010) Molecular mechanisms of perineural invasion, a forgotten pathway of dissemination and metastasis. Cytokine Growth Factor Rev 21(1):77–82

Martin TA, Jiang WG (2009) Loss of tight junction barrier function and its role in cancer metastasis. Biochim Biophys Acta 1788(4):872–891

Martin TA, Mansel RE, Jiang WG (2002) Antagonistic effect of NK4 on HGF/SF induced changes in the transendothelial resistance (TER) and paracellular permeability of human vascular endothelial cells. J Cell Physiol 192(3):268–275

Martin TA, Watkins G, Mansel RE, Jiang WG (2004) Hepatocyte growth factor disrupts tight junctions in human breast cancer cells. Cell Biol Int 28(5):361–371

Masieri L, Lanciotti M, Nesi G, Lanzi F, Tosi N, Minervini A, Lapini A, Carini M, Serni S (2010) Prognostic role of perineural invasion in 239 consecutive patients with pathologically organ-confined prostate cancer. Urol Int 85(4):396–400

Michl P, Barth C, Buchholz M, Lerch M, Rolke M, Holzmann KH et al (2003) Claudin-4 expression decreases invasiveness and metastatic potential of pancreatic cancer. Cancer Res 63:6265–6271
Miyamori H, Takino T, Kobayashi Y, Tokai H, Itoh Y, Seiki M, Sato H (2001) Claudin promotes activation of pro-matrix metalloproteinase-2 mediated by membrane-type matrix metalloproteinases. J Biol Chem 276(30):28204–28211
Morin PJ (2005) Claudin proteins in human cancer: promising new targets for diagnosis and therapy. Cancer Res 65:9603–9606
Mullin JM, Laughlin KV, Ginanni N, Marano CW, Clarke HM, Peralta Soler A (2000) Increased tight junction permeability can result from protein kinase C activation/translocation and act as a tumor promotional event in epithelial cancers. Ann N Y Acad Sci 915:231–236
Mullin JM, Valenzano MC, Trembeth S, Allegretti PD, Verrecchio JJ, Schmidt JD, Jain V, Meddings JB, Mercogliano G, Thornton JJ (2006) Transepithelial leak in Barrett's esophagus. Dig Dis Sci 51(12):2326–2336
Oku N, Sasabe E, Ueta E, Yamamoto T, Osaki T (2006) Tight junction protein claudin-1 enhances the invasive activity of oral squamous carcinoma cells by promoting cleavage of laminin-5 gamma2 chain via matrix metalloproteinase (MMP-2) an membrane-type MMP-1. Cancer Res 66:5251–5257
Oliveira SS, Morgado-Díaz JA (2007) Claudins: multifunctional players in epithelial tight junctions and their role in cancer. Cell Mol Life Sci 64(1):17–28
Ouban A, Ahmed AA (2010) Claudins in human cancer: a review. Histol Histopathol 25(1):83–90
Paschoud S, Bongiovanni M, Pache JC, Citi S (2007) Claudin-1 and claudin-5 expression patterns differentiate lung squamous cell carcinomas from adenocarcinomas. Mod Pathol 20(9):947–954
Pfeiffer F, Kumar V, Butz S et al (2008) Distinct molecular composition of blood and lymphatic vascular endothelial cell junctions establishes specific functional barriers within the peripheral lymph node. Eur J Immunol 2008(38):2142–2155
Rescigno M, Rotta G, Valzasina B, Ricciardi-Castagnoli P (2001) Dendritic cells shuttle microbes across gut epithelial monolayers. Immunobiology 204:572–581
Sakaguchi T, Suzuki S, Higashi H, Inaba K, Nakamura S, Baba S, Kato T, Konno H (2008) Expression of tight junction protein claudin-5 in tumor vessels and sinusoidal endothelium in patients with hepatocellular carcinoma. J Surg Res 147(1):123–131
Seo KW, Kwon YK, Kim BH, Kim CI, Chang HS, Choe MS, Park CH (2010) Correlation between claudins expression and prognostic factors in prostate cancer. Korean J Urol 51(4):239–244
Singh AB, Sharma A, Dhawan P (2010) Claudin family of proteins and cancer: an overview. J Oncol 2010:541957
Sobel G, Németh J, Kiss A, Lotz G, Szabó I, Udvarhelyi N, Schaff Z, Páska C (2006) Claudin 1 differentiates endometrioid and serous papillary endometrial adenocarcinoma. Gynecol Oncol 103(2):591–598
Soini Y (2004) Claudins 2, 3, 4 and 5 in Paget's disease and breast carcinoma. Hum Pathol 35:1531–1536
Soini Y (2005) Expression of claudins 1, 2, 3, 4, 5 and 7 in various types of tumours. Histopathology 46:551–560
Soini Y, Talvensaari-Mattila A (2006) Expression of claudins 1, 4, 5, and 7 in ovarian tumors of diverse types. Int J Gynecol Pathol 25(4):330–335
Soini Y, Kinnula V, Kahlos K, Pääkkö P (2006a) Claudins in differential diagnosis between mesothelioma and metastatic adenocarcinoma of the pleura. J Clin Pathol 59(3):250–254
Soini Y, Tommola S, Helin H, Martikainen P (2006b) Claudins 1, 3, 4 and 5 in gastric carcinoma, loss of claudin expression associates with the diffuse subtype. Virchows Arch 448:52–58
Soler AP, Miller RD, Laughlin KV, Carp NZ, Klurfeld DM, Mullin JM (1999) Increased tight junctional permeability is associated with the development of colon cancer. Carcinogenesis 20(8):1425–1431
Sung SS, Fu SM, Rose CE Jr, Gaskin F, Ju ST, Beaty SR (2006) A major lung CD103 (alphaE)-beta7 integrin-positive epithelial dendritic cell population expressing Langerin and tight junction proteins. J Immunol 176:2161–2172
Szasz AM, Tokes AM, Micsinai M, Krenacs T, Jakab C, Lukacs L, Nemeth Z, Baranyai Z, Dede K, Madaras L, Kulka J (2011) Prognostic significance of claudin expression changes in breast cancer with regional lymph node metastasis. Clin Exp Metastasis 28(1):55–63

Takala H, Saarnio J, Wiik H, Soini Y (2007) Claudins 1, 3, 4, 5 and 7 in esophageal cancer: loss of claudin 3 and 4 expression is associated with metastatic behavior. APMIS 115:838–847

Tobioka H, Sawada N, Zhong Y, Mori M (1996) Enhanced paracellular barrier function of rat mesothelial cells partially protects against cancer cell penetration. Br J Cancer 74(3):439–445

Tsukita S, Furuse M (1999) Occludin and claudins in tight-junction strands: leading or supporting players? Trends Cell Biol 9:268–273

Turksen K, Troy TC (2004) Barriers built on claudins. J Cell Sci 117:2435–2447

Turksen K, Troy TC (2011) Junctions gone bad: claudins and loss of the barrier in cancer. Biochim Biophys Acta 1816(1):73–79

Turunen M, Talvensaari-Mattila A, Soini Y, Santala M (2009) Claudin-5 overexpression correlates with aggressive behavior in serous ovarian adenocarcinoma. Anticancer Res 29(12):5185–5189

Väre P, Soini Y (2010) Twist is inversely associated with claudins in germ cell tumors of the testis. APMIS 118(9):640–647

Väre P, Loikkanen I, Hirvikoski P et al (2008) Low claudin expression is associated with high Gleason grade in prostate adenocarcinoma. Oncol Rep 19(1):25–31

Wirtz D, Konstantopoulos K, Searson PC (2011) The physics of cancer: the role of physical interactions and mechanical forces in metastasis. Nat Rev Cancer 11(7):512–522

Chapter 12
Signaling Pathways Regulating Endothelial Cell-Cell Junctions as a Barrier to Tumor Cell Metastasis

Shigetomo Fukuhara and Naoki Mochizuki

Abstract Tumor metastasis, the spread of cancer cells from primary tumor sites to distant organs, is the main cause of mortality in cancer patients. The key steps leading to metastasis are tumor cell intravasation and extravasation, the processes by which tumor cells penetrate the junctions between endothelial cells in blood vessels and lymphatic vessels. Endothelial cell-cell junctions are organized by adherens junctions and tight junctions, and dynamically regulated by various types of extracellular stimuli to sustain vascular homeostasis and to control the transendothelial migration of inflammatory cells. Tumor cells need to weaken endothelial cell-cell junctions to successfully penetrate the endothelial barrier. Thus, transendothelial migration and metastasis of tumor cells are tightly controlled by endothelial cell-cell junctions. This article describes the molecular mechanisms that positively and negatively regulate endothelial cell-cell junctions and their impact on transendothelial migration of tumor cells and metastasis.

Keywords Endothelial cell-cell junctions • Transendothelial migration • Extravasation • Intravasation adherens junction • Tight junction • VE-cadherin

12.1 Introduction

Metastasis, the spread of cancer cells from primary tumor sites to distant organs, is the main cause of mortality in cancer patients. Tumor metastasis consists of a series of complex processes, which include tumor cell invasion from the primary tumor, intravasation, survival in the circulation, and extravasation and colonization at distinct sites (Joyce and Pollard 2009). Tumor cells metastasize either via blood vessels to

S. Fukuhara (✉) • N. Mochizuki
Department of Cell Biology, National Cerebral and Cardiovascular Center Research Institute, 5-7-1 Fujishirodai, Suita, Osaka 565-8565, Japan
e-mail: fuku@ri.ncvc.go.jp

T.A. Martin and W.G. Jiang (eds.), *Tight Junctions in Cancer Metastasis*, Cancer Metastasis - Biology and Treatment 19, DOI 10.1007/978-94-007-6028-8_12,

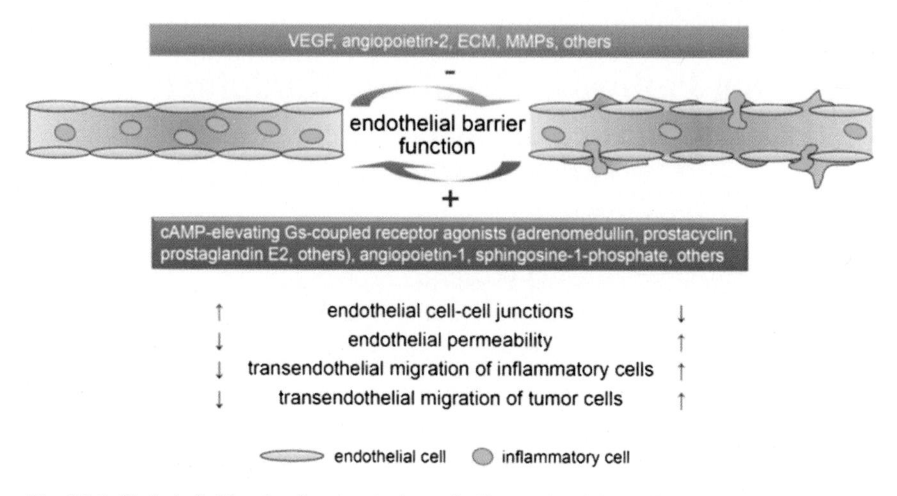

Fig. 12.1 Endothelial barrier function is dynamically regulated by various signaling molecules. Barrier stabilizing factors potentiate endothelial cell-cell junctions, reduce endothelial permeability, and inhibit transendothelial migration of inflammatory cells and tumor cells. On the other hand, barrier destabilizing factors weaken endothelial cell-cell junctions, increase endothelial permeability, and facilitate transendothelial migration of inflammatory cells and tumor cells

remote organs (remote metastasis) or via lymphatic vessels to regional lymph nodes (regional metastasis). Therefore, transmigration of tumor cells through blood and lymphatic vessels is an essential step in intravasation and extravasation, and is tightly controlled by endothelial cell-cell junctions (Fig. 12.1) (Joyce and Pollard 2009). Tumor cells interact with endothelial cells and disrupt endothelial cell-cell junctions through various mechanisms, leading to their penetration across the endothelium. In contrast, the transendothelial migration of tumor cells is thought to be suppressed by potentiation of endothelial cell-cell junctions.

Endothelial cell-cell junctions act as a barrier which restricts the passage of plasma proteins and circulating cells across the endothelial cells. Compromising vascular integrity leads to an increase in vascular permeability, which is associated with not only tumor metastasis but also chronic inflammation, edema and tumor angiogenesis (Dejana et al. 2009). Similar to epithelial cells, endothelial cells have two types of specialized junctional domains, adherens junctions (AJs) and tight junctions (TJs). Endothelial cell-cell junctions need to be highly dynamic to sustain vascular homeostasis and to control the transendothelial migration of leukocytes during inflammation (Dejana et al. 2009). Therefore, the AJs and TJs are intermingled throughout cell-cell contacts in endothelial cells, although TJs are located apically with respect to AJs in epithelial cells (Dejana et al. 2009).

The endothelial cell-cell adhesions are positively and negatively regulated by a variety of extracellular stimuli (Fig. 12.1) (Vestweber et al. 2009). Inflammatory mediators such as thrombin and histamine induce intercellular gap formation leading to an increase in endothelial permeability (Gavard 2009). Inter-endothelial cell-cell junctions are also weakened by vascular endothelial growth factor (VEGF), which is thought to be required for the initiation of angiogenesis (Dejana et al. 2008; Weis et al. 2004). In addition, tumor cells also induce disruption of endothelial cell-cell junctions to successfully undergo intravasation and extravasation (Joyce and Pollard 2009). In contrast, endothelial barrier integrity is stabilized by various factors such as angiopoietin (Ang)-1 and sphingosine-1-phosphate (S1P) (Adams and Alitalo 2007; Fukuhara et al. 2009; Obinata and Hla 2012). Furthermore, an increase in intracellular cyclic AMP (cAMP) level in endothelial cells strengthens barrier function and attenuates endothelial permeability both in vitro and in vivo (Kooistra et al. 2005; Fukuhra et al. 2006; Fukuhara et al. 2005; Cullere et al. 2005).

In this article, we highlight the recent progress in elucidating the signaling pathways regulating endothelial cell-cell junctions as a barrier to metastatic tumor cells by focusing mainly on remote metastasis. Role of cell-cell junctions between lymphatic endothelial cells in regional metastasis has been reviewed previously (Alitalo 2011).

12.2 Cell-Cell Adhesion Molecules in Endothelial Cells

12.2.1 VE-cadherin and N-cadherin

Endothelial AJs are constituted by vascular endothelial (VE)-cadherin (also known as cadherin-5 and CD144) and nectins. VE-cadherin is a single-span transmembrane protein that belongs to the classical cadherin superfamily, and exhibits *cis*- and *trans*-homophilic association via its extracellular domains in a Ca^{2+}-dependent manner (Dejana et al. 2008). Its cytoplasmic region binds three armadillo-family proteins, β-, γ- and p120-catenins. p120-catenin associates with the juxtamembrane region, while β-/γ-catenins bind the distal cytoplasmic region. Since β-/γ-catenins also bind α-cetenin that is connected to actin cytoskeleton, it is believed that cadherin-mediated cell-cell adhesions are stabilized by the actin cytoskeleton. Although the Weiss and Nelson groups have previously challenged this model by showing that α-catenin does not stably connect actin to cadherin (Drees et al. 2005; Yamada et al. 2005), several recent reports have also shown that cadherin stabilizes at cell-cell junctions by anchoring to circumferential actin bundles through α-catenin and β-catenin (Abe and Takeichi 2008; Noda et al. 2010; Yonemura et al. 2010).

VE-cadherin-mediated cell-cell junctions act as a barrier to tumor cell intravasation and extravasation. Thus, tumor cells disrupt VE-cadherin-dependent cell adhesions to penetrate the endothelial barrier. Consistently, disrupting the endothelial cell-cell junctions by anti-VE-cadherin antibody amplifies tumor cell metastasis (Weis et al. 2004).

In addition to VE-cadherin, N-cadherin (also known as cadherin-2) is also expressed in endothelial cells. Although N-cadherin appears to be involved in interactions between endothelial cells and pericytes, its role in formation and maintenance of endothelial cell-cell junctions remains controversial (Luo and Radice 2005; Giampietro et al. 2012).

12.2.2 Nectins

Nectins are Ca^{2+}-independent cell-cell adhesion molecules of the immunoglobulin superfamily, which comprise four members (Takai et al. 2008). Each nectin homophilically and heterophilically *trans*-associates to establish cell-cell adhesions. Similar to cadherins, nectins are also linked to actin cytoskeleton through afadin (Takai et al. 2008). It has been suggested that nectins regulate the initial step of cell-cell junction formation and then recruit cadherins to the nectin-based adhesion sites to form AJs (Takai et al. 2008). However, the role of nectins in formation of endothelial cell-cell junctions is still largely unknown.

12.2.3 Tight Junction Proteins

TJs regulate the passage of ions and solutes through the paracellular route. Endothelial TJs are organized by several cell adhesion molecules which include claudins, occludins, junctional adhesion molecule (JAM) family members and endothelial cell-specific adhesion molecule (ESAM) as well as intracellular adaptors such as the zonula occludens (ZO) proteins ZO-1 and ZO-2 (Dejana et al. 2009; Fukuhra et al. 2006). Endothelial TJs play a key role in the function of the blood-brain barrier (BBB), which is a selective interface separating circulating blood from the brain extracellular fluid in the central nervous system (Zlokovic 2008). Therefore, metastatic tumor cells need to disrupt the BBB TJs to invade the brain. Claudin-5 is a major cell adhesion molecule involved in formation of BBB TJs (Furuse 2009; Nitta et al. 2003). Consistently, claudin-5-deficient mice exhibit size selective opening of the BBB (Furuse 2009; Nitta et al. 2003).

12.3 Disruption of Endothelial Cell-Cell Junctions and Promotion of Tumor Cell Metastasis

12.3.1 VEGF

VEGF is unique among angiogenic growth factors, since it not only regulates vascular network formation but also induces increase in endothelial permeability by disrupting endothelial cell-cell adhesions (Weis and Cheresh 2005; Gavard 2009). Indeed, the original designation for VEGF is vascular permeability factor (VPF). VEGF is produced by tumor cells via hypoxia-inducible factor-1, and promotes tumor cell extravasation and metastasis by disrupting VE-cadherin-dependent cell-cell adhesions (Weis et al. 2004). VEGF also induces BBB breakdown by inducing down-regulation of claudin-5 (Argaw et al. 2009). VEGF-induced endothelial barrier disruption occurs through VEGF receptor 2 (VEGFR2)-mediated activation of Src tyrosine kinase (Weis et al. 2004; Weis and Cheresh 2005; Criscuoli et al. 2005). Consistently, genetic inhibition of Src or pharmacological blockade of VEGFR2 and Src inhibit not only VEGF-induced vascular permeability but also metastasis and extravasation of VEGF-expressing tumor cells in vivo (Weis et al. 2004; Weis and Cheresh 2005).

Until now, several mechanisms have been suggested to account for the disruption of VE-cadherin-mediated cell-cell junctions by VEGF. Gavard and Gutkind have reported that VEGF induces p21-activated kinase (PAK)-mediated phosphorylation and internalization of VE-cadherin (Gavard and Gutkind 2006). This process is initiated by the activation of Rac by VEGFR2 through the Src-dependent phosphorylation of Vav2, a guanine nucleotide-exchange factor (GEF) for Rac (Gavard and Gutkind 2006). It has also been reported that VEGF induces activation of focal adhesion kinase (FAK), which leads to dissociation of β-catenin from VE-cadherin and breakdown of endothelial cell-cell junctions through direct phosphorylation of β-catenin at tyrosine 142 (Chen et al. 2012). Recently, the crucial role of T cell-specific adaptor (TSAd) in VEGF-induced vascular permeability has been reported (Sun et al. 2012). VEGF-induced tyrosine phosphorylation of VEGFR2 facilitates binding of VEGFR2 to the Src homology (SH) 2-domain of TSAd, which in turn leads to Src activation and increases vascular permeability (Sun et al. 2012).

12.3.2 Angiopoietin-2 and Angiopoietin-Like Proteins

Ang family consists of Ang1, Ang2, Ang3 and Ang4. Among them, Ang1 and Ang2 are well characterized (Fukuhara et al. 2010; Adams and Alitalo 2007). Ang1 binds to Tie2 and stimulates its kinase activity to promote vascular quiescence and angiogenesis, whereas Ang2 is known to act as a context-dependent agonist and antagonist for Tie2 (Fukuhara et al. 2010; Adams and Alitalo 2007). Ang1/Tie2 signaling maintains quiescence of mature blood vessels by enhancing vascular integrity and endothelial survival. In contrast, Ang2 induces disruption of vascular integrity and facilitates VEGF-induced angiogenesis by blocking Ang1-induced Tie2 activation (Fiedler and Augustin 2006).

Ang2 expression is induced in hypoxic vascular endothelial cells in tumors, where it promotes tumor angiogenesis and growth (Nasarre et al. 2009; Fiedler and Augustin 2006). Consistently, Ang2-deficient mice exhibit inhibition of early stages of tumor growth and angiogenesis (Nasarre et al. 2009). In addition, the blockade of Ang2 by neutralizing antibody and inhibitory Ang2 binding peptide fused to immunoglobulin Fc results in reduction of primary tumor growth and angiogenesis (Brown et al. 2010; Falcon et al. 2009; Nasarre et al. 2009; Oliner et al. 2004). Recently, it has also been reported that Ang2-blocking antibody promotes vascular integrity, thereby suppressing tumor metastasis (Holopainen et al. 2012). These results suggest that Ang2 facilitates tumor metastasis by disrupting endothelial cell-cell junctions.

Recently, Ang-like proteins that do not bind to Tie2 but contain motif structurally conserved in Ang have been identified. Among them, the role of Ang-like protein-4 (ANGPTL-4) in tumor metastasis has been reported (Padua et al. 2008). Transforming growth factor-β induces expression of ANGPTL4 via the SMAD signaling pathway in tumor cells (Padua et al. 2008). Tumor cell-derived ANGPTL4 disrupts endothelial cell-cell junctions and increases the vascular permeability, thereby facilitating the transendothelial migration of tumor cells (Huang et al. 2011). Recently, it has also been reported that tumor-secreted C-terminal fibrinogen-like domain of ANGPTL4 (cANGPTL4) binds integrin α5β1 to activate Rac1/PAK-signaling pathways which weakens endothelial cell-cell junctions (Huang et al. 2011). In addition, cANGPTL4 associates with and declusters VE-cadherin and claudin-5, leading to disruption of endothelial cell-cell junctions (Huang et al. 2011). In addition to ANGPTL4, Ang-like protein-2 (ANGPTL2) is also expressed by tumor cells, and promotes tumor cell metastasis (Aoi et al. 2011; Endo et al. 2012). Very recently, the immune-inhibitory receptor leukocyte immunoglobulin-like receptor B2 (LILRB2) and its mouse orthologue paired immunoglobulin-like receptor (PIRB) have been identified as receptors for several ANGPTLs (Zheng et al. 2012). Thus, it is interesting to investigate whether these receptors are involved in regulation of tumor cell metastasis.

12.3.3 Extracellular Matrix and Matrix Metalloproteinases

Extracellular matrix (ECM) proteins secreted by tumor cells and stromal cells affect the transendothelial migration of tumor cells by modulating the interaction between tumor cells and endothelial cells (Bernstein and Liotta 1994). Consistently, large-scale analyses of gene expression profiles of human cancers have revealed that aberrant expression patterns of ECM proteins are observed in various types of cancers (Buckhaults et al. 2001). βig-h3/TGFBI (transforming growth factor, β-induced) is a secreted ECM protein with a domain structure identical to periostin. βig-h3/TGFBI contains an N-terminal signal peptide, followed by a cyctein-rich domain, four internal homologous repeats (FAS1 domain), and a C-terminal RGD motif. This ECM protein is known to be highly expressed in various types of cancers which include colon cancer (Argani et al. 2001; Zhang et al. 1997). Overexpression of βig-h3/TGFBI in colon cancer cells leads to a more aggressive phenotype of metastasis (Ma et al. 2008). βig-h3/TGFBI secreted by colon cancer cells induces disruption of VE-cadherin-mediated endothelial cell-cell junctions through the αvβ5 integrin-Src signaling pathway, which leads to enhanced tumor cell extravasation and metastasis (Ma et al. 2008).

Matrix metalloproteinases (MMPs) are a family of zinc-dependent endopeptidases, and are known to mediate many aspects of tumorigenesis (Kessenbrock et al. 2010). Among them, MMP-2 and MMP-9 are secreted from leukemic cells, which frequently invade from the primary site to distinct parts of the body which include the brain (Feng et al. 2011). MMP-2 and MMP-9 produced by leukemic cells increase the permeability of BBB by inducing degradation of TJ proteins claudin-5, occludin and ZO-1, which facilitates the extravasation of leukemia cells into the brain (Feng et al. 2011). In addition, it has also been shown that tissue inhibitor of metalloproteinase-2 (TIMP-2) released by metastatic breast cancer cells induces activation of endothelial MMP2 in a MMP-14-dependent manner, which leads to disruption of endothelial cell-cell junctions and transendothelial migration of the tumor cells (Shen et al. 2010).

12.4 Signaling Pathways Which Potentiate Endothelial Cell-Cell Junctions and May Suppress Transendothelial Migration of Tumor Cells

12.4.1 Cyclic AMP and Rap1

cAMP, a second messenger downstream of Gs-coupled receptor, improves endothelial cell barrier function. Consistently, cAMP-elevating G protein-coupled receptor agonists such as adrenomedullin, prostacyclin, prostaglandin E2 and β-adrenergic agonists, reduce endothelial hyperpermeability induced by inflammatory stimuli (Langeler and van Hinsbergh 1991; Hippenstiel et al. 2002; Farmer et al. 2001). cAMP regulates diverse cellular functions via two downstream effectors, protein kinase A (PKA) and exchange protein directly activated by cAMP (Epac), a GEF for Rap1 (Bos 2003). Previously, we and others have revealed that cAMP potentiates VE-cadherin-dependent cell-cell adhesions by inducing activation of a Rap1 small GTPase through Epac, thereby enhancing endothelial cell-cell junctions (Kooistra et al. 2005; Fukuhra et al. 2006; Fukuhara et al. 2005; Cullere et al. 2005). Rap1 is a small GTPase which belongs to the Ras superfamily and is thought to antagonize Ras functions by competing Ras effector molecules such as c-Raf, RalGDS, and phosphatidylinositol 3-kinase (PI3 kinase) (Bos et al. 2001; Bos 2003). However, recent data have revealed that Rap1 functions, not only as a Ras competitor, but also as a regulator of cell-ECM and cell-cell adhesions (Bos et al. 2001; Bos 2003). Interestingly, VE-cadherin engagement also induces Rap1 activation at nascent cell-cell contacts by inducing junctional recruitment of MAGUK with an inverted domain structure-1 (MAGI-1) through β-catenin (Sakurai et al. 2006). MAGI-1 subsequently recruits PDZ-GEF, a GEF for Rap1, into the cell-cell junctions, thereby inducing activation of Rap1. Rap1 activation by VE-cadherin engagement is responsible for maturation of VE-cadherin-mediated cell adhesions (Sakurai et al. 2006). Thus, Rap1 and VE-cadherin reciprocally influence each other to regulate endothelial cell-cell junctions.

Activation of Rap1 leads to stabilization of VE-cadherin-mediated cell-cell contacts by inducing formation of circumferential actin bundles along the cell-cell junctions (Noda et al. 2010). Circumferential actin bundles anchor VE-cadherin to cell-cell contacts through α- and β-catenins, thereby stabilizing VE-cadherin-dependent

Printed by Printforce, the Netherlands